QED STRUCTURE FUNCTIONS

CONFERENCE PROCEEDINGS NO. **201**

PARTICLES AND FIELDS SERIES 39

QED STRUCTURE FUNCTIONS

ANN ARBOR, MI 1989

EDITOR:
GIOVANNI BONVICINI
UNIVERSITY OF MICHIGAN

American Institute of Physics   New York

Authorization to photocopy items for internal or personal use, beyond the free copying permitted under the 1978 US Copyright Law (see statement below), is granted by the American Insitute of Physics for users registered with the Copyright Clearance Center (CCC) Transactional Reporting Service, provided that the base fee of $2.00 per copy is paid directly to CCC, 27 Congress St., Salem, MA 01970. For those organizations that have been granted a photocopy license by CCC, a separate system of payment has been arranged. The fee code for users of the Transactional Reporting Service is: 0094-243X/87 $2.00.

Copyright 1990 American Institute of Physics.

Individual readers of this volume and non-profit libraries, acting for them, are permitted to make fair use of the material in it, such as copying an article for use in teaching or research. Permission is granted to quote from this volume in scientific work with the customary acknowledgment of the source. To reprint a figure, table or other excerpt requires the consent of one of the original authors and notification to AIP. Republication or systematic or multiple reproduction of any material in this volume is permitted only under license from AIP. Address inquiries to Series Editor, AIP Conference Proceedings, AIP, 335 E. 45th St., New York, NY 10017.

L.C. Catalog Card No. 90-80229
ISBN 0-88318-671-3
DOE CONF 8905213

Printed in the United States of America.

Contents

Preface vii

HIGHER ORDER CALCULATIONS IN QED AND QCD

Coherent State and Structure Functions in QED 3
    Mario Greco
The Structure Function Method in QED 12
    Luca Trentadue and Oreste Nicrosini
Exponentiation of Large Corrections in QCD 36
    George Sterman
Exponentiation at the Edges of the Phase Space 51
    Luca Trentadue and Stefano Catani
The Helicity Method: A Review 58
    R. Gastmans
Next-To-Leading Factorization in QED 73
    O. Nicrosini
The Structure of Gluon Radiation in QCD 91
    Stephen Parke and Michelangelo Mangano
Renormalization Group Improved Calculation of Radiative Corrections
in Annihilation and Scattering Channels 105
    E. A. Kuraev
Application of Structure Function Method to Polarized Particles 118
    V. S. Fadin
Renormalization in the Standard Model of the Electroweak Interactions ... 132
    Giampiero Passarino
Radiative Corrections to WW Scattering in the Standard Model 148
    R. Bouamrane

ALGORITHMIC SOLUTIONS IN QED AND QCD

Algorithmic Solutions in QED and QCD 161
    Michelangelo Mangano
Review of QCD Generators 175
    Kiyoshi Kato
The Renormalization Group Improved YFS Method in QED 201
    B. F. L. Ward and S. Jadach
An Inclusive and Exclusive Algorithm for QED Evolution 216
    Giovanni Bonvicini and Luca Trentadue
Helicity Amplitudes Calculation 243
    R. Miquel, M. Martinez, and C. Mana

Noodles and Stars Allow a Precise and Efficient Calculation
of the Z-line Shape and the Polarization Asymmetry 259
    Carl Jung-Choon Im

THEORY AND EXPERIMENTAL DATA

Radiative Corrections at SLC and LEP 287
    Patricia Rankin

Status Report on Radiative Correction at KEK 311
    J. Fujimoto

Heavy Flavor Resonances and QED Radiative Corrections..................................... 326
    G. Bonvicini et al.

An Introduction About Precise Measurement of QED γ Structure Functions 347
    A. Courau

Neutrino Counting at e^+e^- Colliders 360
    R. J. Wilson

Radiative Corrections to the Neutrino Counting Experiment................................... 395
    C. Mana, M. Martinez, and R. Miquel

Study of the All Neutral Final State of $f_2(1270)$ Produced in Two-Photon
Collisions as a Background to Radiative Neutrino Counting
on the Z^0 Resonance 428
    K. Riles et al.

Radiative Corrections to Polarized Compton Scattering.. 441
    H. Veltman

PREFACE

The Workshop on QED Structure Functions was held at the University of Michigan, Ann Arbor, May 22–25, 1989, under the sponsorship of the Department of Energy, the National Science Foundation, and the University of Michigan. Forty participants from all major laboratories in the world attended the three-and-a-half-day plenary sessions and the four working sessions. Twenty five talks were given, covering a broad range of topics from theoretical to experimental.

The Workshop was organized in view of the upcoming high-precision experiments at LEP/SLC, to review and further discuss recently developed techniques to calculate higher order processes, experimental needs, and use of techniques popular in QCD but still not used much in QED. Four major working groups were formed: (1) Experimental aspects of QED Structure Functions; (2) Experiments at LEP/SLC; (3) Algorithmic solutions in QED and QCD; and (4) Evolution equations in QED and QCD.

This Workshop was successful in exposing the participants to the whole spectrum of ideas and developments of the last 3–4 years (1985–1989), in singling out directions in which to proceed to improve the present techniques and make them more suitable for experiments. These Proceedings collect most of the talks, and the reader who is researching on topics close to those presented here will find this a useful source book.

I should mention that the enthusiasm and active participation of the people present was fundamental and maintained the scientific debate (as well as the social activity) very alive throughout the Workshop. P. Rankin, M. Mangano, R. Akhoury, and G. Sterman coordinated the working groups and their help was determinant. L. Phillips almost single-handedly organized the non-scientific part of the Workshop, ran it smoothly until the end, and amazed everyone with her energy. Finally, I am indebted to my former supervisor R. Thun for having taken on himself bureaucratic duties of this Workshop, and for having given me academic freedom, encouragement, and excellent advice throughout my post-doctoral education.

G. Bonvicini
December 1989

Higher Order Calculations in QED and QCD

COHERENT STATES AND STRUCTURE FUNCTIONS IN QED

Mario Greco

INFN - Laboratori Nazionali di Frascati
P.O. Box 13
00044 Frascati, Italy.

ABSTRACT

The methods of coherent states and of structure functions in QED are considered in detail for a precise evaluation of the radiative effects at LEP/SLC. They are explicitly shown to give identical results for the exponentiated infrared factors and the $O(\alpha)$ terms corresponding to the initial and final state radiation. Furthermore interference effetcs from initial and final state radiation are introduced within the formalism of structure functions in a way which is suitable for Monte Carlo applications. The final formulae improve to $O(0.1\%)$ the evaluation of the e.m. effects.

The important role played by QED radiative corrections at LEP/SLC energies for precision tests of the standard model is well known[1,2]. The understanding and the detailed description of the multiphoton effects closely follow the pioneering work of Touschek and collaborators[3-5] at Frascati, more than twenty years ago, when a similar problem of precision had to be faced in testing Quantum Electrodynamics at "large" transverse momenta, i.e. at $Q^2 \sim 0 \, (1 \, \text{GeV}^2)$.

The strategy was very clear: (i) to sum to all orders the large double and single soft and collinear logarithms of perturbation theory; (ii) to calculate exactly to one loop all remaining terms of $O(\alpha)$. The discovery of the J/ψ and other narrow resonances introduced further complications into the problem, due to the very nature of the resonant process and the subtle interference effects between initial and final leptonic states. Those were successfully described[6] along the same lines, in particular by exploiting the technique of the coherent states, introduced earlier[4,7] with the aim of having a realistic QED S-matrix.

The study[8] of genuine weak radiative effects at LEP/SLC energies clearly indicated the necessity of controlling e.m. corrections to a high degree of accuracy. Then the earlier analytical results were generalized to the case of Z_0 production to exact one-loop accuracy[9,10] and to all orders[9] in the leading logarithmic approximation, reaching a level of $O(1\%)$ precision. More recently initial state radiation effects have been evaluated[11] to two loops, pushing the theoretical accuracy - for the line shape measurements - to $O(0.1\%)$.

© 1990 American Institute of Physics

To this aim the method of the structure functions[12], extended also to final states[13] in the reaction $e^+e^- \to \mu^+\mu^-$, has been shown to be quite powerful, suggesting a systematic approach to other processes as, for example, Bhabha scattering. However, only numerical solutions have been obtained so far in the case of a resonant cross section, leaving unclear the connection to the previous analytical approach of the problem. Furthermore the full extension of this technique to leptonic final states clearly demands an appropriate treatment of initial - and final - state interference effects which, on the other hand, are described to all orders in the coherent state formalism.

In this talk I will address the above questions, discussing analytical solutions of the structure functions method, including initial-final state interference effects, which are explicitly introduced in the formalism. More in detail I will explicitly show that the method of structure functions coincides with the coherent state approach with an accuracy of $O(1\%)$, giving explicitly in addition the $O(\alpha^2)$ corrections needed to improve further the theoretical precision.

The basic formula which describes the reaction $e^+e^- \to \mu^+\mu^-$ in the approach of structure functions without interference effects is the following[13]

$$d\sigma(s) = \int_0^\varepsilon dx \, d\sigma_0((1-x)s) \, H_e(x,s) \, F_\mu(\varepsilon-x, (1-x)s), \tag{1}$$

where $\varepsilon \equiv \Delta E/E$ is the energy resolution, and the initial and final state radiation kernels are given in terms of the electron and muon structure functions as

$$H_e(x,s) = \int_{1-x}^1 \frac{dz}{z} D_e(z,s) D_e\left(\frac{1-x}{z}, s\right), \tag{2}$$

$$F_\mu(x,s) = \int_0^x dy \, H_\mu(y, (1-y)s), \tag{3}$$

with $H_\mu(x,s)$ defined as in eq. (2). Eq. (1) describes the factorizable corrections only, corresponding to real and virtual photon emission from the initial and final legs, with no relative interference. By taking into account the effect of the soft radiation to all orders and of the hard one up to $o(\alpha^2)$ one obtains[13]

$$H_e(x,s) = \Delta_e(s) \beta_e x^{\beta_e - 1} - \frac{1}{2} \beta_e (2-x)$$
$$+ \frac{1}{8} \beta_e^2 \left\{ (2-x)[3 \ln(1-x) - 4 \ln x] - 4 \frac{\ln(1-x)}{x} + x - 6 \right\}, \tag{4}$$

with $\beta_e = \frac{2\alpha}{\pi}(L_e - 1)$, $L_e = \ln\left(\frac{s}{m_e^2}\right)$ and

$$\Delta_e(s) = 1 + \frac{\alpha}{\pi}\left[\frac{3}{2}L_e + 2(\zeta(2) - 1)\right] + \left(\frac{\alpha}{\pi}\right)^2 \left\{\left[\frac{9}{8} - 2\zeta(2)\right]L_e^2\right.$$
$$\left. + \left[3\zeta(3) + \frac{11}{2}\zeta(2) - \frac{45}{16}\right]L_e + \left[-\frac{6}{5}\zeta^2(2) - \frac{9}{2}\zeta(3) - 6\zeta(2)\ln 2 + \frac{3}{8}\zeta(2) + \frac{57}{12}\right]\right\} \quad (5)$$
$$\equiv 1 + \frac{\alpha}{\pi}\Delta_e^{(1)} + \left(\frac{\alpha}{\pi}\right)^2 \Delta_e^{(2)}$$

By insertion of (4) in eqs. (2) and (1) one easily obtains

$$\sigma(s) = \int_0^\varepsilon dx \, \sigma_0(s(1-x))\{\Delta_e(s)\Delta_\mu(s)\beta_e x^{\beta_e - 1}(\varepsilon - x)^{\beta_\mu} + R(x, \ldots)\} \quad (6)$$

where the first term in the r.h.s. of eq. ((6) is proportional to the leading soft contribution, while $R(x, \ldots)$ give further correction terms of order (β^2) and β_ε in the final cross section. We will assume the fractional energy resolution $\varepsilon \equiv \Delta E/E$ of order $10^{-1} - 10^{-2}$.

By splitting the Born cross section $\sigma_0(s)$ as $\sigma_0 = \sigma_0^{QED} + \sigma_0^{INT} + \sigma_0^{RES}$, with

$$\sigma_0^{QED}(s) = A\frac{1}{s}$$

$$\sigma_0^{INT}(s) = B\,\text{Re}\left\{\frac{1}{s - M^2 + i\Gamma M}\right\} \quad (7)$$

$$\sigma_0^{RES}(s) = C\frac{s}{(s - M^2)^2 + \Gamma^2 M^2}$$

the corresponding radiatively corrected cross sections are found[14] to be

$$\sigma^{QED}(s) = \sigma_0^{QED}(s)\Delta_e(s)\Delta_\mu(s)\varepsilon^{\beta_e + \beta_\mu} + \ldots \quad (8)$$

$$\sigma^{INT}(s) = \sigma_0^{INT}(s)\Delta_e(s)\Delta_\mu(s)\varepsilon^{\beta_\mu}\frac{\Gamma(1+\beta_e)\Gamma(1+\beta_\mu)}{\Gamma(1+\beta_e+\beta_\mu)}\frac{1}{\cos\delta_R}$$
$$\cdot \text{Re}\left\{e^{i\delta_R}\left[\frac{\varepsilon}{1 + \left(\frac{\varepsilon s}{M\Gamma}\right)\sin\delta_R e^{i\delta_R}}\right]^{\beta_e}\right\} + \quad (9)$$

$$\sigma^{RES}(s) = \sigma_0^{RES}(s)\Delta_e(s)\Delta_\mu(s)\varepsilon^{\beta_\mu}\frac{\Gamma(1+\beta_e)\Gamma(1+\beta_\mu)}{\Gamma(1+\beta_e+\beta_\mu)}$$
$$\cdot \left|\frac{\varepsilon}{1 + \left(\frac{\varepsilon s}{M\Gamma}\right)\sin\delta_R e^{i\delta_R}}\right|(\cos\beta_e\phi - \cot\delta_R\sin\beta_e\phi) + \ldots \quad (10)$$

where $(M_R^2 - s)^{-1} \equiv \dfrac{\sin\delta_R \, e^{i\delta_R}}{M\Gamma}$, $\tan\delta_R = \dfrac{M\Gamma}{(M^2 - s)}$, $\phi = \arctan\left[\dfrac{\varepsilon s + M^2 - s}{M\Gamma}\right] - \arctan\left[\dfrac{M^2 - s}{M\Gamma}\right]$

and the dots indicate next to leading terms corresponding to R (x, ...) in eq. (6).

Comparing with the analogous expressions obtained in the framework of coherent states[2,9], one finds that exponentiated infrared and the $O(\alpha)$ factors coincide with those in eqs. (8-10). On the other hand extra terms are contained in eqs. (8-10) of $O(\beta^2)$ - in particular the factor $-\beta_e^2 (\phi^2/2)$ in eq. (13) arising from the expansion of $\cos \phi\beta_e$ - and $O(\varepsilon\beta)$, not written explicitly[15].

Concerning the remaining terms, including initial-final states interference, box diagrams, etc. they can also be included to $O(\alpha)$ in the approach of structure functions. However a general treatment of interference effects, to all orders, can be obtained as follows. For pure QED processes the simple rescaling[16] $s \to s (t/u)$, where s, t and u are the Mandelstam variables, gives the usual result

$$d\sigma^{QED} \approx d\sigma_0^{QED} \, e^{\beta_e + \beta_\mu + 2\beta_{int}}, \tag{11}$$

where $\beta_{int} = (4\alpha/\pi) \ln \text{tg } \theta/2$. However this simple rule does not work for a resonant proces. Then, more generally, a full account of all radiative effects can be simply obtained by replacing the Born cross section $d\sigma_0 [(1 - x) s]$ in eq. (1) by $d\sigma_0 [(1 - x) s] \cdot K(x)$, where the K-factor includes all not-factorizable corrections, which can be determined to all orders for the soft contribution and to $O(\alpha)$ for the nonleading terms.

Indeed one obtains[17] the soft contributions, corresponding to initial-final state interference, the following expressions for the $K^{(i)}(x)$ factors

$$K^{QED}(x) = \dfrac{(\beta_e + \beta_{int})}{\beta_e} [x(\varepsilon - x)]^{\beta_{int}}$$

$$K^{INT}(x) = \dfrac{(\beta_e + \beta_{int})}{\beta_e} \left[\dfrac{sx}{s(1-x) - M_R^2} (\varepsilon - x)\right]^{\beta_{int}} \tag{12}$$

$$K^{RES}(x) = \dfrac{(\beta_e + 2\beta_{int})}{\beta_e} \left|\dfrac{sx}{s(1-x) - M_R^2}\right|^{2\beta_{int}}$$

The above result is based on the observation[6] that in the pure QED process the virtual matrix element M_V^{QED} scales as $\{\lambda^2/s\}^{(\beta_e+\beta_\mu+2\beta_{int})/4}$, while for a resonant process one has $M_V^{RES} \sim \{\lambda^2/s\}^{(\beta_e+\beta_\mu)/4} \cdot \{\lambda^2/(s - M_R^2)\}^{\beta_{int}/2}$, where λ is the minimum energy cutoff. Eqs. (12) allow us to obtain the complete analytical solution within the method of the structure functions, generalizing the results of ref. [14].

Then, after the substitution $d\sigma_0^{(i)} [(1 - x)s] \to d\sigma_0^{(i)} [(1 - x) s] \cdot K^{(i)}(x)$ in eq. (1), using eq. (12), and following ref. [14], one then finds[17] for the leading terms - corresponding to the resummation of the soft contributions:

$$d\sigma^{QED}(s) = d\sigma_0^{QED}(s) \left\{ \Delta_e(s) \Delta_\mu(s) \varepsilon^{\bar{\beta}_e + \bar{\beta}_\mu} \right.$$

$$\left. \cdot \frac{\Gamma(1+\bar{\beta}_e)\Gamma(1+\bar{\beta}_\mu)}{\Gamma(1+\bar{\beta}_e+\bar{\beta}_\mu)} {}_2F_1(1, \bar{\beta}_e; 1+\bar{\beta}_e+\bar{\beta}_\mu; \varepsilon) + \ldots \right\}, \tag{13}$$

$$d\sigma^{INT}(s) = d\sigma_0^{INT}(s) \Delta_e(s) \Delta_\mu(s) \varepsilon^{\bar{\beta}_\mu} \frac{\Gamma(1+\bar{\beta}_e)\Gamma(1+\bar{\beta}_\mu)}{\Gamma(1+\bar{\beta}_e+\bar{\beta}_\mu)} \frac{1}{\cos\delta_R} \cdot$$

$$\cdot \operatorname{Re} \left\{ e^{i\delta_R} \left[\frac{\varepsilon}{1+\left(\frac{\varepsilon s}{M\Gamma}\right)\sin\delta_R \, e^{i\delta_R}} \right]^{\beta_e} \left[\frac{\varepsilon}{\varepsilon + \left(\frac{M\Gamma}{s}\right)\frac{e^{-i\delta_R}}{\sin\delta_R}} \right]^{\beta_{int}} \right\} + \ldots, \tag{14}$$

$$d\sigma^{RES}(s) = d\sigma_0^{RES}(s) \Delta_e(s) \Delta_\mu(s) \varepsilon^{\bar{\beta}_\mu} \frac{\Gamma(1+\bar{\bar{\beta}}_e)\Gamma(1+\bar{\beta}_\mu)}{\Gamma(1+\bar{\bar{\beta}}_e+\bar{\beta}_\mu)} \cdot \left| \frac{\varepsilon}{\varepsilon + \left(\frac{M\Gamma}{s}\right)\frac{e^{-i\delta_R}}{\sin\delta_R}} \right|^{2\beta_{int}}$$

$$\left| \frac{\varepsilon}{1+\left(\frac{\varepsilon s}{M\Gamma}\right)\sin\delta_R \, e^{i\delta_R}} \right|^{\beta_e} (\cos\beta_e\phi - \cot\delta_R \sin\beta_e\phi) + \ldots, \tag{15}$$

where $\bar{\beta}_{e,\mu} = \beta_{e,\mu} + \beta_{int}$, $\bar{\bar{\beta}}_e = \beta_e + 2\beta_{int}$.

A few comments are in order here. First of all the main β_{int}-dependence in eqs. (13-15) appears through exponentiated factors, which coincide with those found in refs. [6, 9], using the method of coherent states. Of course they also reproduce the exact one-loop calculations[9, 10]. Furthermore the β_{int}-dependence drops out completely in the limiting case of narrow resonance production ($\Gamma \ll \Delta\omega$), as for example the J/ψ.

Physically this can be understood through the observation that the initial and final state can no longer interfere since the long time delay ($\tau \sim 1/\Gamma$) due to the resonance formation and decay.

Grouping together all non-infrared factors coming from $\Delta_e(s)$, $\Delta_\mu(s)$ and the Γ-functions in eqs. (13-15), as well as those coming from the non-soft terms of the electron and muon radiators, one can then write, as in ref. [14],

$$\frac{d\sigma}{d\Omega}^{QED, INT, RES} = \sum_i \frac{d\sigma_0^{(i)}}{d\Omega} \left\{ C_{infra}^{(i)} (1 + \bar{C}_F^{(i)}) + C_F^{'(i)} \right\} \tag{16}$$

where the infrared factors $C_{infra}^{(i)}$ are simply obtained from eqs. (8-10), and

$$\bar{C}_F^{(QED)} = \left(\frac{\alpha}{\pi}\right)\left[\Delta_e^{(1)} + \Delta_\mu^{(1)}\right] - \beta_\mu \varepsilon$$
$$+ \left(\frac{\alpha}{\pi}\right)^2 \left[\Delta_e^{(2)} + \Delta_\mu^{(2)} + \Delta_e^{(1)}\Delta_\mu^{(1)}\right] - \frac{\pi^2}{6}\bar{\beta}_e\bar{\beta}_\mu - \frac{1}{4}\beta_e\bar{\beta}_e \varepsilon^{1-\bar{\beta}_e} \tag{17}$$

$$\bar{C}_F^{(INT)} = \left(\frac{\alpha}{\pi}\right)\left[\Delta_e^{(1)} + \Delta_\mu^{(1)}\right] - \beta_\mu \varepsilon + \left(\frac{\alpha}{\pi}\right)^2 \left[\Delta_e^{(2)} + \Delta_\mu^{(2)} + \Delta_e^{(1)}\Delta_\mu^{(1)}\right] - \frac{\pi^2}{6}\bar{\beta}_e\bar{\beta}_\mu$$

$$- \bar{\beta}_e \frac{\cos\phi\,(\beta_e + 1) + \tan\delta_R \sin\phi\,(\beta_e + 1)}{\cos\phi\,\beta_e + \tan\delta_R \sin\phi\,\beta_e} \left|\frac{\varepsilon}{1 + \frac{\varepsilon s}{M_R^2 - s}}\right|$$

$$- \frac{1}{4}\beta_e \bar{\beta}_e \frac{\cos\phi + \tan\delta_R \sin\phi}{\cos\phi\,\beta_e + \tan\delta_R \sin\phi\,\beta_e} \left|\frac{\varepsilon}{1 + \frac{\varepsilon s}{M_R^2 - s}}\right|^{1-\bar{\beta}_e} \tag{18}$$

$$\bar{C}_F^{(RES)} = \left(\frac{\alpha}{\pi}\right)\left[\Delta_e^{(1)} + \Delta_\mu^{(1)}\right] - \beta_\mu \varepsilon + \left(\frac{\alpha}{\pi}\right)^2 \left[\Delta_e^{(2)} + \Delta_\mu^{(2)} + \Delta_e^{(1)}\Delta_\mu^{(1)}\right] - \frac{\pi^2}{6}\bar{\beta}_e\bar{\beta}_\mu$$

$$- 2\bar{\bar{\beta}}_e \frac{\cos\phi\,(\beta_e + 1) - \cot\delta_R \sin\phi\,(\beta_e + 1)}{\cos\phi\,\beta_e - \cot\delta_R \sin\phi\,\beta_e} \left|\frac{\varepsilon}{1 + \frac{\varepsilon s}{M_R^2 - s}}\right| \tag{19}$$

$$- \frac{1}{4}\beta_e \bar{\bar{\beta}}_e \frac{\cos\phi - \cot\delta_R \sin\phi}{\cos\phi\,\beta_e - \cot\delta_R \sin\phi\,\beta_e} \left|\frac{\varepsilon}{1 + \frac{\varepsilon s}{M_R^2 - s}}\right|^{1-\bar{\beta}_e}$$

Finally the factors $C_F^{(0)}$ contain other $O(\alpha)$ finite terms, coming from bremsstrahlung and box diagrams, odd in the exchange $\theta \leftrightarrow \pi - \theta$, etc., and can be obtained from refs. [2,14].

The above equations represent our final result, which describes the radiative correction factors to an accuracy better than (1%). The effect of the new terms of $O(\beta^2, \varepsilon\beta)$ in eqs. (18) is shown[14] in figs. (1, 2), where we plot the ratios

$$R^{(i)} = \frac{d\sigma^{(i)}\left[1+O(\alpha) + 1 + O(\beta\epsilon) + O(\alpha^2)\right]}{d\sigma^{(i)}\left[1+O(\alpha)\right]}$$

for i = QED, INT, RES, for ϵ = 0.01 and ϵ = 0.05.

Notice that the factors $C^{(i)}_{infra}$ do not appear in the ratios $R^{(i)}$. We have taken the scattering angle $\theta = \pi/2$ in the factors $C^{(i)}_F$ to authomatically cancel the box diagram and other non factorizable contributions.

As is clear from figs. (1,2) the QED and INT corrections are practically constant in the resonant region to a value of about -0.01, while the RES correction is modulated essentially by the term $1 - (\beta_e^2 \phi^2/2)$, with an extra factor of $O(0.01 - 0.02)$.

The extension of our results to the case of the Z line shape is straightforward. It simply corresponds to take the limit β_μ $\beta_{int}= 0$, $\Delta_\mu = 1$ and $\epsilon = 1 - (4\mu^2/s)$ in eq. (15). Then one simply obtains for $s \approx M_R^2$

$$\sigma(s) \approx \sigma_0(s) \left|\frac{M_R^2 - s}{M_R^2}\right|^{\beta_e} \frac{\sin(1-\beta_e)\delta_R}{\sin \delta_R} \frac{\pi \beta_e}{\sin \pi \beta_e} \Delta_e(s) \tag{20}$$

which agrees with refs [6, 9, 18] up to constant factors of $O(\beta^2)$.

To conclude, we have explicitly shown that the approach of the coherent states and that of the structure functions offer two complementary methods in QED to achieve the theoretical accuracy required for precision measurements at LEP/SLC energies. They give identical results for the exponentiated and finite $O(\alpha)$ factors relative to the initial and final states radiation. Moreover we have also shown how to include the interference effects coming from initial and final state radiation within the approach of structure functions in QED. The additional $O(\alpha^2)$ corrections improve the theoretical accuracy to 0 (0.1%). The overall picture provides simple analytical formulae which can be easily extended to e^+e^- reactions other than $e^+e^- \to \mu^+\mu^-$, and in addition the method is suitable for MonteCarlo applications.

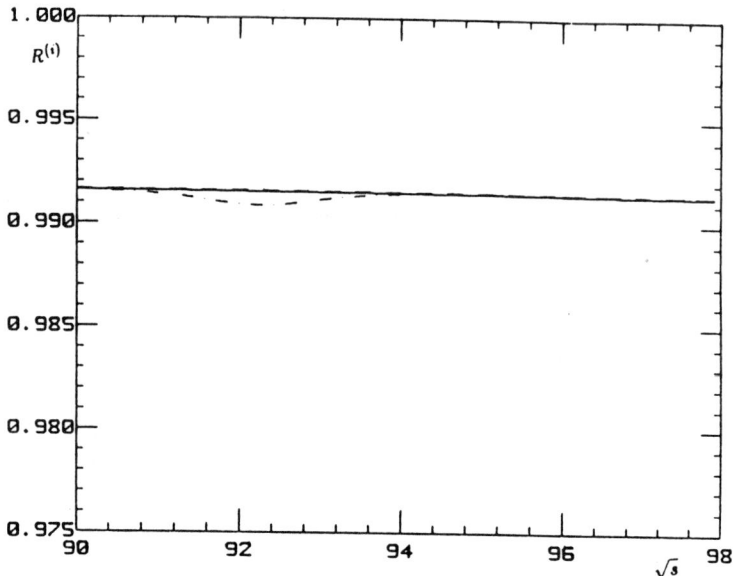

Fig. 1 Ratio $R^{(i)} = \frac{d\sigma^{(i)}[1+o(\alpha)+o(\beta\epsilon)+o(\alpha^2)]}{d\sigma^{(i)}[1+o(\alpha)]}$ for $e^+e^- \to \mu^+\mu^-$, with i=QED (solid line), INT (dashed line), RES (dotted-dashed line) for $\epsilon = 0.01$. The values of M and Γ are taken to be 92 Gev and 2.6 Gev respectively.

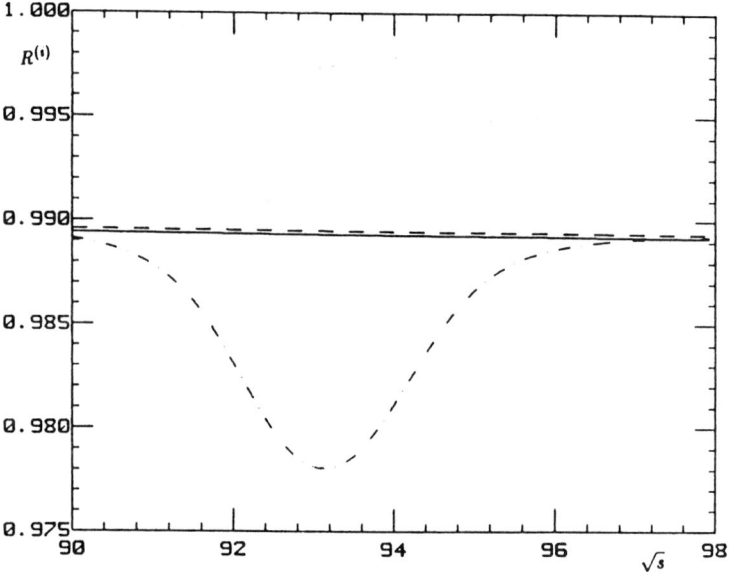

Fig. 2 Same as in fig. (1) for $\epsilon = 0.05$.

REFERENCES

[1] For recent reviews of LEP/SLC physics see, for example: Physics at LEP, edited by J. Ellis and R. Peccei, CERN 86-02 (1986);
Polarization at LEP, edited by G. Alexander et al., CERN 88-02 (1988).

[2] For a review of radiative corrections at LEP/SLC see, for example, M. Greco, La Rivista del Nuovo Cimento, Vol. 11, n. 5 (1988).

[3] E. Etim, G. Pancheri and B. Touschek, Nuovo Cimento 51B, 276 (1967).

[4] M. Greco and G. Rossi, Nuovo Cimento 50, 168 [1967].

[5] G. Pancheri, Nuovo Cimento 60, 321 (1969).

[6] M. Greco, G. Pancheri and Y. Srivastava, Nucl. Phys. B101, 234 (1975).

[7] V. Chung, Phys. Rev. B140, 1110 (1965).

[8] G. Passarino and M. Veltman, Nucl. Phys. B160, 151 (1979);
M. Consoli, Nucl. Phys. B160, 268 (1979).

[9] M. Greco, G. Pancheri and Y. Srivastava, Nulc. Phys. B171, 118 (1980).

[10] F.A. Berends, R. Kleiss and S. Jadach, Nucl. Phys. B202, 63 (1982);
M. Bohm and W. Hollik, Nucl. Phys. B204, 45 (1982).

[11] F.A. Berends, W.L. van Neerven and G.J.H. Burgers, Nucl. Phys. B297, 429 (1988).

[12] E.A. Kuraev and V.S. Fadin, Sov. J. Nucl. Phys. 41, 41, 466 (1985);
G. Altarelli and G. Martinelli, in Physics at LEP, ref. (1);
O. Nicrosini and L. Trentadue, Phys. Lett. 196B, 551 (1987).

[13] O. Nicrosini and L. Trentadue, Z. Phys. C39, 479 (1988).

[14] F. Aversa and M. Greco, LNF-89/025 (PT), May 1989.

[15] Soft corrections of $O(\beta^2)$ to the coherent state method are also discused by H. Spiesberger and M. Böhm Würzburg preprint, 1989.

[16] O. Nicrosini, talk given at the "Workshop on QED Structure Functions", Ann Arbor, May 1989.

[17] M. Greco, LNF-89/042 (PT), June 1989.

[18] R.N. Cahn, Phys. Rev. D36, 2666 (1987).

THE STRUCTURE FUNCTION METHOD IN QED[*]

Oreste Nicrosini
Istituto Nazionale di Fisica Nucleare, Sezione di Pavia, and
Dipartimento di Fisica Nucleare e Teorica dell'Universitá, Pavia, Italy
and
Luca Trentadue
CERN, Geneva, Switzerland
and
Dipartimento di Fisica, Universitá di Parma, Parma, Italy, and
Istituto Nazionale di Fisica Nucleare,
Gruppo Collegato di Parma, Sezione di Milano

ABSTRACT

The use of the structure function formalism to describe the electromagnetic radiative corrections in e^+e^- colliders is briefly discussed. Specific applications of the method to physical quantities are shown. The total cross-section with both initial and final state radiative corrections is derived. We define here p_t-dependent structure functions and their application to the radiative neutrino counting problem. Expressions are derived for the forward-backward asymmetry A_{FB}. Some more recent uses of the structure function method are also listed.

INTRODUCTION

Precision tests of the electroweak theory around the Z^0 are among the goals of the new electron-positron colliders LEP and SLC [1]. A significant role is played by the electromagnetic radiative corrections.

Starting from the first developments of the electron-positron accelerators this subject has been actively investigated [2]. Recently, the formalism of the structure functions, widely applied to describe the interactions of partons within the Quantum Chromodynamics (QCD), has been used [3-5] for a new series of applications also in the framework of the Quantum Electrodynamics (QED) theory in order to systematically evaluate electromagnetic radiative corrections to leptonic processes.

Structure function evolution equations have been applied to describe the interactions of fermions within vector theories [6]. The more complex case of the interaction within non-Abelian gauge theories has been also considered [7]. This formalism has been proposed as a new tool [3-5] to deal with the problems of describing the dynamics also of electron and photon states by using the corresponding evolution equations. At high energies, particularly around peaked

---

[*] Talk given by L. Trentadue

resonances, has the infra-red and collinear singularity structure of the radiative corrections dominates the dynamics of the reaction [2,3-5]. Structure functions are in this respect able to deal with the mass singularity sector in a simple and effective way.

GENERAL FORMALISM

In order to introduce the formalism of the structure functions for the QED case, it is useful to follow the analogy with the corresponding QCD case. A parton 'electron' can be defined as consisting of a cloud of real and virtual particles, quarks and gluons 'electrons and photons'. The probability of finding within this given state a an electron or a photon b at a given scale or virtualness $k^2 = s$ with a fraction of longitudinal momentum $x = \frac{k_l}{E}$ with k_l the longitudinal momentum of the electron and $E = \sqrt{s}/2$ can be given by defining the distribution $D_{ab}(x, s)$.

In general 'matrix' coupled equations should be considered connecting the different branching probabilities among the various channels. In fact if we call $D_{ab}(x,s)$ the probability distribution we may have that $a, b = e^+, e^-, \gamma$ and $D_{ab} = D_{ee}, D_{e\gamma}, D_{\gamma e}, D_{\gamma\gamma}$, as well as $P(z) = P_{ee}, P_{e\gamma}, P_{\gamma e}, P_{\gamma\gamma}$. The corresponding coupled 'matrix' equations are:

$$s\frac{\partial D_{ee}(x,s)}{\partial s} = \frac{\alpha}{2\pi}\int_x^1 \frac{dz}{z}[D_{ee}(\frac{x}{z},s)\, P_{ee}(z) + D_{e\gamma}(\frac{x}{z},s)\, P_{e\gamma}(z)]$$

$$s\frac{\partial D_{e\gamma}(x,s)}{\partial s} = \frac{\alpha}{2\pi}\int_x^1 \frac{dz}{z}[D_{e\gamma}(\frac{x}{z},s)P_{\gamma\gamma}(z)+(D_{ee^+}(\frac{x}{z},s)+D_{ee^-}(\frac{x}{z},s))P_{e\gamma}(z)] \quad (1)$$

with

$$P_{ee}(z) = P_{e^+e^-}(z) = P_{\gamma e}(1-z) = P_{\gamma e^+}(1-z)$$

and

$$P_{e^+\gamma}(z) = P_{e\gamma}(z) = P_{e\gamma}(1-z) = \frac{1}{2}(z^2 + (1-z)^2)$$

We will be mainly interested in the radiative corrections to the cross-section for photon emission. From a QCD analogy these contributions correspond to the Non-Singlet (NS) channel in the evolution equations where the electron or positron line is continuously connected to the annihilation vertex and all the coupled contributions of the singlet type are neglected. The corresponding evolution equations can be therefore written with the electron and positron distributions that satisfy the 'master' evolution equation:

$$D_{ee}(x,s) = D(x,s) = \delta(1-x) + \int_{m_e^2}^{s} \frac{dk^2}{k^2} \frac{\alpha(k^2)}{2\pi} \int_x^1 \frac{dz}{z} P^{NS}(z) D_{ee}(\frac{x}{z}, k^2) \quad (2)$$

with

$$\alpha(k^2) = \frac{\alpha}{1 - \frac{\alpha}{3\pi}\ln(k^2/m_e^2)}$$

where

$$P^{NS}(z) = P(z) = \frac{1+z^2}{1-z} - \delta(1-z) \int_0^1 dx \frac{1+x^2}{1-x}$$

is the regularized *electron → electron + photon* vertex where the first term represents the 'real' photon radiation and the second the 'virtual' corrections. The $\delta(1-x)$ represents a Born source term within the evolution equation and the regularization is obtained with the inclusion of the 'virtual' self-energy type contributions.

INITIAL STATE RADIATIVE CORRECTIONS IN e^+e^- AROUND THE Z^0

The physics involved in the problem of the radiative corrections to the initial state in e^+e^- annihilation is related to the emission of quanta from the annihilating electron and positron states and to their effect on the size and shape of the resonance peak. In the production of a resonance of mass M with colliding beams of total centre-of-mass energy $\sqrt{s} = 2E$. For $\sqrt{s} > M$, in the energy region above the resonance mass, this has the effect of decreasing the total center-of-mass energy to M. These states, therefore, effectively contribute to the resonance production. As a result a radiative tail arises on the right-hand side of the resonance peak and, the area under the peak not being affected by radiative corrections, a lowering of the maximum of the peak follows [2].

The determination of the Z^0 resonance cross-section requires the evaluation of the radiative corrections to the electron-positron vertex. At LEP/SLC it is expected that the accuracy in the determination of the mass and width of the Z^0 will be of the order of a few dozen MeV [1]. The size of the systematic errors can therefore be reduced by obtaining a comparable accuracy in the estimates of the radiative electromagnetic corrections to the initial state.

For an electron or positron state the radiative corrections are characterized in the perturbative series by terms of the type $(\alpha/\pi)^n \ln^p(s/m_e^2) \ln^q(E/\lambda)$ with m_e the electron mass and λ a scale on which there is a sizeable variation of the cross-section. In our case $\lambda = \delta E$ is the difference $\sqrt{s} - M$, i.e. the energy radiated away by the annihilating states. At the mass of the Z^0 $L = \ln(s/m_e^2) \simeq 24$ and the possibility that also the soft logarithms $l = \ln(E/\delta E)$ are also large implies that the effective expansion parameter $(\alpha/\pi)Ll$ is large too, and therefore these terms in the perturbative series must be taken into account and summed.

The process we are considering is the e^+e^- annihilation into the Z^0. According to a QCD analogy this scattering might be seen as a Drell-Yan process and, taking into account the well known theorems on the factorization of the mass and infra-red singularities [8], its cross-section can be written in the following form :

$$\sigma(s) = \int dx_1 \int dx_2 \, D_{e^-}(x_1,s) \, D_{e^+}(x_2,s) \, \sigma_0(x_1 x_2 s) \qquad (3)$$

$D_{e^-(+)}(x,s)$ represents the electron (positron) structure function. $\sigma_0(x_1 x_2 s)$ represents the resonance-cross section at the reduced energy $x_1 x_2 s = s'$.

Various solutions can be obtained for $D(x,s)$:

a) One possible method is simply to iterate eq.(2) and write the distribution in terms of a series which for the non-singlet case is:

$$D_{e^-(+)}(x,s) = \delta(1-x) + \int \frac{dk^2}{k^2} \frac{\alpha(k^2)}{2\pi} \left(P(x) + \int \frac{dt^2}{t^2} \frac{\alpha(t^2)}{2\pi} \int \frac{dz}{z} P(z) P(\frac{x}{z}) \right) + ...$$

The series can be truncated to the required level of approximation [5]. The perturbative expansion contains terms of the form : $(\alpha/\pi)^n L^n l^n$ (dominant) ; $(\alpha/\pi)^n L^n l^{n-j}$ (L−dominant) ; $(\alpha/\pi)^n L^{n-j} l^n$ (l−dominant) ; $(\alpha/\pi)^n L^{n-j} l^{n-i}$ (non−dominant) ; with $n \geq 1$ and $n \geq i,j \geq 1$. A first classification of the various terms according to their decreasing importance can be made by considering that the largest contribution will be given by those containing at least a logarithm $L = \ln(s/m_e^2)$ for each power of α/π, i.e. $(\alpha L)^n l^m$, $n \geq m \geq 0$.

b) A second solution can be obtained in the soft, large-x, limit. Here the large logarithm can be taken into account and summed by using the method developed by Gribov and Lipatov [9]. With the use of the running coupling constant the solution to eq.(2) reads :

$$D(x,s) = \frac{e^{\frac{\eta}{4}(\frac{3}{2} - 2\gamma_E)}}{\Gamma(\frac{\eta}{2})} (1-x)^{\frac{\eta}{2}-1} \qquad (4)$$

with $\eta = -6\ln(1 - \frac{\alpha}{3\pi} \ln(\frac{s}{m_e^2}))$ where γ_E is the Euler constant.

By taking the product $D(x_1,s) \, D(x_2,s)$ in the expression for the cross-section in eq.(3) and substituting the solution of eq.(4) and expanding in series in α/π one has, with the substitution $1 - x = E/\delta E$, that eq.(3) becomes in the large-x soft limit:

$$\sigma(s) = \sigma_0(s)(1 + \sum_{n=1}^{2}\sum_{m=0}^{n} g_{nm}(l)(\frac{\alpha}{\pi})^n L^m) \qquad (5)$$

with

$$g_{11}(l) = -2l + \frac{3}{2}; \quad g_{22}(l) = 2l^2 - \frac{10}{3}l + \frac{11}{8} - \frac{\pi^2}{3}$$

It is now interesting to compare this result with a finite second-order result. We use the results of the work of Barbieri, Mignaco and Remiddi [10] on the electron form factor.

The cross-section, with emission of only photons by the electron-positron state, is :

$$\sigma^{(\gamma)}(s) = \sigma_R(s)(1 + \sum_{n=1}^{2}\sum_{m=0}^{n} a_{nm}(l)(\frac{\alpha}{\pi})^n L^m) \qquad (6)$$

with

$$a_{10}(l) = 2l + \frac{\pi^2}{3} - 2; \quad a_{11}(l) = -2l + \frac{3}{2};$$

$$a_{20}(l) = 2l^2 + (\frac{2}{3}\pi^2 - 4)l - \frac{6}{5}(\zeta(2))^2 - \frac{9}{2}\zeta(3) - 6\zeta(2)\ln 2 + \frac{3}{8}\zeta(2) + \frac{57}{12};$$

$$a_{21}(l) = -4l^2 - \frac{45}{16} + (7 - \frac{2}{3}\pi^2)l + \frac{11}{2}\zeta(2) + 3\zeta(3);$$

$$a_{22}(l) = 2l^2 - 3l + \frac{9}{8} - 2\zeta(2);$$

This expression contains all the logarithmic contributions and in $a_{20}(l)$ the constant terms. By comparing eq.(6) with eq.(5) it appears that the coefficients of the large $(\alpha L)^n l^n$ leading logarithmic terms are exactly reproduced by the Gribov-Lipatov solution. In eq.(4), however, other terms are not reproduced which do appear in the second-order solution eq.(6). These contributions, which are of the type $(\alpha/\pi)^n L^j$ with $n > j$, can be included by applying the iterative method to solve eq.(2). Here we use a fixed coupling constant and factorize the summed Gribov-Lipatov solution. From the second-order result [3-5,13] one can also extract in fact the remaining virtual terms. This procedure gives:

$$D^{NS}_{e^-(+)}(x,s) = \frac{\beta}{2}(1-x)^{\frac{\beta}{2}-1}\Delta^{\frac{1}{2}} - (1+x)\frac{\alpha L}{2\pi} + (1+x)\frac{\alpha}{2\pi}$$

$$+\frac{1}{2}(\frac{\alpha L}{2\pi})^2[(1+x)(-4\ln(1-x)+3\ln x) - \frac{4}{1-x}\ln x - 5 - x]$$

$$-\frac{1}{2}(\frac{\alpha}{2\pi})^2(2L-1)[(1+x)(-4\ln(1-x)+3\ln x) - \frac{4}{1-x}\ln x - 5 - x]$$
(7)

where $\beta = \frac{2\alpha}{\pi}(L-1)$ and Δ is given by the expression [3,5]:

$$\Delta = 1 + \frac{\alpha}{\pi}(\frac{3}{2}L + \frac{\pi^2}{3} - 2) + (\frac{\alpha}{\pi})^2[(\frac{9}{8} - 2\zeta(2))L^2 + (-\frac{45}{16}$$
(8)

$$+\frac{11}{2}\zeta(2) + 3\zeta(3))L - \frac{6}{5}(\zeta(2))^2 - \frac{9}{2}\zeta(3) - 6\zeta(2)\ln 2 + \frac{3}{8}\zeta(2) + \frac{57}{12}].$$

By substituting the result for $D_{e^-(e^+)}(x,s)$ into eq.(1) we have by defining $(1-\chi)s = s'$ that the cross-section becomes:

$$\sigma(s) = \int d\chi\, \sigma_0((1-\chi)s)\, H(\chi,s)$$
(9)

with

$$H(\chi,s) = \Delta\,(\beta\chi^{\beta-1} - \frac{\beta^2\chi^\beta}{4}) - \frac{\beta\chi^{\frac{\beta}{2}}\Delta_{(1)}^{\frac{1}{2}}}{4}\left[(2-\chi)(1+(1-\chi)^{-\frac{\beta}{2}})\right.$$

$$\left.-\frac{\beta}{(2+\beta)}\chi(1-(1-\chi)^{-\frac{\beta}{2}})\right] + \frac{\beta^2}{16}\left[(\chi-2)(4\ln\chi - 3\ln(1-\chi))\right.$$

$$\left.-\frac{4}{\chi}\ln(1-\chi) - 6 + \chi + 2\chi - (2-\chi)\ln(1-\chi)\right],$$

where $\Delta_{(1)} = 1 + \frac{\alpha}{\pi}(\frac{3}{2}L + \frac{\pi^2}{3} - 2)$.

In the expression above not only the dominant $(\alpha L)^n$ terms are summed but also the less dominant $\alpha^n L^m$ with $n \geq m \geq 0$ are taken into account up to $n = 2$. In eq.(7) the first term $\frac{\beta}{2}(1-x)^{\frac{\beta}{2}-1}$ corresponds to the soft photon approximation. The terms proportional to $(1+x)$ and the last term contain contributions which modify the result accounting for emission of hard photons. The Δ factor contains terms of the type $(\frac{\alpha}{\pi})^n L^m, n \geq m \geq 0$.

This expression differs [5] from the one obtained in ref.[3] for the inclusion of the terms $(\frac{\alpha}{\pi})^2 L^2, (\frac{\alpha}{\pi})^2 L, (\frac{\alpha}{\pi})^2 constant$ in the factor Δ. It agrees with the result obtained in ref. [11] by an exact $O(\alpha^2)$ calculation apart from terms that are relevant only within the 'hard' $x \to 0$ limit.

These facts all show that, $O(\alpha^3)$ corrections being really negligible, the accuracy reached in the cross-section in eq.(9) is below the one per cent level.

FINAL STATE RADIATIVE CORRECTIONS

As we have seen before radiation from the initial states strongly affects the behaviour of the production cross-section around the Z^0 peak by modifying both its shape and size. Also final state corrections [12] together with initial-final states interference, are needed to reproduce physical quantities such as acollinearities and asymmetries. Any separation between radiation arising from initial states as compared to the one from final states is, in processes involving charged particles, of no physical content. Radiative corrections should be considered therefore as a whole set of corrections for any given process.

Since finite $O(\alpha)$ results are highly inadequate to describe the many-particle nature of the electromagnetic radiation [2], expressions to all orders of the perturbative expansion are necessary. In ref.[13] the general case of including the electromagnetic radiative corrections to the entire process has been considered. To this purpose the structure function formalism developed in [3-5] has been used and extended to both initial and final states. This is done by performing an $O(\alpha^2)$ calculation which takes into account the resummation to all orders of the dominant and next-to-dominant [3] logarithmic contributions and uses finite order expressions [12] for the box and the interference contributions. In order to describe the annihilation process $e^+e^- \to \mu^+\mu^-$ in an inclusive sense, i.e. with all the radiated photons summed over.

As shown before we can factorize initial state radiation and write the cross section as a convolution of a "bare" cross-section with a "radiator" representing initial state radiation eq.(9). By applying the Kinoshita-Lee-Nauenberg factorization theorem [9], the radiatively corrected cross-section can be written as:

$$\sigma(s) = \int dx \, \sigma_0((1-x)s) \, H_e(x,s)$$

According to the theorem on the factorization of infra-red and mass singularities, the evaluation of the radiator gives a quantity independent from the process itself. Analogously also for the final state, a radiator can be defined that does factorize like the initial one. If one calls σ_k the set of the factorizable (Born) plus non-factorizable (box diagrams and initial-final state interference) elementary cross-sections, the totally radiatively corrected process can be effectively decomposed. The calculation of the total cross-section can be performed. The accuracy that might be obtained depends on the various dominant and non-dominant contributions that are properly included within the three factors $H_e(x,s)$, $H_\mu(x,s)$ and σ_k [13]. Initial, final states and the kernel σ_k can be evaluated with any desired accuracy, provided this is done consistently order by order in the perturbation expansion.

The radiator $H_e(x,s)$ is defined as :

$$H_e(x,s) = \int_{1-x}^{1} \frac{dz}{z} D_e(z,s) D_e(\frac{1-x}{z},s). \qquad (10)$$

H_e contains contributions at all orders in perturbation theory at dominant and non-dominant levels.

The generalization of eq.(9) to take into account also final state radiation is, for the factorized part of the cross-section,

$$\sigma_f(s) = \int dx_1 dx_2 dy_1 dy_2 \, \sigma_0(s') \, D_e(x_1,s) D_e(x_2,s) D_\mu(y_1,s'') D_\mu(y_2,s'') \qquad (11)$$

where $s' = x_1 x_2 s$ and $s'' = y_1 y_2 s'$, with $x_{1,2}$ and $y_{1,2}$ fractions of longitudinal momentum of the electrons and muons respectively. $D_\mu(x,s)$ is the structure function for the muon, obtained from $D_e(x,s)$ with the substitution $m_e \to m_\mu$. Note that the scale s'' at which D_μ are evaluated is the invariant mass squared of the final real muon pair. By making the substitution $1 - x = x_1 x_2$ and $1 - y = y_1 y_2$ and recalling eq.(2) for the initial state radiator one has:

$$\sigma_f(s) = \int_0^{r_{max}} dx \, \sigma_0(s') \, H_e(x,s) \, F_\mu(r_{max} - x, s') \qquad (12)$$

where the upper limit r_{max} is the maximum fraction of radiation emitted and can be properly choosen and $F_\mu(z,s) = \int_0^z dx \, H_\mu(x,(1-x)s)$. The final state radiation kernel $H_\mu(x,s)$ is defined as:

$$H_\mu(x,s) = \int_{1-x}^{1} \frac{dy}{y} D_\mu(y,s) D_\mu(\frac{1-x}{y},s).$$

H_μ contains the same set of contributions as $H_e(x,s)$. eq.(12) is in a factorized form. σ_0 corresponds to the Born cross-section. To take into account also the non-factorizable corrections having photon lines connecting the initial with the final state, box diagrams and initial-final state interference contributions should be also included. Let us define the effective kernel σ_k. It contains all these non-factorizable diagrams. Special care must be devoted to the cancellation of infra-red singularities among initial, final and non-factorizable contributions. By closely following the recipe sketched above, the cross-section becomes:

$$\tilde{\sigma}(s) = \sigma_f(s) + \sigma_{box}(s) + \sigma_{int}(s) =$$

$$= \int_0^{r_{max}} dx\, \sigma_0(s')\, H_e(x,s)\, F_\mu(r_{max} - x, s') + \sigma_{box}(s) + \sigma_{int}(s) \qquad (13)$$

where x is the energy fraction radiated away.

$H_e(x,s)$, the initial state radiator as calculated by means of eq.(11), is [3]

$$H_e(x,s) = \Delta_e(s)\, \beta_e\, x^{\beta_e-1} - \frac{1}{2}\beta_e\,(2-x)$$

$$+ \frac{1}{8}\beta_e^2 \left[(2-x)\,[3\ln(1-x) - 4\ln x] - \frac{4\ln(1-x)}{x} - 6 + x \right] \qquad (14)$$

$F_\mu(t,s)$, the final state radiator, can be written as:

$$F_\mu(t,s) = \int_0^t dy\, H_\mu(y,(1-y)s) \approx$$

$$\approx \int_0^t dy\, H_\mu(y,s) = \Delta_\mu(s)\, t^{\beta_\mu(s)} - \beta_\mu(s)(t - \frac{1}{4}t^2)$$

$$-\frac{1}{2}\beta_\mu^2(s)\left[Li_2(1-t) + \ln(1-t)\ln t + \frac{3}{2}\ln(1-t)\left[(1-t) + \frac{1}{4}(t^2-1)\right]\right.$$

$$\left. + 2\ln t(t - \frac{1}{4}t^2) - \frac{1}{16}t^2 + \frac{5}{8}t - \zeta(2) \right] \qquad (15)$$

The second approximate equality corresponds to neglecting the y dependence of $L_\mu((1-y)s) = \ln \frac{(1-y)s}{m_\mu^2}$ in Δ_μ and β_μ. With this approximation we neglect hard $y \approx 1$ radiation from the final state and this corresponds to setting the scale of the final state to be s' instead of s''.

$\sigma_{box}(s)$ and $\sigma_{int}(s)$ respectively represent the contributions of non-factorizable virtual and real interference terms.

Consistently to the O(α) in the corrections, one can rewrite eq.(13) as

$$\sigma(s) = \int_0^{r_{max}} dx\, \sigma_k(s')\, H_e(x,s)\, F_\mu(r_{max} - x, s') \qquad (16)$$

where $\sigma_k(s)$ is the effective integration kernel as defined above and given by

$$\sigma_k(s) = \sigma_0(s) + \sigma_{box}(s) + \sigma_{int}(s).$$

By comparing eq. (16) with eq.(13) we see that further corrections are taken into account:

$$\sigma(s) = \sigma_f(s) + \int_0^{r_{max}} dx\, \left[\sigma_{box}(s') + \sigma_{int}(s')\right] H_e(x,s)\, F_\mu(r_{max} - x, s')$$

$\sigma(s)$ reduces to $\tilde\sigma(s)$ only if the $O(\alpha^0)$ in H_e and F_μ in the second term are taken into account [13].

To better understand the content of eq.(16), let us focus our attention on the first term on the r.h.s. of eq.(13). By substituting in eq.(13) the definition of $F_\mu(t,s)$ of eq.(15) and by properly splitting the integration domain in the $x - y$ plane by using a cut-off ϵ, three contributions can be defined.

a) The *soft* cross-section

$$\sigma_{soft}(s) = \int_0^{\epsilon} dx \int_0^{\epsilon-x} dy\, \sigma_0((1-x)s)\, H_e(x,s)\, H_\mu(y, (1-y)(1-x)s)$$

reduces to the following expression:

$$\sigma_{soft}(s) \approx \sigma_0(s) \int_0^{\epsilon} dx\, H_e(x,s) \int_0^{\epsilon} dy\, H_\mu(y,s) =$$

$$\sigma_0(s)\left[1 + 2\frac{\alpha}{\pi}\left[[-1 + \frac{\pi^2}{6} + \frac{3}{4}L_e + (L_e - 1)\ln\epsilon] + [-1 + \frac{\pi^2}{6} + \frac{3}{4}L_\mu + (L_\mu - 1)\ln\epsilon]\right]\right]$$

which represents the factorized part of the contribution of soft radiation from both initial and final states.

b) The *hard − initial* cross-section

$$\sigma_e(s) = \int_\epsilon^{r_{max}} dx \int_0^{r_{max}-x} dy\, \sigma_0((1-x)s)\, H_e(x,s)\, H_\mu(y,(1-y)(1-x)s)$$

by freezing the final state radiation ($H_\mu(y,s) = \delta(y)$) and expanding to $O(\alpha)$ reduces to

$$\sigma_e(s) \approx \int_\epsilon^{r_{max}} dx\, \frac{d\sigma_e}{dx}$$

with

$$\frac{d\sigma_e}{dx} = \sigma_0((1-x)s)\, \frac{\alpha}{\pi}(L_e(s) - 1)\, \frac{1+(1-x)^2}{x}$$

which is precisely the initial state bremsstrahlung spectrum.

c) The *hard − final* cross-section

$$\sigma_\mu(s) = \int_0^\epsilon dx \int_{\epsilon-x}^{r_{max}-x} dy\, \sigma_0((1-x)s)\, H_e(x,s)\, H_\mu(y,(1-y)(1-x)s)$$

by freezing initial state radiation ($H_e(x,s) = \delta(x)$) and expanding to $O(\alpha)$ as for the previous case becomes

$$\sigma_\mu(s) \approx \int_\epsilon^{r_{max}} dy\, \frac{d\sigma_\mu}{dy}$$

where

$$\frac{d\sigma_\mu}{dy} = \sigma_0(s)\, \frac{\alpha}{\pi}\left[L_\mu\big((1-y)s\big) - 1\right]\, \frac{1+(1-y)^2}{y}$$

is nothing but the final state hard bremsstrahlung spectrum. The result in eq.(16) is nothing but a factorized correction to initial and final states of an effective one-loop kernel. It reproduces the bulk of second-order contributions to the Born cross-section, resumming in addition the leading and next-to-leading singularities.

Recalling the definition of $F_\mu(z,s)$, one can see that the integrand in eq.(12) is the double differential cross-section

$$\frac{d^2\sigma}{dx\, dy} = \sigma_k((1-x)s)\, H_e(x,s)\, H_\mu(y,(1-x)(1-y)s) \qquad (17)$$

with x and y fractions of energy radiated away from initial and final states respectively. From eq.(17) one can derive the initial, final and total radiation spectra, resummed to all orders:

$$\frac{d\sigma_e}{dx} = \sigma_0((1-x)s)H_e(x,s) \qquad (18)$$

$$\frac{d\sigma_\mu}{dy} = \sigma_0(s)H_\mu(y,(1-y)s) \qquad (19)$$

$$\frac{d\sigma}{dz} = \int_0^z dx\, \sigma_k((1-x)s)\, H_e(x,s)\, H_\mu(z-x,(1-z+x)(1-x)s) \qquad (20)$$

Eq.(18) and (19) are obtained by freezing final and initial radiation respectively. eq.(20) is obtained by integrating eq.(17) over x and y with the constraint $x+y=z$.

The effect of the full set of the radiative corrections on the total cross-section in eq.(11) is to lower the peak, to shift its position and to raise a radiative tail at high energy with respect to the Born approximation $\sigma_0(s)$ [13].

Recently Aversa and Greco [14] by using the structure function formalism and the expression of the total cross section in eq.(16) have been able to derive an analytical solution of the same equation by explicitely including box and interference contributions.

TRANSVERSE MOMENTUM STRUCTURE FUNCTIONS

In order to deal with the problem of taking into account also the angles and the transverse momentum of the emitted photons, some extended distributions can be defined. Structure functions in eq.(1) and (2) can be generalized to take into account also the transverse degrees of freedom [15]. The evolution equation for $D(x, p_t; s)$ in the case of space-like kinematics, which is appropriate for the annihilation process, has the form:

$$D(x, p_t; s) = \delta(1-x)\delta^{(2)}(p_t) + \frac{\alpha}{2\pi} \int_{m^2}^{s} \frac{dk^2}{k^2+m^2} \int_x^1 \frac{dz}{z} P(z)$$

$$\int \frac{d^2 q_t}{\pi} \delta((1-z)k^2 + z(1-z)m^2 - q_t^2) D(\frac{x}{z}, p_t - \frac{x}{z}q_t; k^2). \qquad (21)$$

$D(x, p_t; s)$ represents the probability of finding inside a parent electron, at the scale s, an electron with fraction of longitudinal momentum x and transverse momentum p_t with respect to the initial beam direction. eq.(21) obeys transverse momentum conservation.

Let us solve eq.(21) by iterating the first term on the r.h.s., analogously to what can be done for the integrated $D(x,s)$ distribution. With one iteration of the source term, the solution at order α gives the following form for the distribution function:

$$D(x,p_t;s) = \delta(1-x)\delta^{(2)}(p_t) \\ + \frac{\alpha}{2\pi}P(x)\frac{1}{\pi}\frac{1}{p_t^2+(1-x)^2m^2}\Theta\left((1-x)s-p_t^2\right) + O\left(\frac{p_t^2}{E_\gamma^2}\right), \quad (22)$$

where Θ is the step function. $D(x,p_t;s)$ is normalized in such a way that, neglecting the electron mass m in the denominator, one has, by integrating over the transverse momentum degrees of freedom:

$$\int d^2p_t D(x,p_t;s) = D(x,s)$$

consistently at order α, so that the original longitudinal momentum distribution is recovered.

As for the x-dependent distributions, we define a p_t-dependent radiator $H(x,p_t;s)$ as:

$$H(x,p_t;s) = \int_{1-x}^{1} \frac{dz}{z} \int d^2k_t D_{e^-}(z,k_t;s) D_{e^+}(\frac{1-x}{z},p_t-k_t;s),$$

where the indices e^- and e^+ label radiation from the electron and the positron respectively. The expression for $H(x,p_t;s)$ is, at order α,

$$H^{(\alpha)}(x,p_t;s) = \delta(x)\delta^{(2)}(p_t) + \frac{\alpha}{2\pi}P(1-x) \\ \frac{1}{\pi}\left[\frac{1}{p_{te+}^2+x^2m^2} + \frac{1}{p_{te-}^2+x^2m^2}\right]\Theta\left(xs-p_t^2\right) + O\left(\frac{p_t^2}{E_\gamma^2}\right). \quad (23)$$

An angle-dependent radiator $H^{(\alpha)}(x,\cos\theta;s)$ can be also defined. The transverse momentum and the angle of the emitted photon, measured with respect to the incoming electron, are linked by the relations

$$p_{te-}^2 = 2EE_\gamma(1-\cos\theta) \quad p_{te+}^2 = 2EE_\gamma(1+\cos\theta).$$

One obtains, for the angle-dependent radiator, the expression:

$$H^{(\alpha)}(x,\cos\theta;s) = \frac{\alpha}{\pi}\frac{1+(1-x)^2}{x}\frac{1}{1+\frac{4m^2}{s}-\cos^2\theta} + O\left(\frac{m^2}{s}\right).$$

The matrix element in the last equation contains the leading soft and collinear singularities respectively represented by the $\frac{1}{x}$ and $\frac{1}{1-\cos^2\theta}$ poles. We can include also other less dominant (less singular) contributions by using the exact matrix element in the radiator, i.e. $H(x, \cos\theta; s)$ can be rewritten as [16]:

$$H^{(\alpha)}(x, \cos\theta; s) = \frac{\alpha}{2\pi} \frac{1}{x} \left[2 \frac{1 + (1-x)^2}{1 + \frac{4m^2}{s} - \cos^2\theta} - x^2 \right] + O\left(\frac{m^2}{s}\right).$$

This expression is more appropriate in the case when very hard photons ($x \approx 1$) are emitted at large angles ($\theta \approx \pi/2$). Moreover, for photons of energy $E_\gamma \approx 10$ GeV at $\sqrt{s} = 100$ GeV, the difference between the exact matrix element and the one given in eq.(23) is less than the 1% level, so that, to this accuracy, one can use the following form for the radiator:

$$H^{(\alpha)}(x, \cos\theta; s) = \frac{\alpha}{\pi} \frac{1 + (1-x)^2}{x} \frac{1}{1 - \cos^2\theta} \Theta\left(1 - \frac{4m^2}{s} - |\cos\theta|\right). \quad (24)$$

RADIATIVE NEUTRINO COUNTING

It has been proposed [17] that the number of neutrino families is a quantity that can be determined by counting photons produced in electron-positron annihilation around the Z^0 resonance peak.

As for the total cross-section case this formalism has the main advantage that it is able to take into account and resum, in a compact and straightforward way, initial as well as final state configurations that contain soft and collinear radiation.

The neutrino counting problem is related to different final states, i.e. those containing one or more isolated photons [17]. These, radiated only by the initial electrons, are detected. Those configurations define a quantity which is less inclusive than the total cross-section and, for the neutrino counting problem, states with one or more radiated photons should also be accounted for.

The problem is to determine the energy and the angle of emission of photons radiated by initial electrons and positrons accompanying the production of Z^0's. When a Z^0 decays into a $\nu\bar{\nu}$ pair, the photon is the only observed state. In the average event, however, together with this process also hard photons radiated in the very forward direction and soft ones radiated all over the solid angle are likely to be produced. The former photons are lost in the beam pipe or rejected according to the geometrical set-up of the apparatus; the second ones will not be observed for energies below the detector threshold.

Let us consider the cross-section for producing a single photon by an electron-positron pair, $e^+e^- \to Z^0, W \to \gamma\nu\bar{\nu}$. The bare spectrum has been computed by many authors [17]. The corresponding cross-section can be written as follows:

$$\frac{d^2\sigma_0}{dx\,dy} = \frac{G_F^2 \alpha s (1-x)\left[(1-\frac{x}{2})^2 + \frac{x^2 y^2}{4}\right]}{6\pi^2 x (1-y^2)}$$

$$\cdot \left(2 + \frac{N_\nu\left(g_v^2 + g_a^2\right) + 2\left(g_v + g_a\right)\left[1 - \frac{s(1-x)}{M_Z^2}\right]}{\left[1 - \frac{s(1-x)}{M_Z^2}\right]^2 + \Gamma_Z^2/M_Z^2}\right), \qquad (25)$$

where G_F is the Fermi coupling constant, $x = 2E_\gamma/\sqrt{s}$ is the fraction of energy carried away by the emitted photon, $y = \cos\theta$, with θ the angle of the photon, N_ν is the number of neutrinos and $g_v = -\frac{1}{2} + 2\sin^2\theta_W$ and $g_a = -\frac{1}{2}$. In eq.(25) the first term in the last factor comes from the square of the W exchange amplitude, the term containing N_ν from the square of the Z amplitude and the last one from the $Z - W$ interference.

Radiative corrections to $e^+e^- \to \gamma\nu\bar{\nu}$ have been evaluated with standard Feynman-diagram techniques [18,19] to the order α. The corresponding cross-section can be also computed by using the method of the structure functions for the initial fermions [20].

By using the radiator obtained in eq.(24), the cross-section for the process under consideration is

$$\frac{d^2\sigma_0}{dx\,dy} = H^{(\alpha)}(x,y;s)\,\sigma_0\left((1-x)s\right), \qquad (26)$$

where σ_0 is the "reduced" cross-section for the process $e^+e^- \to Z, W \to \nu\bar{\nu}$:

$$\sigma_0(s) = \frac{G_F^2 s}{12\pi}\left(2 + \frac{N_\nu\left(g_v^2 + g_a^2\right) + 2\left(g_v + g_a\right)\left[1 - \frac{s}{M_Z^2}\right]}{\left[1 - \frac{s}{M_Z^2}\right]^2 + \Gamma_Z^2/M_Z^2}\right). \qquad (27)$$

In order to analyze the various contributions to the radiative corrections, let us first consider the virtual and soft ones. By using the bare spectrum defined in eq.(26), we obtain the total bare cross-section for producing a real photon by integrating x from $x_{min} = 2E_{\gamma min}/\sqrt{s}$ to 1, where $E_{\gamma min}$ is the minimum detectable energy, and y from minus $\cos\theta_{min}$ to plus $\cos\theta_{min}$, where θ_{min} is the minimum detectable angle. The total bare cross-section $\sigma^{(\gamma)}$ is then given by the expression:

$$\sigma^{(\gamma)}(s) = \int_{x_{min}}^{1} dx \int_{-\cos\theta_{min}}^{\cos\theta_{min}} dy \, H^{(\alpha)}(x,y;s) \, \sigma_0\left((1-x)s\right). \tag{28}$$

The integral over y of $H^{(\alpha)}(x,y;s)$ is easily performed to give:

$$\int_{-\cos\theta_{min}}^{\cos\theta_{min}} dy \, H^{(\alpha)}(x,y;s) \approx$$

$$\frac{\alpha}{\pi} \frac{1+(1-x)^2}{x} \left[\ln\left(\frac{1+\cos\theta_{min}}{1-\cos\theta_{min}}\right) \cdot \Theta\left(1 - \frac{4m^2}{s} - \cos\theta_{min}\right) + \right.$$

$$\left. + \ln\frac{s}{m^2} \cdot \Theta\left(\cos\theta_{min} - 1 + \frac{4m^2}{s}\right) \right] = H^{(\alpha)}_{\theta_{min}}(x;s). \tag{29}$$

In order to evaluate the $O(\alpha)$ virtual and soft radiative corrections to the bare process, it is sufficient to correct the cross section $\sigma^{(\gamma)}$ by the $O(\alpha)$ radiator $H_{(1)}(x,s)$ as given by:

$$H_{(1)}(x,s) = \frac{\beta}{x} - \frac{\beta}{2}(2-x)$$

in the following way:

$$\sigma^{(\gamma)}_{(1)}(s) = \int_0^{x_{min}} d\xi \, H(\xi,s) \, \sigma^{\gamma}\left((1-\xi)s\right)|_{(\alpha)} =$$

$$= \int_0^{x_{min}} d\xi \, H_{(1)}(\xi,s) \left[\sigma^{(\gamma)}((1-\xi)s) - \sigma^{(\gamma)}(s)\right]$$

$$+ \sigma^{(\gamma)}(s) \int_0^{x_{min}} d\xi \, H(\xi,s)|_{(\alpha)}, \tag{30}$$

where $\sigma^{(\gamma)}_{(1)}$ is the $O(\alpha)$ corrected cross section. Since $H_{(1)}(x,s)$ is the $O(\alpha)$ expression of the radiator $H(x,s)$ of eq.(7), the all-orders soft and virtual radiative corrections [1-3] are taken into account by substituting, in the previous expression, $H(x,s)$ to $H_{(1)}(x,s)$. The fully corrected cross-section $\sigma_c^{(\gamma)}$ is given by:

$$\sigma_c^{(\gamma)}(s) = \int_0^{x_{min}} d\xi \, H(\xi, s) \, \sigma^{(\gamma)}((1-\xi)s). \tag{31}$$

By interchanging the order of integration over ξ and x and dropping the integral over x, one obtains the differential spectrum by using both the single bremstrahlung spectrum of eq.(24) and the radiator H_e:

$$\frac{d\sigma^{soft}}{dx} = H_{\theta_{min}}^{(\alpha)}(x,s) \int_0^{x_{min}} d\xi \, H(\xi, s) \, \sigma_0\left((1-x)(1-\xi)s\right). \tag{32}$$

eq.(34) describes the x spectrum of a real photon accompanied by soft and virtual radiation. Let us now consider the case of hard corrections to the bare process. By hard corrections we mean those due to photons that are detectable but that are lost for various reasons. These photons effectively contribute to the bare process and must be taken into account. The differential spectrum is given by:

$$\frac{d\sigma}{dx_1 dx_2 dy_1 dy_2} = H^{(\alpha)}(x_1,y_1;s) H^{(\alpha)}(x_2,y_2;(1-x_1)s) \sigma_0((1-x_1)(1-x_2)s).$$

By substituting to $H^{(\alpha)}(x,y;s)$ the matrix element as given in eq.(26), the double bremsstrahlung matrix element is recovered.

It is necessary to distinguish two kinds of hard corrections: a) hard photons lost in the pipe and b) hard photons parallel (within an apparatus-dependent resolution angle) to the observed one.

a) These photons are lost, being emitted at angles smaller than a *veto* angle. The collinear singularity gives rise to logarithmic contributions of the form $\frac{\alpha}{\pi}\ln(\frac{s}{m^2})$. As for the infra-red singularity, one should sum them to all orders. The production spectrum must be integrated from the forward 0 angle direction to the veto angle.

To $O(\alpha)$, the contribution coming from one photon lost in the pipe is given by:

$$\frac{d\sigma_{(1)}^{pipe}}{dx} = H_{\theta_{min}}^{(\alpha)}(x;s) \int_{\frac{x_{min}}{\sqrt{1-x}}}^{\sqrt{1-x}} dx_1 \cdot 2$$

$$\int_{\cos\theta_v}^{1-\frac{m^2}{s}} dy_1 H^{(\alpha)}(x_1,y_1;(1-x)s) \sigma_0\left((1-x_1)(1-x)s\right),$$

where $H^{(\alpha)}(x,y;s)$ is the angle-dependent radiator of eq.(24) and θ_v is the veto angle. The factor 2 takes into account the backward collinear singularity.

The expression in eq.(24) can be generalized in order to handle the collinear singularity resummation problem. By defining

$$H(x,y;s) = \Delta(s) \frac{2\alpha}{\pi} \frac{1}{1-y^2} x^{\beta(y)-1} [1+\beta(y)\ln(x)]$$

$$-\frac{\alpha}{\pi}\frac{1}{1-y^2}(2-x) + \left(\frac{\alpha}{\pi}\right)^2 \frac{1}{1-y^2} \ln\left(\frac{s(1-y^2)}{2m^2}\right) \cdot$$

$$\cdot \left[(2-x)\left[3\ln(1-x) - 4\ln x\right] - \frac{4\ln(1-x)}{x} - 6 + x\right], \qquad (33)$$

where $\beta(y)$ is given by

$$\beta(y) = \frac{2\alpha}{\pi} \ln\left(\frac{s(1-y^2)}{2m^2}\right).$$

The integral over y of $H(x,y;s)$ gives the resummed x-dependent radiator. The contribution to the spectrum, by integrating over the angular region as discussed before, is

$$\frac{d\sigma^{pipe}}{dx} = H^{(\alpha)}_{\theta_{min}}(x;s) \int_{\frac{x_{min}}{\sqrt{1-x}}}^{\sqrt{1-x}} dx_1 \cdot 2$$

$$\cdot \int_{\cos\theta_v}^{1-\frac{m^2}{s}} dy_1 H(x_1,y_1;(1-x)s)\sigma_0\left((1-x_1)(1-x)s\right).$$

The integral over y_1 is easily performed [20] to give

$$\frac{d\sigma^{pipe}}{dx} = H^{(\alpha)}_{\theta_{min}}(x;s) \cdot$$

$$\cdot \int_{\frac{x_{min}}{\sqrt{1-x}}}^{\sqrt{1-x}} dx_1 H^{pipe}(x_1;(1-x)s)\sigma_0\left((1-x_1)(1-x)s\right), \qquad (34)$$

where H^{pipe} is given by

$$H^{pipe}(x,s) \equiv 2 \cdot \int_{\cos\theta_{veto}}^{1-\frac{m^2}{s}} dy\, H(x,y;s) = \Delta(s)\,\beta_v\, x^{\beta_v-1} - \frac{1}{2}\beta_v(2-x)$$

$$+ \frac{1}{8} \beta_v^2 \left[(2-x) \left[3\ln(1-x) - 4\ln x \right] - \frac{4\ln(1-x)}{x} - 6 + x \right], \qquad (35)$$

with $\beta_v = \frac{2\alpha}{\pi} \ln \left(\frac{s(1-c_v)}{m^2} \right)$ and $c_v = \cos\theta_v$. b) photons parallel to the observed one. The contribution to the cross-section give by photons parallel to the observed one can also be computed. One has [20]:

$$\frac{d\sigma^{par}}{dx} = 2 \cdot \int_0^{c_m} dy H^{(\alpha)}(x,y;s) \int_{\frac{x_{min}}{\sqrt{1-x}}}^{\sqrt{1-x}} dx_1$$

$$\int_{yc_r - \sqrt{1-y^2} s_r}^{yc_r + \sqrt{1-y^2} s_r} dy_1 H^{(\alpha)}(x_1, y_1; (1-x)s) \, \sigma_0((1-x)(1-x_1)s), \qquad (36)$$

where $c_r = \cos\theta_r$, $s_r = \sin\theta_r$, θ_r being half of the resolution angle.

The contribution can be written as:

$$\frac{d\sigma^{par}}{dx} = 2 \cdot \int_0^{c_m} dy H^{(\alpha)}(x,y;s)$$

$$\int_{\frac{x_{min}}{\sqrt{1-x}}}^{\sqrt{1-x}} dx_1 H_{par}^{(\alpha)}(x_1, y; (1-x)s) \sigma_0((1-x)(1-x_1)s). \qquad (37)$$

The spectrum of the observed photon $\frac{d\sigma}{dx}$ is in fact given by the following sum:

$$\frac{d\sigma}{dx} = \frac{d\sigma^{soft}}{dx} + \frac{d\sigma^{pipe}}{dx} + \frac{d\sigma^{par}}{dx}, \qquad (38)$$

where the three contributions of the r.h.s. are respectively given by eqs. (32), (34) and (37).

FORWARD-BACKWARD ASYMMETRIES

Let us confine ourselves to the process $e^+e^- \to \gamma, Z^0 \to \mu^+\mu^-$. The zeroth order differential cross-section is given by

$$\frac{d\sigma_0}{d\Omega_\mu} = \frac{\alpha^2}{4s} \left[W_1(s)(1+c^2) + W_2(s)c \right], \qquad (39)$$

where $c = \cos\vartheta$, ϑ being the angle between the momenta of e^- and μ^-, and the functions $W_1(s)$ and $W_2(s)$ are given by

$$W_1(s) = 1 + \frac{2(s-M^2)sc_v^2}{|Z(s)|^2} + \frac{s^2(c_v^2+c_a^2)^2}{|Z(s)|^2},$$
$$W_2(s) = \frac{4(s-M^2)sc_a^2}{|Z(s)|^2} + \frac{8s^2c_v^2c_a^2}{|Z(s)|^2}, \qquad (40)$$
$$Z(s) = s - M^2 + iM\Gamma.$$

In eq.(40) M and Γ are the mass and width of the Z^0 respectively. c_a and c_v are given by

$$c_a = -\frac{1}{2\sin(2\vartheta_W)}, \qquad c_v = -c_a(4\sin^2\vartheta_W - 1).$$

The forward-backward asymmetry is defined as

$$A_0(s) = \frac{\sigma_0^{(+)}(s) - \sigma_0^{(-)}(s)}{\sigma_0(s)}, \qquad (41)$$

where $\sigma_0^{(+)}(s)$ and $\sigma_0^{(-)}(s)$ are the forward and backward hemisphere cross sections respectively, and $\sigma_0(s)$ is the total cross-section. An explicit calculation gives

$$\sigma_0^{(+)}(s) = \frac{\pi\alpha^2}{4s}\left[\frac{8}{3}W_1(s) + W_2(s)\right],$$
$$\sigma_0^{(-)}(s) = \frac{\pi\alpha^2}{4s}\left[\frac{8}{3}W_1(s) - W_2(s)\right], \qquad (42)$$
$$\sigma_0(s) = \frac{4\pi\alpha^2}{3s}W_1(s).$$

Inserting eq.(42) into (41) one obtains the expression for the Born approximation asymmetry:

$$A_0(s) = \frac{3}{8}\frac{W_2(s)}{W_1(s)}. \qquad (43)$$

The calculation of the QED radiative corrections is reduced to the evaluation of the structure function for the initial states [3-5]. In particular the corrected cross-section can be written as a convolution of a "bare" cross-section and a "radiator" H. The structure function approach can be generalized to include final state radiative corrections too [13]. The corrected cross-section can be written in the form [21]

$$\sigma(s) = \int_0^\varepsilon dx\, H_e(x,s) F_\mu(\varepsilon - x, (1-x)s)\, \sigma_0((1-x)s) + \int d\Omega_\mu \frac{d\sigma_B}{d\Omega_\mu}, \quad (44)$$

where $\frac{d\sigma_B}{d\Omega_\mu}$ is the box and interference contribution (see for example ref.[12]). H_e and F_μ represent initial and final state radiation respectively [13]. Radiative corrections to the Born approximation asymmetry are implemented by simply applying eq.(44) to the forward and backward hemisphere Born approximation cross-sections $\sigma_0^{(+)}$ and $\sigma_0^{(-)}$ of eq.(42)

$$\sigma^{(+)}(s) = \int_0^\varepsilon dx\, H_e(x,s) F_\mu(\varepsilon - x, (1-x)s)\, \sigma_0^{(+)}((1-x)s) + \int_0^1 d\Omega_\mu \frac{d\sigma_B}{d\Omega_\mu},$$

$$\sigma^{(-)}(s) = \int_0^\varepsilon dx\, H_e(x,s) F_\mu(\varepsilon - x, (1-x)s)\, \sigma_0^{(-)}((1-x)s) + \int_{-1}^0 d\Omega_\mu \frac{d\sigma_B}{d\Omega_\mu}, \quad (45)$$

and then combining the results according to the definition

$$A(s) = \frac{\sigma^{(+)}(s) - \sigma^{(-)}(s)}{\sigma(s)}. \quad (46)$$

eq.(45) takes into account the exact $O(\alpha^2)$ real and virtual photonic radiation from initial and final state, box and interference contributions and takes into account the resummation to all orders of the leading and part of next-to-leading logarithms both from initial and final state radiation. At the peak soft photon contributions represent the largest effect of the radiation. As for the line shape case the factorized form in eq.(46) is justified [21]. This result has been recently confirmed by an independent calculation [22].

By using the structure functions formalism it has been observed that initial-final interference contributions diagrams become important if photon-energy cuts are considered [21,23]. The contribution of the interference is, on the contrary, negligible when the radiation is inclusively integrated [21,23].

MONTE CARLO

Structure functions formalism has been applied to develop a Monte Carlo code for electromagnetic processes [24]. The structure and the results of the corresponding algorithm are discussed in detail elsewhere in these proceedings [25] and will not be mentioned here. It is only worth saying here that at the basis of this algorithm are the main features of the structure function approach as described above. The radiatively corrected cross-sections have the simple, factorized structure given by the convolution of a radiator with the corresponding Born cross-section. p_t-dependent structure functions are used to describe both the longitudinal and the transverse degrees of freedom in the electron evolution. Multiphoton exponentiation and next-to-leading factorization are naturally implemented [25].

FURTHER APPLICATIONS

The structure function approach to radiative corrections has shown the ability to deal with the infrared structure of the electromagnetic radiation in a simple and effective way. Among the possible further applications of this same formalism is the inclusion of the effects due to the interference between initial and final states [26]. The relevance of such extension of the structure function method is related to the evaluation, within this same formalism, of the radiative Bhabha process [27].

Acknowledgements

We would like to thank G. Bonvicini, S. Catani, M. Greco and G. Montagna for many useful discussions. I would also thank G. Bonvicini for the efforts devoted to organize this workshop and Laura Phillips for the efficient help given to us.

References

[1] G. Altarelli, in "Physics at LEP", CERN - Yellow Report, 86-02, J. Ellis and R. Peccei Editors, Geneva, February 1986; F. Gilman, SLAC-PUB-4002, June 1986, Talk presented at the Seventh Vanderbilt Conference on High Energy Physics, Nashville, Tennessee, May 15-17, 1986.

[2] E. Etim, G. Pancheri and B. Touschek, Nuovo Cimento 51B (1967) 276; G. Bonneau and F. Martin, Nucl. Phys. B27 (1971) 381; M. Greco, G. Pancheri-Srivastava and Y. Srivastava, Nucl. Phys. B101 (1975) 11, B171 (1980), 118; J. D. Jackson and D. L. Scharre, Nucl. Instruments and Methods 128 (1975) 13; F. A. Berends and R. Kleiss, Nucl. Phys. B178 (1981) 141; V. Baier, V. S. Fadin, V. Khoze and E. A. Kuraev, Phys. Rep. 78, 294 (1981); Y. S. Tsai, SLAC-PUB-3129 (1983), Presented at the Asia Pacific Conference, Singapore, June 12-18 1983; V. S. Fadin and V. A. Khoze, Yad. Fiz., 47 (1988) 1693.

[3] E. A. Kuraev and V. S. Fadin, Yad. Fiz. 41, 753 (1985) [Sov. J. Nucl. Phys. 41 (3), 1985, 466]. E. Kuraev, these proceedings; V. Fadin, these proceedings.

[4] G. Altarelli and G. Martinelli, in "Physics at LEP", CERN-Yellow Report, 86-02, J. Ellis and R. Peccei Editors, Geneva, February 1986.

[5] O. Nicrosini and L. Trentadue - Phys. Lett. 196B (1987) 551.

[6] L. Lipatov, Yad. Fiz. 20, 181 (1974)[Sov. J. of Nucl. Phys. 20, 94 (1975)]; V. Baier, V. S. Fadin and V. A. Khoze, Nucl. Phys. B65 (1973) 381; J. Kogut and L. Susskind, Phys. Rev. D9 (1974) 693, 3391.

[7] G. Altarelli and G. Parisi, Nucl. Phys. B126 (1977) 298.

[8] T. Kinoshita and J. Math. Phys., 3, 650 (1962); T. D. Lee and M. Nauenberg, Phys. Rev. 133 (1964) 1549.

[9] V. Gribov and L. Lipatov, Yad. Fiz. 15 (1972), 781, 1218 [Sov. J. Nucl. Phys, 15 (1972), 938, 675].

[10] R. Barbieri, J. A. Mignaco and E. Remiddi, Nuovo Cimento 11A (1972) 824.

[11] F. A. Berends, G. J. H. Burgers, and W. L. van Neerven, Phys. Lett. 185B (1987) 395; G. J. Burgers, Phys. Lett. 164B (1985) 167; F. A. Berends, G. Burgers, W. Hollik and W. L. van Neerven, Phys. Lett. B203 (1988) 177; J. P. Alexander, G. Bonvicini, P. S. Drell and R. Frey, Phys. Rev. D37 (1988).

[12] M. Greco, G. Pancheri and Y. Srivastava, Nucl. Phys. B101 (1975) 11, B171 (1980) 118, B197 (1982) 543; M. Greco - "Physics at LEP", CERN - Yellow Report, 86-02, J. Ellis and R. Peccei Editors, Geneva, February 1986; F. A. Berends, R. Kleiss and S. Jadach, Nucl. Phys. B202 (1982) 63.

[13] O. Nicrosini and L. Trentadue, Zeitsch. Phys. C39 (1988) 479.

[14] F. Aversa and M. Greco, LNF-Preprint 89/125 1989; M. Greco these proceedings.

[15] A. Bassetto, M. Ciafaloni and G. Marchesini, Phys. Rep. 100 (4) 1983; O. Nicrosini, L. Trentadue, Universitá di Pavia, preprint 1989.

[16] G. Bonneau and F. Martin, Nucl. Phys. B27 (1971) 381; see also F. A. Berends and R. Kleiss, Nucl. Phys. B260 (1985) 32.

[17] A. D. Dolgov, L. B. Okun and V. I. Zacharov, Nucl. Phys. B41 (1972) 197; V. S. Fadin and V. Khoze, ZhETF, Pis. Red. 17, 8, 438 (1973); E. Ma and J. Okada, Phys. Rev. Lett. 41 (1978) 287; K. J. F. Gaemers, R. Gastmans and F. M. Renard, Phys. Rev. D19 (1979) 1605; G. Barbiellini, B. Richter and J. L. Siegrist, Phys. Lett. 106B (1981) 414.

[18] M. Igarashi and N. Nakazawa, Nucl. Phys. B288 (1987) 301.

[19] F. A. Berends, G. J. H. Burgers, C. Mana, M. Martinez and W. L. van Neerven, Nucl. Phys. B301 (1988) 583.

[20] O. Nicrosini and L. Trentadue, Nucl. Phys. B318 (1989) 1.

[21] O. Nicrosini, L. Trentadue, Proceedings of the "Ringberg Workshop on Electroweak Radiative Corrections", Ringberg Castle, Germany, April 3-7, 1989, J.H. Kuehn ed., Springer Verlag 1989; G. Montagna, O. Nicrosini and L. Trentadue, CERN-TH 5445/89, Phys. Lett. B in press.

[22] D. Bardin, M. Bilenky, A. Chizov, A. Sazonov, Yu. Sedykh, T. Ricmann and M. Sachwitz, Phys. Lett. B229 (1989) 405;

[23] J. E. Campagne and R. Zitoun, LPNHE preprints 88-06, 88-08 (1988).

[24] G. Bonvicini and L. Trentadue, Nucl. Phys. B323 (1989) 253.

[25] G. Bonvicini, these proceedings.

[26] O. Nicrosini, these proceedings.

[27] F. Aversa, G. Bonvicini, M. Greco, O. Nicrosini, L. Trentadue, work in progress.

EXPONENTIATION OF LARGE CORRECTIONS IN QCD

George Sterman
Institute for Theoretical Physics
State University of New York, Stony Brook, NY 11794-3840

ABSTRACT

Logarithms in moments of the Drell Yan cross section with respect to the variable $\tau = Q^2/s$ exponentiate into a power which includes integrals of the running coupling. This result gives insight into the sensitivity of hard QCD cross sections to soft physics.

I. INTRODUCTION

Most applications of perturbative QCD depend on factorization theorems, in which hard scattering cross sections are written as the convolution of perturbatively calculable short distance coefficient functions with nonperturbative long distance parton densities. Schematically, the cross section for the production of a system of mass Q^2 in the collision of two hadrons $A(p_A)$ and $B(p_B)$ with $(p_A + p_B)^2 = s = Q^2/\tau$ is given by

$$\frac{d\sigma}{dQ^2} = \sigma_0 \sum_{ab} \int_\tau^1 \frac{dz}{z}\, \omega^{(f)}_{ab}(z, \alpha_s(Q^2))\, \mathcal{F}^{(f)}_{ab,AB}(\tau/z)\,, \qquad (1)$$

where the sum goes over all parton types, including quarks, antiquarks and gluons. The function $\mathcal{F}$ is a "parton luminosity",

$$\mathcal{F}^{(f)}_{ab,AB}(\tau/z) = \int dx_a dx_b\, \delta(x_a x_b - \tau/z)\, f_{a/A}(x_a, Q^2) f_{b/B}(x_b, Q^2)\,, \qquad (2)$$

with $f_{a/A}(x_a, Q^2)$ the density for partons of type a and fractional momentum x_a in hadron A. These parton densities must be detemined from experiment. Once this is done in one process, however, the factorization scheme gives absolute predictions for all other processes which involve the same parton densities. The factor σ_0 is conveniently defined as the Born cross section for the corresponding partonic process. Then, at lowest order in the QCD coupling, $\omega_{ab}^{(f)} = \delta(1-z)$, and to this approximation perturbative QCD is equivalent to the parton model. At higher orders, it is often referred to as the "QCD improved" parton model, and to the extent that the parton model is in semi-quanititative agreement with experiment, so is perturbative QCD.

Ideally, of course, one would like to see that QCD really does improve the parton model, and supplies quantitative theoretical predictions. This possibility depends on the size of low-order calculations for $\omega_{ab}^{(f)}$, and here one runs into immediate problems. Lowest order nontrivial corrections to the parton model are not in general small, while still higher order corrections are prohibitively difficult to calculate exactly, at least with present techniques. This is not the end of the story, however. The possibility remains that we may identify the origin of large corrections to perturbative cross sections, and use this knowledge to resum the perturbation series taking these large corrections into account. The model for this approach is the Bloch-Nordseick procedure for infrared divergences in QED.

In this talk, I will try to summarize how such a resumation can be carried out to all orders in perturbation theory for "large-moment" corrections to the Drell Yan (DY) process[1,2], when the cross section is normalized in terms of deeply inelastic scattering (DIS) structure functions. This is, in fact, a classic case, where large corrections are present at first order in perturbative QCD[3], and are also required by the comparison of the parton model to experiment[4]. Thus, I shall describe an attempt to calculate the "K-factor", which is to be understood as the ratio of the true DY cross section $d\sigma/dQ^2$ to the parton model result.

To be specific, consider the moments of eq. (1). Defining, for any function g,

$$\tilde{g}(n) = \int_0^1 dx\, x^{n-1} g(x) , \qquad (3)$$

we have

$$\int_0^1 d\tau\, \tau^{n-1} \frac{d\sigma}{dQ^2} = \sigma_0 \sum_{ab} \tilde{\omega}_{ab}^{(f)}(n) \tilde{f}_{a/A}(n, \alpha_s(Q^2)) \tilde{f}_{b/B}(n, \alpha_s(Q^2)) . \qquad (4)$$

For the Drell Yan cross section[5],

$$\tilde{\omega}_{ab}^{(f)}(n) = 1 + \frac{\alpha_s}{2\pi} C_F [2\ln^2(ne^{\gamma_E}) - 3\ln(ne^{\gamma_E}) + \frac{3\pi^2}{2} + 1 + f(n)] \,, \quad (5)$$

where $f(n)$ vanishes as $n \to \infty$, and γ_E is the Euler constant.

I will begin with a review of the origin of large corrections in the Drell Yan cross section, emphasizing the role of phase space. I will then describe a method of summation based on modified parton densities, tailored to the phase space of the Drell Yan and deeply inelastic scattering cross sections. These densities satisfy evolution equations, whose solution leads to an explicit expression for the large-moment behavior of the Drell Yan short distance function of eq. (4), in which all logarithms of n are exponentiated. I will briefly discuss some consequences of this result.

II. ORIGIN OF LARGE CORRECTIONS

Large corrections involve only[1] $\omega_{q\bar{q}}^{(f)}$, and the function $\omega_{q\bar{q}}^{(f)}(z, \alpha_s(Q^2))$ in eq. (1) is of the form

$$\omega_{q\bar{q}}^{(f)}(z, \alpha_s(Q^2)) = \delta(1-z) + \sum_{n=1}^{\infty} \left(\frac{\alpha_s(Q^2)}{\pi}\right)^n \left(\sum_{m=1}^{2n-1} a_{q\bar{q}}^{(mn)} \mathcal{D}_m(z)\right. \\
\left. + a_{q\bar{q}}^{(n0)} \delta(1-z) + b_{q\bar{q}}^{(n)}(z)\right) , \quad (6)$$

where the functions $b^{(n)}(z)$ are smooth functions of z, and the $\mathcal{D}_k(z)$ are plus distributions, defined by their integral with any smooth function $g(z)$,

$$\int_\tau^1 \mathcal{D}_k(z) g(z) = \int_\tau^1 \left(\frac{\ln^k(1-z)}{1-z}\right)_+ g(z)$$
$$\equiv \int_\tau^1 \left(\frac{\ln^k(1-z)}{1-z}\right)(g(z) - g(1)) - g(1)\frac{1}{k+1}\ln^{k+1}(1-\tau). \quad (7)$$

In the short-distance function, the positive term in the plus distribution comes from real gluon emission, the negative term from virtual gluon corrections.

At one loop, the large corrections to $\omega^{(f)}$ arise (i) from a large coefficient $a_{q\bar{q}}^{(n0)}$ for the $\delta(1-z)$ term and (ii) from the plus distributions themselves. In the following, I will discuss only the latter corrections, since our understanding of the former is much less complete. The two types of terms may in principle be separated experimentally, because τ dependence comes primarily from the plus distributions.

The contribution of plus distributions grows with $Q^2/s = \tau$, because as τ increases the lepton pair gets heavier, and there is less phase space available for the emission of gluons. This is only part of the story, however. The short-distance cross section ω in eq. (1) is calculated with $Q^2/\hat{s} = z$, where $\hat{s}$ is the invariant mass of the partons which initiate the hard scattering. The phase space for the perturbative cross section thus vanishes whenever $z = 1$, whatever the size of τ.

This effect is greatly magnified by the behavior of the parton distributions in eq. (2), in which the delta function fixes

$$z = \frac{\tau}{x_a x_b}. \tag{8}$$

In general the parton distributions behave as $x^a(1-x)^b$, where b is about 4 for a valence quark in a nucleon, and may be as large as 8 or 9 for sea quarks. This means that the distributions decay very rapidly for moderate x, even of the order 0.3 or 0.4. Because z must be less than unity, (8) enforces τ as a lower bound for the product $x_a x_b$, which forces, roughly speaking, the x's to be of the order of the square root of τ. When τ of 0.36, for example, the typical x must be of order 0.6, where the structure functions are both small and rapidly decreasing. In this way, the z integral in eq. (1) then gets most of its contribution from z of order unity, even at quite modest values of τ. Thus the plus distributions, which are singular at $z = 1$ can give large corrections even for τ quite far from unity. This accounts for the fact[6] that a "leading logarithm" aproximation in τ, in which $\mathcal{F}$ is approximated by a constant in (1) generally gives a poor approximation to the integral. The point is simply that $\mathcal{F}$ is not a constant.

Now let us discuss specifically how the plus distributions occur at one loop in the Drell Yan cross section. The one loop hard scattering function may be expressed as the difference between the one loop corrections to the DY and DIS cross sections, normalized to $\delta(1-x)$ to lowest order. For instance, for the quark-antiquark correction

$$\omega_{q\bar{q}}^{(F,1)}(z) = \frac{1}{\sigma_0}\frac{d\sigma}{dQ^2}^{(1)}(Q^2 = zs) - 2F_{qq}^{(1)}(z) , \qquad (9)$$

where $F_{qq} = F_{\bar{q}\bar{q}}$ is a normalized structure function for a quark in a quark. Fig. 1 illustrates how leading logarithm contributions cancel in Feynman gauge.

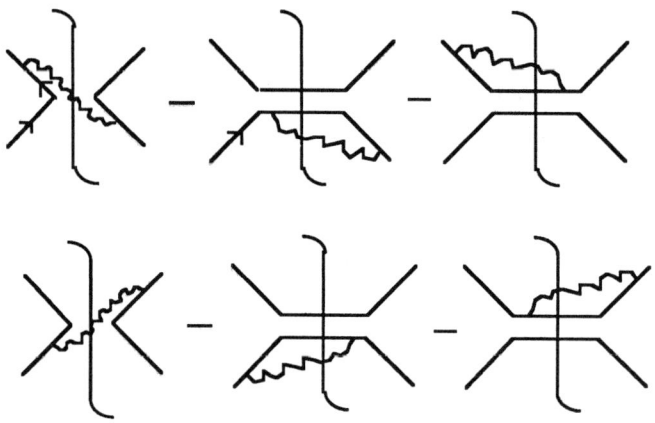

FIGURE 1

Real gluon diagrams which give rise to leading divergences in the DY and DIS cross sections. The vertical line represents the final state.

The leading contributions in these diagrams come about when the gluon becomes soft and nearly parallel to one of the quarks to which it attaches. In the (two) DY diagrams, these are the two incoming quarks. In the (four) DIS diagrams we have the same two incoming quarks, but also the two outgoing quarks. The divergences associated with the incoming quarks exactly cancel between the two sets of diagrams, but they leave over divergences associated with outgoing quarks in DIS. These divergences cancel against contributions from the two final state virtual diagrams of fig. 2, which have no analog in the DY process.

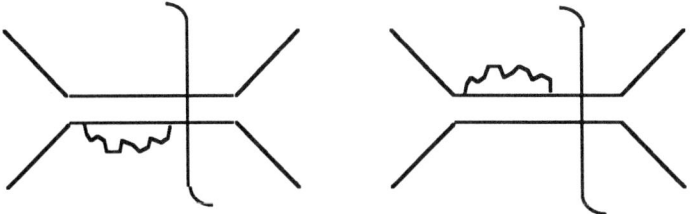

FIGURE 2

Virtual diagrams in DIS needed to cancel leftover divergences from fig. 1.

The cancellation of real and virtual final state contributions in DIS, however, is not exact, and it fails in the $x \to 1$ limit. Large corrections thus arise from regions of momentum space in DIS which are not present in DY[7]. This is the observation that makes it possible to organize these corrections, using the method sketched below.

III. ALTERNATE PARTON DENSITIES

To organize large corrections, we will use the important fact that the parton density $f_{i/J}(x)$ in eqs. (1) and (2) is not unique. Indeed, any two choices $f_{i/J}$ and $f'_{i/J}$ related by

$$f'_{a/J}(x) = \int_x^1 \frac{dy}{y} C_{ac}^{(f'f)}(x/y, Q^2) f_{c/J}(y, Q^2) \tag{10}$$

are equally acceptable, so long as $C_{ac}^{(f'f)}(x, Q^2)$ is itself a short-distance function. The short-distance coefficients corresponding to the two parton densities will be related by

$$\omega_{ab}^{(f')}(z) = \int_0^1 \frac{d\xi_1}{\xi_1} \frac{d\xi_2}{\xi_2} C_{ac}^{(f'f)}(\xi_1) \omega_{cd}^{(f)}(z\xi_1\xi_2) C_{db}^{(f'f)}(\xi_2) . \tag{11}$$

I will exhibit below a new parton density, $\psi_{a/A}(x)$ which incorpretates the correct phase space for the DY process. $\psi_{a/A}(x)$ absorbs the leading corrections $\mathcal{D}_n(z)$ to $\omega^{(F_2)}$, in the sense that both real and virtual contributions to the corresponding short distance functions are individually infrared finite, and their sum is nonsingular as $z \to 1$.

The use of this strategy may be seen most easily in terms of moments. Applying eq. (4) to $f = F_2$, and $f = \psi$, we find

$$\tilde{\omega}^{(F)}(n) = \tilde{\omega}^{(\psi)}(n)[\tilde{\psi}(n)/\tilde{F}(n)]^2 \ . \qquad (12)$$

If ψ is constructed as above, $\tilde{\omega}^{(\psi)}(n)$ will be well-behaved as $n \to \infty$. The probelm of computing large corrections in $\tilde{\omega}^{(F)}(n)$ then reduces to computing the ratio $\tilde{\psi}(n)/\tilde{F}(n)$, which, as we shall see below, can be carried out by analyzing the DIS cross section in a similar fashion.

To identify the function ψ which does the job, we only need to note that near threshold in the center of mass of the partonic system, the phase space for the DY process is given by the restriction

$$1 - z = \frac{2\ell_0}{\sqrt{x_a x_b s}} \ , \qquad (13)$$

where $\ell^\mu = [x_a p_a + x_b p_b - Q]^\mu$ is the total momentum available for the emission of soft particles. This suggests that the appropriate parton density is defined at measured energy, rather than, as is more customary[8], measured fractional momentum. We thus take for the parton density the definition

$$\psi_{q/A}(x) = \frac{1}{6\pi 2^{5/2}} \int_{-\infty}^{\infty} dy_0 e^{-ixp_0 y_0} < A(p)|\bar{q}(x_0, 0)\gamma^+ q(0)|A(p) > , \qquad (14)$$

where $|A(p)>$ represents the hadronic state (in this case, spin 1/2) of momentum p^μ. This parton density is normalized to $\delta(1-x)$ at zeroth order. The "measured" quark has a fixed energy xp^0, while its spatial momenta are integrated over. Note that ψ is neither Lorentz nor gauge invariant. We evaluate it in $A_0 = 0$ gauge and, as suggested above, in the center of mass of the partonic subsystem. With these definitions, it absobs the leading plus distributions of the DY cross section[1].

In fact, to organize the full large-n behavior in $\tilde{\omega}^{(F)}(n)$ - or equivalently all plus distributions - in $\omega^{(F)}(z)$, it is necessary to use a slightly modified version of the parton luminosity, eq. (2). To this end, we introduce an eikonal distribution $U(x\sqrt{s})$ for "central" particles, defined to include soft gluons (and their interactions) without collinear divergences. In physical gauges, the corresponding infrared logarithms are associated with gluons which connect the two incoming particles and their fragments. At lowest order, $U(x\sqrt{s})$ is given by the graphs shown in fig. 3. At higher order it may be constructed by the "tulip-garden" method of separating infrared and collinear divergences introduced in ref. 9.

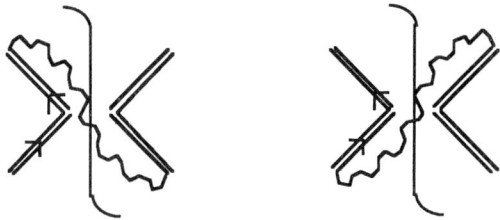

FIGURE 3

Lowest order contributions to the soft gluon function $U(x\sqrt{s})$.

The double lines represent eikonal propagators and vertices, of the form[9] $\pm v^\mu/(v \cdot q \pm i\epsilon)$, with v^μ an appropriate lightlike vector. In operator notation, the function $U(x\sqrt{s})$ is defined by the relation

$$\sum_n \delta(E_n - x\sqrt{s})| < 0|T(P\exp[ig\int_0^\infty d\lambda v_a \cdot A(\lambda v_a)]$$

$$\times P\exp[ig\int_0^\infty d\lambda v_b \cdot A(\lambda v_b)])|n > |^2 =$$

$$\int dx_1 dx_2 dx_3 U(x_1\sqrt{s})\delta(x - x_1 - x_2 - x_3)$$

$$\times \sum_{m_a} \delta(E_{m_a} - x_2\sqrt{s})| < 0|T(P\exp[ig\int_0^\infty d\lambda v_a \cdot A(\lambda v_a)])|m_a > |^2$$

$$\times \sum_{m_b} \delta(E_{m_b} - x_2\sqrt{s})| < 0|T(P\exp[ig\int_0^\infty d\lambda v_b \cdot A(\lambda v_b)])|m_b > |^2 .$$

(15)

Here $v_{a,b}$ are unit vectors in the direction of the incoming hadrons A and

B.

With these ingredients, the quark-quark DY cross section, which enables us to compute the function $\omega_{q\bar{q}}$, may be written in the form

$$\frac{d\sigma_{q\bar{q}}}{dQ^2} = \sigma_0 \int_\tau^1 \frac{dz'}{z'}\, \omega_{q\bar{q}}^{(\psi,U)}(z',\alpha_s(Q^2))\, \mathcal{F}_{qq,qq}^{(\psi,U)}(z/z') \,, \tag{16}$$

where now

$$\mathcal{F}_{qq,qq}^{(\psi,U)}(z/z') = \int dx_a dx_b dw\, \delta(x_a x_b(1-w) - z/z') \\ \times \psi_{q/q}(x_a,Q^2)\psi_{q/q}(x_b,Q^2)\, U(w\sqrt{\hat{s}}) \,, \tag{17}$$

which is analogous to eqs. (1) and (2). This approach may be tested by computing $\omega^{(\psi,U)}$ to one loop, and checking that it is free of plus distributions,

$$\omega_{q\bar{q}}^{(\psi,U)}(z',\alpha_s(Q^2)) = \delta(1-z') + \frac{\alpha_s(Q^2)}{\pi}[2C_F\delta(1-z') + (\frac{1+z'^2}{1-z'})\ln z' + 1] \,, \tag{18}$$

which is to be compared with the inverse moment of eq. (5),

$$\omega_{q\bar{q}}^{(F)}(z,\alpha_s(Q^2)) = \delta(1-z) + \frac{\alpha_s(Q^2)}{2\pi}[4C_F \mathcal{D}_1(z) + 3\mathcal{D}_0(z) \\ + [4\pi^2/3 + 1]\delta(1-z) + \tilde{f}(z)] \,, \tag{19}$$

with $\tilde{f}(z)$ a smooth and relatively small function. Up to corrections which vanish in the large-moment limit, $\omega^{(\psi,U)}$ is proportional to $\delta(1-z)$. In Ref. 1, it is treated as an overall factor in eq. (16), and is denoted by $|H_{DY}(Q)|^2$. The introduction of the new parton densities ψ and U thus succeeds in reducing the size of the short distance function drastically. This result is not immediately useful, because the function ψ is itself infrared sensitive. According to our comments after eq. (12), however, we expect that the ratio of moments of ψ to moments of the deeply inelastic scattering cross section are calculable. To implement this observation, we perform a similar decomposition for DIS.

The factorized form for a DIS structure function analogous to eqs. (16) and (17) is

$$F(x,Q^2) = |H(Q)|^2 \int_x^1 \frac{dy}{y}\phi(y) \\ \times \int_0^{y-x} \frac{dw}{1-w} V(wQ) J((y-x-w)Q^2) \,, \tag{20}$$

which is accurate up to terms which vanish in the high-moment limit. Without going into details, one may note that the function H_{DIS} is ultraviolet dominated, ϕ is analogous to ψ, and V to U, differing only in the phase space which defines them. For the DIS functions, the component of momentum parallel to the incoming hadron takes the place of the center-of-mass energy. The new feature in eq. (20) is the "jet" subdiagram J, which absorbs collinear singularities associated with the outgoing scattered quark. As suggested above, the bulk of the large corrections are associated with this jet, since this is the feature of DIS which is lacking altogether in DY.

The generalization of eq. (12), which gives the full large corrections to the moments of $\omega^{(F)}$ is now given by

$$\tilde{\omega}^{(F)}(n) = \frac{\tilde{\omega}^{(\psi,U)}(n)}{|H_{DIS}|^4}[\frac{\tilde{\psi}(n)}{\tilde{F}(n)}]^2 \left(\frac{\tilde{U}(n)}{\tilde{V}(n)^2}\right)\frac{1}{(\tilde{J}(n))^2} . \qquad (21)$$

The large-n dependence of each of the functions in this expression is derived in a similar fashion. Some details on the most important function, the jet distribution J in DIS, will serve to illustrate the method.

IV. BEHAVIOR OF THE JET DISTRIBUTION

The jet distribution can be shown to satisfy an evolution equation, which is analogous to the Altarelli-Parisi evolution equations[10], of the form

$$\left(\frac{\partial}{\partial \ln(1-x)} - \frac{1}{2}\beta(g)\frac{\partial}{\partial g} + 1\right) J((1-x)Q^2)$$
$$= -\frac{1}{2}\int_0^{1-x} dy \left[\frac{K_{J+}(g[(1-y)^2Q^2])}{1-y}\right]_+ J((y-x)Q^2) \qquad (22)$$
$$- \frac{1}{2}[S(g[Q^2]) - 2\gamma_q(g[Q^2])] J((1-x)Q^2) .$$

Here γ_q is the quark anomalous dimension in axial gauge, $-\alpha_s/\pi$, while K_{J+} and S are power series in the coupling, which to one loop are given by

$$K_{J,+}^{(1)} = S^{(1)} = (2\alpha_s/\pi)C_F . \qquad (23)$$

Corrections to eq. (22) are nonsingular as $x \to 1$, so that their moments vanish for large n.

An interative solution to eq. (22) is given by[1]

$$J((1-x)Q^2) = \sum_{n=0}^{\infty} \frac{(-1/2)^n}{n!} \prod_{i=1}^{n} \int_0^{\infty} dy_i t(y_j, Q^2) \, \delta(\sum_{j=1}^{n} y_j - (1-x)), \qquad (24)$$

where

$$t(y_i, Q^2) = \left[\frac{\int_{1-y_i}^{1} dy' [K_{J,+}(\alpha_s(y_i(1-y')Q^2)/(1-y')]_+}{y_i}\right]_+$$
$$+ \left[\frac{S(\alpha_s(y_iQ^2)) - 2\gamma_q(\alpha_s(y_iQ^2)) + \frac{1}{2}\beta(g)(\partial F(\alpha_s(y_iQ^2))/\partial g)}{y_i}\right]_+$$
$$- \delta(y_i)F(\alpha_s(Q^2)) \, . \qquad (25)$$

The function $F(\alpha_s)$ is a constant of integration, which is to be calculated order-by-order, and which ensures that J in eq. (24) gives exactly the (infrared-finite) imaginary part of the quark two-point function at invariant mass squared $(1-x)Q^2$. Note that at higher orders, F influences the distribution structure. At lowest order, it given by

$$F^{(1)} = -\frac{3\alpha_s}{2\pi} C_F \, . \qquad (26)$$

Up to corrections which vanish in the large-n limit, the moments of eq. (24) exponentiate,

$$\tilde{J}(n, Q^2) \equiv \int_0^1 dx \, x^{n-1} J((1-x)Q^2)$$
$$= exp\{\int_0^1 dz \, z^{n-1} t(1-z, Q^2)\} \, . \qquad (27)$$

All the other factors in eq. (21) behave in a similar manner.

V. EXPONENTIATION AND THE K-FACTOR

The complete expression for the moments of the short-distance function which follows from eq. (22) may now be expressed in exponential form as[2]

$$\tilde{\omega}_{q\bar{q}}^{(F)}(n, Q^2) = \sigma_0 (1 + \frac{2\alpha_s(Q^2)}{\pi} C_F) I(n) \, , \qquad (28)$$

where

$$I(n) = exp\{-\int_0^1 dx(\frac{x^{n-1}-1}{1-x}[\int_0^x dy \frac{g_1(\alpha_s[(1-x)(1-y)Q^2])}{1-y} \\ + g_2(\alpha_s[(1-x)Q^2]) + g_3(\alpha_s[(1-x)^2Q^2])])\} \ . \qquad (29)$$

The crucial feature of this result is that the functions g_i are power series in the running couplings of the indicated arguments, which organize all logarithms of Q^2 and other kinematic variables. The exponentiation is thus highly nontrivial. To first order, the g_i are given by

$$g_1(\alpha_s) = g_3(\alpha_s) = \frac{2\alpha_s}{\pi}C_F \ , \ g_2(\alpha_s) = -\frac{7\alpha_s}{2\pi}C_F \ . \qquad (30)$$

Essentially, g_1 and g_2 are associated with the jet function J described above, the quark anomalous dimension appearing in g_2. g_3 is associated with the ratio of central functions $\tilde{U}$ and $\tilde{V}$ in eq. (21). The contributions from the distributions ψ and ϕ cancel in the large-n behavior.

A number of comments may be made concerning eq. (29), what it says about the K-factor, and about the summation of large order corrections in general.

(1) Soft gluon contributions. The quantity $(1-x)Q$ in eq. (29) may be interpreted as the energy of a radiated gluon[1,2]. from this point of view, the exponentiated form includes contributions from gluons of arbitrarily small energy. But since $(1-x)Q$ appears in the argument of the running coupling, (29) is not well defined as it stands. It may, however, be made well-defined by introducing a cutoff which ensures that the integral never goes through the pole in the running coupling. Order-by-order in perturbation theory, the dependence of the resulting expression on the cutoff is higher twist[2], that is, proportional to extra inverse powers of Q. The introduction of a cutoff is necessary because the summation of perturbation theory does not commute with the $Q \to \infty$ limit. Such behavior is typical of perturbative QCD, which at best is expected to provide an asymptotic series[11].

(2) Soft gluons and the self-consistency of the resummation procedure. By inverting eq. (29) with a finite cutoff, it is possible to derive explicit predictions for the K-factor when DY structure functions are taken from

DIS. The resulting dependence on energy, τ and the perturbative cutoff can then be studied. The results have been described in Ref. 2. Cutoff dependence is indeed higher-twist, but is not always negligible for at fixed-target energies, especially for τ's of 0.2 or greater. This soft-gluon sensitivity is, as described above, due to the $x \to 1$ behavior of the parton distributions. For such energies, perturbative predictions for the K-factor, resummed or not, must be interpreted with care. It is gratifying, however, that at collider energies, soft-gluon dependence is quite small, of the order of tenths of a percent for τ's of 0.5 or less.

(3) Implications for two-loop calculations. If we expand the running couplings in eq. (29), it gives explicit predictions for the n-dependence of $\tilde{\omega}^{(F)}$ at order $\alpha_s^2(Q^2)$, due to the one-loop calculation alone. Inverting the moments then gives explicit predictions for plus distributions at two loops. These may be compared with the explicit two-loop calculations described in Ref. 6. It should be noted that this is a much more sensitive test of the resummation procedure than is comparison of the two-loop calculation with an inverted "leading logarithm" short-distance function, because it does not depend on the approximation that parton distributions are constants. Lorenzo Magnea and I are currently making the comparison, and preliminary results are encouraging[12].

(4) Two-loop anomalous dimensions. Catani and Trentedue[5] have recently noted that considerable information on nonleading logarithms may be gleaned from already existing computations of Sudakov effects. In particular, they argue, from a point of view related to the one described above, that next-to-leading logarithms in the exponent of eq. (28) are generated from the combination

$$exp[2 \int_0^1 dx \frac{x^{n-1}-1}{1-x} \{ \int_{Q^2}^{Q^2(1-x)} \frac{dk^2}{k^2} A(\alpha_s((1-x)k^2)) + \frac{3C_F}{2\pi} \alpha_s((1-x)Q^2) \}] , \quad (31)$$

where the function A is known up to two loops[13]. Although eq. (31) is not obviously of the form of eq. (29), it is possible, by relating the functions $K_{J,+}$ and A, to show that their predictions for the explicit two-loop calcualtion of Ref. 6 are the same[12]. For instance, it is interesting to note the correspondence between the variables $(1-y)Q^2$ in eq. (29) and k^2 in eq. (31). The precise relation between the two approaches, however, has not yet been fully explored.

(5) Abelian Theory. The same reasoning may be applied to cross sections and parton densities in QED cross sections such as[14] $e^+e^- \to Z^0$. In this case one is interested not so much in a K-factor as in a direct computation of the cross section. The method described above can be used to derive the singular $z \to 1$ behavior, which is associated entirely with the emisson of soft photons. As such, the leading behavior exponentiates according to the Yennie-Frautschi-Suura formalism[15]. Corrections will also exponentiate, as above, and will be due at order α^2 to the running of the QED coupling, and at order α^4 to photon-photon scattering. These topics are, of course, discussed at length and depth in other contributions to this conference.

ACKNOWLEDGEMENT

This work was supported in part by the National Science Foundation under Grant No. Phys-85-07627.

REFERENCES

1. G. Sterman, Nucl. Phys. B281, 310 (1987).
2. D. Appel, G. Sterman and P. Mackenzie, Nucl. Phys. B309, 259 (1988).
3. G. Altarelli, R. K. Ellis and G. Martinelli, Nucl. Phys. B157, 461 (1979); J. Kubar-Andre and F. E. Paige, Phys. Rev. D19, 221 (1979); B. Humpert and W. L. van Neerven, Phys. Lett. 84B, 327 (1979); K. Harada, T. Kaneko and N. Sakai, Nucl. Phys. B155, 69 (1979); (E) B165, 545 (1980).
4. I. R. Kenyon, Rep. Prog. Phys. 45, 1261 (1982).
5. S. Catani and L. Trentedue, Firenze preprint DFF 93/3/89 (1989) (unpublished).
6. T. Matsuura and W. L. van Neerven, Z. Phys. C38, 623 (1988), T. Matsuura, S. C. van der Marck and W. L. van Neerven, Phys. Lett. B211, 171 (1988).

7. G. Parisi, Phys. Lett. 90B, 295 (1980); G. Curci and M. Greco, Phys. Lett. 92B, 175 (1980).
8. G. Curci, W. Furmanski and R. Petronzio, Nucl. Phys. B175, 27 (1980); J. C. Collins and D. E. Soper, Nucl. Phys. B194, 445 (1982).
9. J. C. Collins and D. E. Soper, Nucl. Phys. B193, 381 (1981).
10. L. N. Lipatov, Sov. J. Nucl. Phys. 20, 94 (1975); G. Altarelli and G. Parisi, Nucl. Phys. B126, 298 (1977).
11. G. 't Hooft, in The whys of subnuclear physics, Erice 1977, A. Zichichi ed. (Plenum, New York, 1977); A. H. Mueller, Nucl. Phys. B250, 327 (1985).
12. L. Magnea and G. Sterman, in preparation.
13. J. Kodaira and L. Trentedue, Phys. Lett. 112B, 66 (1982); 123B, 335 (1983).
14. E. A. Kuraev and V. S. Fadin, Sov. J. Nucl. Phys. 41, 466 (1985); B. F. L. Ward, Phys. Rev. D36, 939 (1987); J. P. Alexander, G. Bonvicini, P. S. Drell and R. Frey, Phys. Rev. D37, 56 (1988); O. Nicrosini and L. Trentadue, Z. Phys. C39, 479 (1988).
15. D. R. Yennie, S. C. Frautschi and H. Suura, Ann. Phys. (N.Y.) 13, 379 (1961); G. Grammer, Jr. and D. R. Yennie, Phys. Rev. D8, 4332 (1973).

EXPONENTIATION AT THE EDGES OF THE PHASE SPACE

Stefano Catani
Istituto Nazionale di Fisica Nucleare
Sezione di Firenze, I-50125 Firenze, Italy

Luca Trentadue *
CERN, Geneva, Switzerland
and
Dipartimento Fisica, Università di Parma, Parma, and
INFN, Gruppo Collegato di Parma, Sezione di Milano, Italy

ABSTRACT

A general method to exponentiate leading and next-to-leading logarithms to all orders in perturbation theory is presented for the Quantum Chromodynamics (QCD) theory. Within this method that uses the structure function formalism and the eikonal approximation the cases of Deep Inelastic Scattering and Drell-Yan annihilation are explicitly considered. A complete formula for the large N-moments is given. The extension to the case of the Quantum Electrodynamics (QED) is presented for the radiative corrections to the initial state in the e^+e^- annihilation.

THE GENERAL METHOD

The perturbative QCD evaluation of physical quantities in semi-inclusive processes is characterized by the presence of large logarithmic corrections. We outline here a general method [1] to resum these contributions.

In semi-inclusive processes, i.e. inclusive hard processes with more than one large kinematical momentum scale, cross sections are given in terms of a large transferred momentum Q^2 and one or more ratios $z_i = Q^2/Q_i^2$ of the large scales Q_i^2. We choose z_i in such a way that the limit $z_i \to 1$ defines the semi-inclusive region corresponding to the boundary of the phase space allowed by the kinematics.

There are many examples of these processes which are physically relevant. The most familiar ones are Deep Inelastic Scattering (DIS) when $z = x_{\text{Bjorken}} \to 1$ and the Drell-Yan (DY) process when $z = \tau = Q^2/s \to 1$ [2]. We can also consider transverse momentum distributions of colorless bosons of mass Q^2 produced by Drell-Yan type mechanisms [3]. One can have photons or vector bosons produced by $q\bar{q}$ annihilation as well as Higgs bosons and heavy flavour resonances produced by gluon fusion.

For all these processes, in the semi-inclusive limit, the associate parton

---

* Talk given by L. Trentadue

radiation is strongly inhibited and large logarithmic corrections $\log(1-z)$ appear in the perturbative computation. The QCD perturbative expansion in the strong coupling constant α_s has a double logarithmic structure: besides the usual collinear logarithms $\log Q^2$ we get extra $\log(1-z)$ of infrared (IR) origin due to the depletion of soft gluon real emission. In order to get reliable theoretical predictions the double logarithms must be resummed to all orders in the QCD coupling constant. To this purpose the standard renormalization group analysis are no longer applicable in a straightforward way and we need different techniques to deal with both soft and collinear logarithmic singularities. In the following we shall show how our method applies to the DIS and DY processes. Similar analysis have been previously used for the evaluation of transverse momentum form factors [4] with single logarithmic accuracy.

According to the mass singularity factorization theorem the DY cross section can be written as follows

$$W(Q^2,\tau) = \frac{Q^2}{\sigma_0}\frac{d\sigma}{dQ^2} = \int_0^1 \frac{dx_1}{x_1}\frac{dx_2}{x_2} F(x_1,Q^2)F(x_2,Q^2)\Delta(\frac{\tau}{x_1 x_2},Q^2) \quad , \quad (1)$$

where σ_0 is the Born cross section, $\tau = Q^2/s$ and $F(x_i,Q^2)$ are the parton structure functions as measured in DIS. The coefficient function $\Delta(z,Q^2)$ is perturbatively calculable as a series in powers of the QCD coupling constant $\alpha_s(Q^2)$. By introducing the N-moments via the Mellin transform, eq.(1) factorizes and we get

$$\Delta_N(Q^2) = \int_0^1 dz\, z^{N-1}\Delta(z,Q^2) = \frac{W_N(Q^2)}{[F_N(Q^2)]^2} \quad . \quad (2)$$

The perturbative expansion for $\Delta_N(Q^2)$ in the large N limit is the following

$$\Delta_N(Q^2) \underset{N \to \infty}{=} 1 + \sum_{n=1}^{\infty}\sum_{m=0}^{2n} a_{n,m}\alpha_s^n(Q^2)\log^m N \quad (3)$$

The logN terms in eq.(3) are due to the Mellin transform of the singular contributions $\frac{\alpha_s^n}{1-z}\log^m(1-z)$ which appears in $\Delta(z,Q^2)$ when the semi-inclusive limit $z \simeq \tau \to 1$ is considered. In eq.(3) we define as *leading* the terms with $2n \geq m > n$. The *next-to-leading* ones are those with $m = n$, whilst for $m \leq n-1$ we have the next-to-dominant terms. In the following we shall show how the *leading* and *next-to-leading* contributions for Δ_N can be explicitly resummed. This is achieved by evaluating both F_N and W_N in eq.(3) with the following method.

(i) Firstly we make an appropriate gauge choice in order to decouple soft from collinear partons. (ii) Then we can use eikonal techniques to perform the resummation of the soft contributions and double logarithms. (iii) Finally,

Altarelli-Parisi type evolution equations for structure functions can be used in order to resum collinear logarithms [1].

Let us briefly discuss the application of this technique to DIS. By analyzing the scattering kinematics [1] in the semi-inclusive limit $x \to 1$ we conclude that (i) only soft gluons can be radiated by the initial state parton p and (ii) the leading region for the final state jet $\bar{p}$ is the collinear one. It follows that soft gluon emission can be evaluated by introducing the eikonal approximation [5]. At this point, in order to disentangle soft from collinear contributions, we work in an axial gauge $n \cdot A = 0$ with the gauge vector n_μ aligned along the $\bar{p}_\mu = p_\mu + Q_\mu$ direction. In this way soft gluons decouple from the leg $\bar{p}$ because of the structure of the eikonal vertices in the physical gauges. It follows that the IR singularities are completely under control within the soft contribution and the moments of the total cross section take the factorized form

$$F_N(Q^2) = F_N^{\text{soft}}(Q^2) F_N^{\text{collinear}}(Q^2) . \tag{4}$$

Now we can evaluate F_N^{soft} and $F_N^{\text{collinear}}$. F_N^{soft} takes into account soft gluons radiate by p_μ in the axial gauge with $n_\mu = \bar{p}_\mu$. By using the eikonal approximation its computation is quite direct [1] since eikonal cross sections exponentiate [5]. The evaluation of the remaining factor $F_N^{\text{collinear}}$, can be performed with standard structure function techniques [1].

The steps performed to compute the DIS structure function can be simply repeated [1] for the DY normalized cross section $W(\tau, Q^2)$ by taking into account the different kinematics. The final result we get for the coefficient function Δ_N in eq.(2) is the following

$$\Delta_N(Q^2) = \exp\left\{-\int_0^1 dx \frac{x^{N-1}-1}{1-x}\left[2\int_{(1-x)Q^2}^{Q^2}\frac{dk^2}{k^2}A(\alpha_s((1-x)k^2))+\right.\right.$$
$$\left.\left.+B(\alpha_s((1-x)Q^2))\right]\right\} + O(\alpha_s(\alpha_s \log N)^n) . \tag{5}$$

where the functions A and B are given by

$$A(\alpha_s) = \frac{\alpha_s}{\pi}A^{(1)} + \left(\frac{\alpha_s}{\pi}\right)^2 A^{(2)} ; \quad B(\alpha_s) = \frac{\alpha_s}{\pi}B^{(1)} \tag{6}$$

$$A^{(1)} = C_F, \quad A^{(2)} = \frac{1}{2}C_F K \quad K = C_A\left(\frac{67}{18} - \frac{\pi^2}{6}\right) - \frac{5}{9}N_f,$$

$$B^{(1)} = -\frac{3}{2}C_F. \tag{7}$$

Eq.(5) resums all the *leading* and *next-to-leading* contributions $\alpha_s^n(\log N)^m$, $n \leq m \leq 2n$, for Δ_N in the large N limit. This equation agrees with general results [6] about the exponentiation of all log N terms for Δ_N. By putting

$A^{(2)}$ and $B^{(1)}$ equal to zero into eq. (5) we recover the known leading order results [2,7]. The next-to-leading order terms in eq.(5) are obtained by using the coefficients $A^{(2)}$ and $B^{(1)}$ as given by eq.(7). We would like to stress that the pertubative expansion of eq.(5) in powers of $\alpha_s(Q^2)$ agrees with a complete two loop calculation recently performed [8]. We regard this agreement as a sensible check of our result.

The expression in eq.(5) allow us to discuss the range of validity of the perturbation theory in the semi-inclusive region. We see that the momentum index N cannot be extendend to extremely large values (corresponding to $\tau \simeq 1$) without moving the argument of $\alpha_s((1-x)k^2)$ outside from the perturbative regime. Within the context of our method such scale dependence of α_s arises [1] from the transverse momentum of the interacting partons. Therefore the resummation procedure leads to the introduction of the intrinsic transverse momentum q_t of the interacting partons within the hadron as the *natural* upper limit for $N(N \leq Q^2/q_t^2)$. In this way the cross section $W(\tau, Q^2)$, obtained by the inverse Mellin transform of eq.(5), will depend on this non-pertubative scale.

QED RADIATIVE CORRECTIONS TO e^+e^- ANNIHILATION

It is known [9,10] that around the Z^0 resonance the soft and collinear dynamics plays a crucial role in describing the form of the resonance peak. Both structure function techniques [10] and ordinary Feynman diagram calculations [11] have been used to compute electromagnetic radiative corrections to the line shape. The physical motivation is the accurate determination, according to the precise measurements, of the parameters of the Z^0 boson in the LEP/SLC experiments [9]. The contributions coming from soft and collinear photon radiation has been shown to sizeably affect the position and the shape of the peak.

The techniques developed in [1] can be applied to study this problem with an independent method. This approach allows to evaluate the structure of the non-dominant logarithmic terms arising from the mass singularities [1]. The e^+e^- annihilation cross section can be seen as an electrodynamical Drell-Yan process where the electrons, surrounded by the photon radiation, play the role of the annihilating partons. The cross section for e^+e^- annihilation is written as

$$\frac{d\sigma}{ds'} = \frac{1}{s}\sigma_0(s')W(s,\tau) \qquad (8)$$

where s is the invariant energy square, $\tau = s'/s$ and $W(s,\tau)$ is the radiator containing radiation contributions to the elementary Born cross-section $\sigma_0(s')$. The total cross-section is given by $\sigma = \int ds' \, d\sigma/ds'$

The perturbative expansion for $W(s,\tau)$ shows contributions of the type

$$\frac{\alpha^n}{(1-\tau)} ln^{k-1}(1-\tau)L^n, \quad (m+k \leq 2n)$$

with $L = ln(s/m_e^2)$ and m_e the electron mass and terms of the form

$$\alpha^n L^m f(1-\tau), \quad (m \leq n)$$

with $f(1-\tau)$ an integrable function for $\tau \to 1$. The contributions above can be classified as collinear-soft and non-soft respectively.

In order to accomplish the resummation of the logarithmic contributions, it is useful to consider instead of the $\tau \to 1$ limit, the large N behaviour of the N-moment distributions $W_N(s) = \int d\tau \tau^{N-1} W(s,\tau)$. In the N-moment expression energy conservation can be naturally implemented. $W_N(s)$ then reads

$$ln W_N^{IR}(s) = -\int \frac{d^3q}{4\pi\omega_q} \left(\frac{p_1}{p_1 \cdot q} - \frac{p_2}{p_2 \cdot q}\right)^2 A(\alpha(q_t^2))$$

$$[(1 - 2\frac{\omega_q}{\sqrt{s}})^{N-1} - 1]\theta(\sqrt{s}/2 - \omega_q) \tag{9}$$

where $\alpha(q^2)$ is the QED running coupling constant, p_1 and p_2 are the electron and positron momenta and the superscript IR means that eq.(9) takes into account *leading* and *next − to − leading* collinear-soft contributions.

When only the radiation of photons from the interacting leptons is considered we have $A(\alpha) = \alpha/\pi$ with α being the fine structure constant. In this case eq.(2) gives the usual result [9,10,11] of the soft photon exponentiation.

When also the production of real and virtual fermion pairs of mass m_f^2 is taken into account one has

$$A(\alpha(q^2)) = \frac{\alpha}{\pi}(1 + O(\alpha \frac{q^2}{m_f^2})), \quad q^2 \ll m_f^2 \tag{10a}$$

$$A(\alpha(q^2)) = \frac{\alpha(q^2)}{\pi}(1 + K_{QED}\frac{\alpha(q^2)}{2\pi}), \quad q^2 \gg m_f^2 \tag{10b}$$

where $\alpha(q^2) = \alpha/(1 - \frac{\alpha}{3\pi}lnq^2/m_f^2)$ ($q^2 \gg m_f^2$) and in the QED case we have $K_{QED} = -10/9$ for the contribution of each charged lepton pair. By inserting the expressions (10) into (9) we obtain

$$ln W_N^{IR}(s) = \frac{2}{\pi}\int_0^1 dz \frac{z^{N-1} - 1}{1 - z}[\alpha(L - 1) + \theta((1-z)^2 s - m_f^2)$$

$$\int_{m_f^2}^{s(1-z)^2} \frac{dq^2}{q^2}(\alpha(q^2)-\alpha-\frac{5}{9}\frac{\alpha^2(q^2)}{\pi})] \qquad (11)$$

The inverse Mellin transform of the first term in the square brackets gives the well known soft photon exponentiation factor $\beta(1-\tau)^{\beta-1}$ ($\beta=2\frac{\alpha}{\pi}(L-1)$) for $W(s,\tau)$.

The remaining terms prove that also the pair production contributions do exponentiate.

At finite order the coefficients of these terms coincide with the ones given in [11], since by expanding them up to the second order in α we get

$$(\frac{\alpha}{\pi})^2\int_0^1 dz\frac{z^{N-1}-1}{1-z}\theta((1-z)^2 s-m_f^2)[\frac{1}{3}ln^2\frac{(1-z)^2 s}{m_f^2}-\frac{10}{9}ln\frac{(1-z)^2 s}{m_f^2}]+O(\alpha^3) \qquad (12)$$

We stress that the pair production contribution differs qualitatively from the pure photonic first term by showing an explicit logarithmic $(\alpha ln(1-z))^n$ dependence that cannot be reproduced by the ordinary renormalization group equations [10,11].

REFERENCES

[1] S. Catani and L. Trentadue, Phys. Lett. 217B (1989) 539; Università di Firenze preprint DFF-93-3-1989 (Nucl. Phys. B in press.) and references therein.

[2] G. Parisi, Phys. Lett. 90B (1980) 295; G. Curci and M. Greco, Phys. Lett. 92B (1980) 175; Phys. Lett. 102B (1981) 280; M. Ciafaloni and G. Curci, Phys. Lett. 102B (1981) 352.

[3] G. Parisi and R. Petronzio, Nucl. Phys. B154 (1979) 427; Yu. L. Dokshitzer, D. I. Dyakonov and S. I. Troyan, Phys. Lett. 84B (1979) 234.

[4] J. Kodaira, L. Trentadue, Phys. Lett. 112B (1982) 66, 123B (1983) 335; SLAC-PUB-2934 (1982); S. Catani, E. d'Emilio, L. Trentadue, Phys. Lett. 211B (1988) 335.

[5] J. Frenkel and J. C. Taylor, Nucl. Phys. B246 (1984) 231; S. Catani, M. Ciafaloni and G. Marchesini, Nucl. Phys. B264 (1986) 558; G. P. Korchemsky and A. V. Radyushkin, Nucl. Phys. B283 (1987) 342.

[6] G. Sterman, Nucl. Phys. B281 (1987) 310; D. Appel, G. Sterman and P. Mackenzie, Nucl. Phys. B309 (1988) 259; G. Sterman, these proceedings.

[7] D. Amati, A. Bassetto, M. Ciafaloni, G. Marchesini and G. Veneziano, Nucl. Phys. B173 (1980) 429; P. Chiappetta, T. Grandou, M. Le Bellac, J. L. Meunier, Nucl. Phys. B 207 (1982) 251; E. Drukarev and E. M. Levin, Nucl. Phys. B262 (1985) 1.

[8] T. Matsuura and W. L. van Neerven, Z. Phys. C38 (1988) 623; T. Matsuura, S. C. van der Marck and W. L. van Neerven, Phys. Lett. 211B (1988) 171.

[9] E. Etim, G. Pancheri and B. Touschek, Nuovo Cimento 51B (1967) 276; M. Greco, G. Pancheri-Srivastava and Y. Srivastava, Nucl. Phys. B101 (1975) 11.

[10] E. A. Kuraev and V.S. Fadin, Sov. J. Nucl. Phys 41 (1985) 466; G. Altarelli and G. Martinelli, CERN-Yellow Report 86-02 "Physics at LEP", J. Ellis and R. Peccei eds. (1986); O. Nicrosini and L. Trentadue, Phys. Lett. 196B (1987) 551 and these proceedings.

[11] F.Berends, G.Burgers and W. L. van Neerven, Nucl. Phys. B297 (1988) 429; Errata ibidem B304 (1988) 921.

THE HELICITY METHOD: A REVIEW

R. Gastmans[*]

Instituut voor Theoretische Fysica, Universiteit Leuven, B-3030 Leuven, Belgium

ABSTRACT

The helicity method is presented as an efficient way of calculating bremsstrahlung processes in gauge theories at high energies. Recent developments are discussed as well as some unsolved problems.

WHY BREMSSTRAHLUNG?

A bremsstrahlung process is any process in which one or more gauge particles (photons, gluons, etc.) are radiated. In particle physics, they play an important role as can be seen from this nonexhaustive list of examples:

1. $e^+e^- \to \mu^+\mu^-\gamma$, $e^+e^-\gamma$ and $\gamma\gamma\gamma$: these processes are often examined as tests of the SU(2) × U(1) model. The process $e^+e^- \to e^+e^-\gamma$ is also important for the determination of the beam luminosity of e^+e^- colliders.

2. $e^+e^- \to \mu^+\mu^-\gamma\gamma$ and $e^+e^-\gamma\gamma$: these processes are often analyzed in searches for excited leptons.

3. $e^+e^- \to q\bar{q}g, q\bar{q}gg, \ldots$: these processes lead to 3, 4, ..., -jet production in e^+e^- collisions.

4. $qq \to qqg$, $q\bar{q} \to ggg$, $gg \to ggg$, $gggg, \ldots$: these processes are important for testing perturbative QCD, as they describe jet production in hadronic collisions.

5. $gg \to (q\bar{q})g$: this process is responsible for the production of heavy quarkonia at large momentum transfers.

For the study of bremsstrahlung processes, it is useful to have compact cross section formulae which allow more refined analyses. The helicity method [1] provides a convenient way for obtaining such compact formulae, at least in the high energy limit, where the lepton and quark masses can often be neglected in a first approximation. The helicity method is to be contrasted with the standard method of covariant summation over the polarization degrees of freedom. For complicated processes, where many Feynman diagrams contribute, the standard method is not so practical, as I shall show in the coming section.

---

[*]Onderzoeksdirecteur NFWO, Belgium.

PROBLEMS WITH THE STANDARD METHOD

There are four main reason why the standard Feynman techniques are rather cumbersome for applications to bremsstrahlung processes:

1. Usually, one deals with many Feynman diagrams for bremsstrahlung processes. Let n be the number of Feynman diagrams in such a case, then one is led to consider $n(n+1)/2$ interference terms in calculating the cross section, each of which may be fairly complicated. E.g., $e^+e^- \to e^+e^-\gamma\gamma$ is described by 80 Feynman diagrams, if one includes Z-exchange, hence, one has to struggle with 3240 interference terms, which lead to rather lengthy trace calculations.

2. In the high energy limit, one can often neglect fermion masses. Yet, with the standard method, this only leads to a slight simplification. One would like to think that substantial improvements can be made, but this is not so obvious when one uses the standard techniques.

3. Admittedly, with symbolic manipulation programs, it is possible to obtain cross section formulae for complicated processes. But, these formulae are usually very lengthy and it becomes a formidable task to use energy-momentum conservation to simplify the answer. Furthermore, it is by no means obvious what are the most convenient variables in which the answers should be expressed.

4. If one also wants to include the effects of initial state polarization, the standard method becomes even more arduous.

In the coming section, I hope to show that the helicity method offers an attractive alternative, which takes care of the above problems in a rather simple way.

THE HELICITY METHOD

<u>Photon Polarizations</u>

Suppose a photon with four-momentum k is radiated in a process. Let q_+ and q_- be two different light-like four-vectors, both different from k. Then,

$$\epsilon_\mu^{\prime\|} = 2\sqrt{2}N_q[(q_+ \cdot k)q_{-\mu} - (q_- \cdot k)q_{+\mu}],$$

$$\epsilon_\mu^\perp = 2\sqrt{2}N_q \epsilon_{\mu\alpha\beta\gamma}q_+^\alpha q_-^\beta k^\gamma, \qquad [\epsilon_{0123} = +1], \qquad (1)$$

with

$$N_q = \tfrac{1}{4}[(q_+ \cdot q_-)(q_+ \cdot k)(q_- \cdot k)]^{-\tfrac{1}{2}}, \qquad (2)$$

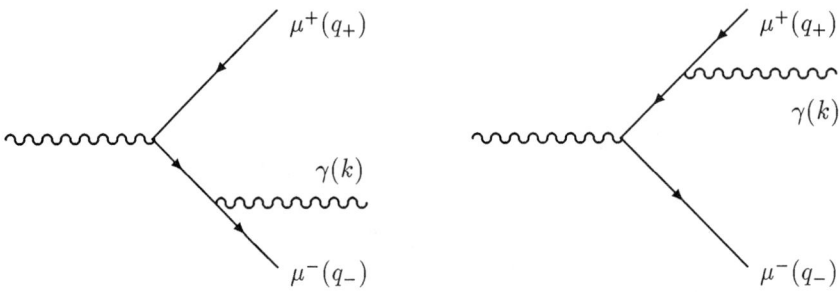

Figure 1: Feynman diagrams for photon radiation by a $\mu^+\mu^-$-pair.

are two orthogonal polarization vectors for the photon: they are orthogonal to k and properly normalized. As we shall see, it is convenient to combine the two linear polarizations into circularly polarized states [1]:

$$\epsilon'^{\pm}_\mu = \frac{1}{\sqrt{2}} \left(\epsilon'^{\parallel}_\mu \pm i\epsilon^{\perp}_\mu \right) . \tag{3}$$

The reason is that, for QED, only the combination $\not{\epsilon} \equiv \epsilon_\mu \gamma^\mu$ occurs in Feynman amplitudes, and, using the identity,

$$i\gamma^\mu \epsilon_{\mu\alpha\beta\gamma} = (\gamma_\alpha \gamma_\beta \gamma_\gamma - \gamma_\alpha g_{\beta\gamma} + \gamma_\beta g_{\alpha\gamma} - \gamma_\gamma g_{\alpha\beta})\gamma_5 , \tag{4}$$

we can write

$$\not{\epsilon}'^{\pm} = -N_q[\not{k} \not{q}_- \not{q}_+(1 \pm \gamma_5) - \not{q}_- \not{q}_+ \not{k}(1 \mp \gamma_5) \mp 2(q_+ \cdot q_-) \not{k}\gamma_5]. \tag{5}$$

Because of axial current conservation for massless fermions, we can omit the last term proportional to $\not{k}\gamma_5$, so that effectively we can use

$$\not{\epsilon}^{\pm} = -N_q[\not{k} \not{q}_- \not{q}_+(1 \pm \gamma_5) - \not{q}_- \not{q}_+ \not{k}(1 \mp \gamma_5)]. \tag{6}$$

Consider now what happens, e.g., in the process where a muon pair in the final state radiates a photon (see Fig. 1).

The Feynman amplitudes then contain the expressions

$$\begin{aligned} M_1 &= \bar{u}(q_-) \not{\epsilon} \frac{\not{q}_- + \not{k}}{2(q_- \cdot k)} \gamma_\mu v(q_+), \\ M_2 &= \bar{u}(q_-)\gamma_\mu \frac{-\not{q}_+ - \not{k}}{2(q_+ \cdot k)} \not{\epsilon} v(q_+), \end{aligned} \tag{7}$$

where γ_μ is the vertex connecting the muon line to the rest of the diagram.

For the case of a positive helicity photon, ϵ^+, we have, upon insertion of Eq.(6) into Eqs.(7),

$$M_1 = -N_q \bar{u}(q_-) \not{k} \not{A}_- \not{A}_+ (1+\gamma_5) \frac{\not{A}_- + \not{k}}{2(q_- \cdot k)} \gamma_\mu v(q_+)$$

$$= -N_q \bar{u}(q_-) \not{A}_+ (\not{A}_- + \not{k}) \gamma_\mu (1 + \gamma_5) v(q_+),$$

$$M_2 = N_q \bar{u}(q_-) \gamma_\mu \frac{-\not{A}_+ - \not{k}}{2(q_+ \cdot k)} \not{A}_- \not{A}_+ \not{k}(1 - \gamma_5) v(q_+)$$

$$= -N_q \bar{u}(q_-) \gamma_\mu (\not{A}_+ + \not{k}) \not{A}_- (1 - \gamma_5) v(q_+). \tag{8}$$

Both expressions were simplified using the Dirac equation for the spinors:

$$\bar{u}(q_-) \not{A}_- = 0, \qquad \not{A}_+ v(q_+) = 0. \tag{9}$$

It is now clear from Eqs.(8) that if we evaluate these expressions for definite muon helicities, only one of the two expressions, M_1 or M_2, can yield a nonvanishing contribution. Indeed, defining the fermion helicities through

$$u_\pm(p) = \tfrac{1}{2}(1 \pm \gamma_5) u_\pm(p),$$

$$v_\pm(p) = \tfrac{1}{2}(1 \mp \gamma_5) v_\pm(p), \tag{10}$$

for all $p^2 = 0$, we see that the helicity projection operators $(1 \pm \gamma_5)/2$ either annihilate the M_1 or M_2 contribution. Furthermore, the μ^+ and the μ^- helicities must necessarily be opposite. We are thus led, in a natural way, to consider helicity amplitudes for bremsstrahlung processes.

A complete calculation of the helicity amplitude $M(+,-;+,-,+)$ for the process

$$e^+(p_+) + e^-(p_-) \to \mu^+(q_+) + \mu^-(q_-) + \gamma(k), \tag{11}$$

with radiation from the muon lines only, then shows that

$$M(+,-;+,-,+) = \frac{e^3}{2s} N_q \bar{v}(p_+) \gamma^\mu (1 - \gamma_5) u(p_-)$$

$$\times \bar{u}(q_-) \gamma_\mu (\not{A}_+ + \not{k}) \not{A}_- (1 - \gamma_5) v(q_+). \tag{12}$$

Here, the helicity labels in $M(+,-;+,-,+)$ correspond respectively to those of the e^+, e^-, μ^+, μ^- and γ, and $s = 2(p_+ \cdot p_-)$.

Elimination of Repeated Indices

Our next task is to simplify the expression for the helicity amplitudes by eliminating the repeated index, which is often present in these expressions. To this end, we observe that, in the helicity method, we can freely insert a polarization sum in expressions of the type

$$(1-\gamma_5)u(p)\bar{u}(p)(1+\gamma_5) = (1-\gamma_5)\left(\sum_{\text{pol}} u(p)\bar{u}(p)\right)(1+\gamma_5)$$

$$= 2(1-\gamma_5)\not{p}. \quad (13)$$

This is so because the helicity projection operators annihilate the contribution of the "wrong" polarization state of the fermion.

Let us illustrate the usefulness of Eq.(13) in the expression (12) for the helicity amplitude $M(+,-;+,-,+)$. We have

$$A = \bar{v}(p_+)\gamma^\mu(1-\gamma_5)u(p_-)\bar{u}(q_-)\gamma_\mu(\not{q}_+ + \not{k})\not{q}_-(1-\gamma_5)v(q_+)$$

$$= \bar{v}(p_+)\gamma^\mu(1-\gamma_5)u(p_-)\frac{\bar{u}(p_-)\not{a}(1-\gamma_5)u(q_-)}{\bar{u}(p_-)\not{a}(1-\gamma_5)u(q_-)}$$

$$\times \bar{u}(q_-)\gamma_\mu(\not{q}_+ + \not{k})\not{q}_-(1-\gamma_5)v(q_+)$$

$$= 4\frac{\bar{v}(p_+)\gamma^\mu \not{p}_- \not{a} \not{q}_- \gamma_\mu (\not{q}_+ + \not{k})\not{q}_-(1-\gamma_5)v(q_+)}{\bar{u}(p_-)\not{a}(1-\gamma_5)u(q_-)}$$

$$= -8\frac{\bar{v}(p_+)\not{q}_- \not{a} \not{p}_- (\not{q}_+ + \not{k})\not{q}_-(1-\gamma_5)v(q_+)}{\bar{u}(p_-)\not{a}(1-\gamma_5)u(q_-)} \quad (14)$$

Choosing $a = p_+$ and using momentum conservation, we have

$$A = -16(p_+ \cdot q_-)\frac{\bar{v}(p_+)\not{p}_-(\not{q}_+ + \not{k})\not{q}_-(1-\gamma_5)v(q_+)}{\bar{u}(p_-)\not{p}_+(1-\gamma_5)u(q_-)}$$

$$= -16(p_+ \cdot q_-)\frac{\bar{v}(p_+)\not{p}_- \not{p}_+ \not{q}_-(1-\gamma_5)v(q_+)}{\bar{u}(p_-)\not{p}_+(1-\gamma_5)u(q_-)}$$

$$= -32(p_+ \cdot q_-)(p_+ \cdot p_-)\frac{\bar{v}(p_+)\not{q}_-(1-\gamma_5)v(q_+)}{\bar{u}(p_-)\not{p}_+(1-\gamma_5)u(q_-)}. \quad (15)$$

Substituting the expression (15) for A in $M(+,-;+,-,+)$, Eq.(12), we obtain

$$M(+,-;+,-,+) = -8e^3 N_q (p_+ \cdot q_-)(p_+ \cdot p_-)\frac{\bar{v}(p_+)\not{q}_-(1-\gamma_5)v(q_+)}{\bar{u}(p_-)\not{p}_+(1-\gamma_5)u(q_-)}, \quad (16)$$

and, if we are only interested in the absolute value of this expression, we have

$$|M(+,-;+,-,+)|^2 = 4e^6 N_q \frac{(p_+ \cdot q_-)^2}{(p_+ \cdot p_-)(q_+ \cdot k)(q_- \cdot k)}$$

$$= 16e^6 s'^2 N_q^2 \frac{u^2}{ss'}. \quad (17)$$

Here, we introduced the notation

$$s = 2(p_+ \cdot p_-), \qquad s' = 2(q_+ \cdot q_-), \qquad u = -2(p_+ \cdot q_-). \tag{18}$$

Radiation from Different Fermion Lines

For a complete calculation of the process (11), one has to include radiation from the electron lines as well. In order to have the same type of simplifications for the evaluation of the diagrams where the photon is radiated from the electron lines, it is now essential to use an expression for ϵ in terms of the external momenta of the electron and positron. More explicitly, one would like to use

$$\not{\epsilon}^\pm(p_+, p_-) = -N_p[\not{k}\,\not{p}_+\,\not{p}_-(1\pm\gamma_5) - \not{p}_+\,\not{p}_-\,\not{k}(1\mp\gamma_5)],$$

$$N_p = \tfrac{1}{4}[(p_+\cdot p_-)(p_+\cdot k)(p_-\cdot k)]^{-\frac{1}{2}}, \tag{19}$$

for those diagrams. Thus, when the photon is radiated from the muon line, it is advantageous to express ϵ in terms of the momenta q_+ and q_- of the muons, but, for radiation from the electron line, one would prefer to express ϵ in terms of the electron momenta, p_+ and p_-.

It is perfectly possible to use $\epsilon(q_+, q_-)$, Eq.(6), for the set of diagrams where the photon is attached to the muon line and to use $\epsilon(p_+, p_-)$, Eqs.(19), for the remaining diagrams provided one takes into account a simple phase factor [2]. To see this, consider the photon to be moving along the z-axis. As the diagrams with radiation from the muon line and from the electron line form separately gauge invariant sets, we can make gauge transformations such that $\epsilon(p_+, p_-)$ and $\epsilon(q_+, q_-)$ only have components in the xy-plane. But then, as these vectors have the same norm, they can at most differ by a phase factor (and terms proportional to the photon momentum arising from the gauge transformations). Thus,

$$\epsilon_\mu^\pm(q_+, q_-) = e^{\pm i\phi}\,\epsilon_\mu^\pm(p_+, p_-) + \beta_\pm k_\mu. \tag{20}$$

Because of gauge invariance, the constants $\beta_\pm$ are irrelevant. The phase ϕ is given by the scalar product of $\epsilon_\mu^\mp(p_+, p_-)$ and $\epsilon_\mu^\pm(q_+, q_-)$:

$$e^{\pm i\phi} = -\epsilon_\mu^\mp(p_+, p_-)\,\epsilon^{\pm\mu}(q_+, q_-)$$

$$= -N_p N_q \text{Tr}[\not{p}_+\,\not{p}_-\,\not{k}\,\not{q}_-\,\not{q}_+\,\not{k}(1\mp\gamma_5)], \tag{21}$$

and must be taken into account.

One then finds for the above helicity amplitude

$$|M(+,-;+,-,+)|^2 = 16e^6|s'N_q + e^{i\phi}sN_p|^2\,\frac{u^2}{ss'}$$

$$= -2e^6\left[\frac{q_+\mu}{(q_+\cdot k)} - \frac{q_-\mu}{(q_-\cdot k)} - \frac{p_+\mu}{(p_+\cdot k)} + \frac{p_-\mu}{(p_-\cdot k)}\right]^2\,\frac{u^2}{ss'}. \tag{22}$$

With our choices for the photon polarization vectors (6) and (19), we can similarly work out the corresponding expressions for the remaining helicity configurations.

Application to QCD Processes

For application to QCD processes, one can in general not drop the $k\gamma_5$ term of Eq.(1) in the expression for the polarization vector $\not{\!\epsilon}$ of the gluon. Instead, one can use [3]

$$\not{\!\epsilon}^\pm = -N_q[\not{k}\,\not{q}_-\,\not{q}_+(1\pm\gamma_5)+\not{q}_+\,\not{q}_-\,\not{k}(1\mp\gamma_5)], \qquad (23)$$

which differs from Eq.(5) by a harmless term proportional to $\not{k}$.

Compared to the QED case, there are additional sources of complications due to the presence of 3- and 4-gluon vertices in QCD. We find it convenient to write the 3-gluon vertex in the following form:

$$V_{\mu\nu\rho}(p,q,k) = (p-q)_\rho g_{\mu\nu} + (q-k)_\mu g_{\nu\rho} + (k-p)_\nu g_{\rho\mu}$$

$$= \tfrac{1}{8}\mathrm{Tr}\{\not{p}_\mu[\gamma_\nu,\gamma_\rho](1\pm\gamma_5)\} + \tfrac{1}{8}\mathrm{Tr}\{\not{q}_\nu[\gamma_\rho,\gamma_\mu](1\pm\gamma_5)\}$$

$$+ \tfrac{1}{8}\mathrm{Tr}\{\not{k}_\rho[\gamma_\mu,\gamma_\nu](1\pm\gamma_5)\}, \qquad (24)$$

where one is free to choose the sign in front of the γ_5-matrix. When the 3-gluon vertex couples to an external gluon, one can always choose this sign so that one of the three terms vanishes upon insertion of the gluon polarization vector. This apparently complicated way of writing the 3-gluon vertex provides then a natural way of separating the diagram with the 3-gluon vertex into two parts which can nicely be combined with other diagrams to form gauge invariant subsets for the process under consideration.

In many cases, the 4-gluon vertex will simplify considerably. Suppose, e.g., that we have three external gluons attached to the 4-gluon vertex and that the three gluons with momenta k_1, k_2 and k_3 have helicities $+$, $-$ and $-$. We can then take

$$\not{\!\epsilon}^+(k_1) = N[\not{k}_1\,\not{k}_2\,\not{k}_3(1+\gamma_5) + \not{k}_3\,\not{k}_2\,\not{k}_1(1-\gamma_5)],$$

$$\not{\!\epsilon}^-(k_2) = N[\not{k}_1\,\not{k}_3\,\not{k}_2(1+\gamma_5) + \not{k}_2\,\not{k}_3\,\not{k}_1(1-\gamma_5)],$$

$$\not{\!\epsilon}^-(k_3) = N[\not{k}_1\,\not{k}_2\,\not{k}_3(1+\gamma_5) + \not{k}_3\,\not{k}_2\,\not{k}_1(1-\gamma_5)]. \qquad (25)$$

With the identity

$$(a\cdot b) = \tfrac{1}{4}\mathrm{Tr}[\not{a}\,\not{b}], \qquad (26)$$

one easily sees that
$$(\epsilon_i \cdot \epsilon_j) = 0, \qquad i,j = 1,2,3. \tag{27}$$

As a result, the 4-gluon vertex vanishes and the entire Feynman diagram with the 4-gluon vertex can be omitted for this helicity configuration.

Phases of Helicity Amplitudes

After the introduction of photon (gluon) polarization vectors and the elimination of the repeated indices, as explained in the previous sections, one obtains expressions for helicity amplitudes which generally contain spinors. To avoid having to perform matrix multiplications for the evaluation of these amplitudes, we rewrite these expressions as ordinary complex functions of the components of the different four-momenta in the process. This can be done as follows [4].

Let us choose a representation of the γ-matrices for which

$$\gamma_5 = \begin{pmatrix} \mathbb{1} & 0 \\ 0 & -\mathbb{1} \end{pmatrix}. \tag{28}$$

The helicity spinors then take the form

$$u_+(k) = v_-(k) = \begin{pmatrix} \sqrt{k_+} \\ \sqrt{k_-} e^{i\phi_k} \\ 0 \\ 0 \end{pmatrix}, \tag{29}$$

$$u_-(k) = v_+(k) = \begin{pmatrix} 0 \\ 0 \\ -\sqrt{k_-} e^{-i\phi_k} \\ \sqrt{k_+} \end{pmatrix}, \tag{30}$$

where, for any four-vector k with $k^2 = 0$, we introduced the notation

$$k_\pm = k_0 \pm k_z, \qquad k_\perp = k_x + ik_y = |k_\perp| e^{i\phi_k}. \tag{31}$$

With the help of Eq.(13), we can now reduce any spinor expression to a product of elementary spinor expressions of the type

$$\bar{u}(k_i)(1 \pm \gamma_5)u(k_j). \tag{32}$$

These expression are easily expressed in terms of the components of the four-vectors k_i and k_j using Eqs.(28) through (30), i.e.,

$$\bar{u}(k_i)(1 - \gamma_5)u(k_j) = \frac{2k_{i\perp}^* Z_{ji}}{k_{i-}\sqrt{k_{i+}k_{j+}}},$$

$$\bar{u}(k_i)(1 + \gamma_5)u(k_j) = \frac{2k_{j\perp} Z_{ij}^*}{k_{j-}\sqrt{k_{i+}k_{j+}}}, \tag{33}$$

with the definition

$$Z_{ij} = k_{i+}k_{j-} - k_{i\perp}^* k_{j\perp} . \qquad (34)$$

In this way, it is possible to express all helicity amplitudes as complex functions of the four-momenta in the process. Moreover, one keeps track of the phases of the helicity amplitudes, which is necessary if one wants to study polarization effects.

FURTHER DEVELOPMENTS

In recent years, several new developments of the helicity method have been worked out. They are either extensions of the method to new types of applications or efforts to improve and streamline the calculations even further.

Mass Corrections for Amplitudes

Special care must be taken for a correct description of collinear bremsstrahlung. This phenomenon occurs when a photon or gluon with four-momentum k is emitted in a direction which is close to the direction of an external fermion with four-momentum p, i.e., when $(p \cdot k) = \mathcal{O}(m^2)$, m being the fermion mass. In the massless fermion limit, we found that the squared absolute value of the helicity amplitude contains terms which are proportional to

$$|M|^2 \sim \frac{1}{(p \cdot k)} , \qquad (35)$$

as can be seen from Eq.(17), e.g. Such terms are in fact of order m^{-2} in the collinear situation. But then, terms proportional to $m^2/(p \cdot k)^2$ are of the same order of magnitude, and must be included in the cross section formula. In Ref. 5, a general method has been developed which allows one to take into account these mass correction terms at the level of helicity amplitudes to leading order in m.

Production of Heavy Quarkonia

The usefulness of the helicity method is by no means restricted to massless or nearly massless fermion theories. In Ref. 6, a study has been made of the helicity amplitudes for $g + g \rightarrow (q\bar{q}) + g$ processes. Here, $(q\bar{q})$ denotes a heavy quarkonium state which can be a $^1S_0$, $^3S_1$, $^1P_1$, $^3P_0$, $^3P_1$ or $^3P_2$ state, like, e.g., the J/ψ or the Υ. For these processes, the mass of the quarkonium state cannot be neglected. Yet, using the helicity formalism for the description of the gluons, it was found that reasonably short and transparent formulae for the cross sections could be obtained in spite of the large number of Feynman diagrams which had to be evaluated.

Supersymmetry

By embedding QCD in a SO(2) supersymmetric extension, Parke and Taylor[7] found several relations between helicity amplitudes for purely gluonic processes and helicity amplitudes involving the supersymmetric partners of the gluons. The main point is that usually helicity amplitudes for lower spin particles are easier to calculate. Supersymmetry then tells one how these amplitudes are related to the higher spin processes in which one is really interested.

Phase Choice of Polarization Vectors

In Eqs.(6) and (23), we introduced expressions for photon and gluon polarization vectors using two reference four-vectors q_+ and q_-. Xu, Zhang and Chang[8] noted that one such reference four-vector is sufficient. They observe that the four-vectors $\epsilon^\pm$, defined by

$$\epsilon^+_\mu(k,q) = \frac{\bar{u}(q)\gamma_\mu(1-\gamma_5)u(k)}{\sqrt{2}\,\bar{u}(q)(1+\gamma_5)u(k)},$$

$$\epsilon^-_\mu(k,q) = \frac{\bar{u}(q)\gamma_\mu(1+\gamma_5)u(k)}{\sqrt{2}\,\bar{u}(k)(1-\gamma_5)u(q)}, \quad (36)$$

satisfy all the requirements

$$(k \cdot \epsilon^\pm) = (\epsilon^+ \cdot \epsilon^+) = (\epsilon^- \cdot \epsilon^-) = 0,$$

$$(\epsilon^+ \cdot \epsilon^-) = -1. \quad (37)$$

Hence, the expressions (36) can also be used in the calculation of helicity amplitudes.

Furthermore, they observe that the transition from one reference four-vector, q, to another one, p, can be achieved with the formula

$$\epsilon^\pm_\mu(k,q) = \epsilon^\pm_\mu(k,p) + \beta_\pm k_\mu. \quad (38)$$

This relation is to be compared with Eq.(20): the main difference is the absence of the a phase factor relating $\epsilon^\pm(k,q)$ to $\epsilon^\pm(k,p)$. As a consequence, the bookkeeping of the phase factors for the different contributions to a specific helicity amplitude is somewhat simplified: it is now done automatically through the normalization factor of $\epsilon^\pm$ in Eqs.(36), which is complex.

Weyl-van der Waerden Formalism

The helicity states for fermions are in fact described by two-component spinors because of the helicity projection operators $(1 \pm \gamma_5)/2$. For this reason, one can nicely reformulate the helicity method using the Weyl-van der Waerden formalism [9], which explicitly refers to the nonvanishing components of the spinors only.

Thus, the helicity spinors of Eqs.(29) and (30) are replaced by

$$u_+(p) = v_-(p) \rightarrow p_A,$$

$$u_-(p) = v_+(p) \rightarrow p^{\dot{A}}, \qquad A, \dot{A} = 1, 2, \qquad (39)$$

where

$$p_A = \frac{1}{\sqrt{p_+}} \begin{pmatrix} p_+ \\ p_\perp \end{pmatrix}, \qquad p^{\dot{A}} = \frac{1}{\sqrt{p_+}} \begin{pmatrix} -p_\perp^* \\ p_+ \end{pmatrix}. \qquad (40)$$

The undotted indices refer to the upper components of the Dirac spinor, whereas the dotted indices refer to the lower components.

The raising and lowering of indices is done with the antisymmetric tensor

$$\epsilon_{\dot{A}\dot{B}} = \epsilon^{AB} = \begin{pmatrix} 0 & 1 \\ -1 & 0 \end{pmatrix}, \qquad (41)$$

i.e.,

$$p^A = p_B \epsilon^{BA} = \frac{1}{\sqrt{p_+}} \begin{pmatrix} -p_\perp \\ p_+ \end{pmatrix},$$

$$p_{\dot{A}} = \epsilon_{\dot{A}\dot{B}} p^{\dot{B}} = \frac{1}{\sqrt{p_+}} \begin{pmatrix} p_+ \\ p_\perp^* \end{pmatrix}. \qquad (42)$$

It then follows from Eqs.(40) and (42) that

$$\bar{u}_+(p) = \bar{v}_-(p) \rightarrow p_{\dot{A}},$$

$$\bar{u}_-(p) = \bar{v}_+(p) \rightarrow p^A, \qquad (43)$$

in this formalism.

One also defines a spinor product

$$<pq> = p_A q^A = \frac{p_\perp q_+ - p_+ q_\perp}{\sqrt{p_+ q_+}}, \qquad (44)$$

which allows one to write, e.g.,

$$\bar{u}(p)(1-\gamma_5)u(q) = 2 <pq>^*, \qquad (45)$$

which is to be compared with our previous formula given in Eq.(33).

With this formalism, one can now reformulate the entire helicity method. For more details, I refer to the article by Berends and Giele [10], where also applications to multi-gluon processes can be found.

Quantum Gravity

In quantum gravity, the gauge theory of gravitational interactions, the graviton is described by a polarization tensor $\epsilon^{\mu\nu}(k)$, which satisfies the following conditions:

$$\epsilon^{\mu\nu}(k) = \epsilon^{\nu\mu}(k), \qquad \epsilon^{\mu}_{\mu} = 0, \qquad k_{\mu}\epsilon^{\mu\nu} = 0. \qquad (46)$$

Using the polarization vectors $\epsilon^{\pm}_{\mu}(k)$, which we introduced for the photons or the gluons, it is possible to construct explicit representations for the graviton polarization tensor. It suffices to note that

$$\epsilon^{\pm}_{\mu\nu}(k) = \epsilon^{\pm}_{\mu}(k)\epsilon^{\pm}_{\nu}(k) \qquad (47)$$

satisfies all the requirements of Eqs.(46). The quantity $\epsilon^{+}_{\mu\nu}$ thus describes a helicity +2 graviton, and $\epsilon^{-}_{\mu\nu}$ a helicity -2 graviton.

An application of the helicity method to quantum gravity can be found in Ref. 11, where Su examined several single bremsstrahlung processes.

Massive Spin-1 Particles

An extension of the helicity method to processes involving massive spin-1 particles has been developed by Passarino [12]. To illustrate the method, consider the process

$$Z(p) \to e^{+}(p_{+}) + e^{-}(p_{-}) + \gamma(k). \qquad (48)$$

As a massive spin-1 particle has three polarization states, let us introduce the following three expressions

$$\begin{aligned}
\slashed{\epsilon}^{*}_{1}(p) &= N\left[\slashed{p}'\,\slashed{p}_{+}\,\slashed{p}_{-}(1+\gamma_{5}) + \slashed{p}_{-}\,\slashed{p}_{+}\,\slashed{p}'(1-\gamma_{5})\right], \\
\slashed{\epsilon}^{*}_{2}(p) &= N\left[\slashed{p}'\,\slashed{p}_{+}\,\slashed{p}_{-}(1-\gamma_{5}) + \slashed{p}_{-}\,\slashed{p}_{+}\,\slashed{p}'(1+\gamma_{5})\right], \\
\slashed{\epsilon}^{*}_{3}(p) &= \frac{1}{M_{Z}}\left[\slashed{p} - \frac{M_{Z}^{2}}{(p\cdot p_{-})}\,\slashed{p}_{-}\right],
\end{aligned} \qquad (49)$$

with

$$p' = p_{+} + k - \frac{(p_{+}\cdot k)}{(p\cdot p_{-})}p_{-}. \qquad (50)$$

The expressions $\slashed{\epsilon}^{*}_{i}(p)$, $i=1,2,3$, satisfy the relations $(p\cdot\epsilon^{*}_{i}) = 0$. Furthermore, they are orthogonal to one another, i.e.,

$$(\epsilon^{*}_{i}\cdot\epsilon_{j}) = 0, \qquad i \neq j, \qquad (51)$$

and they are normalized:

$$(\epsilon_i^* \cdot \epsilon_i) = -1, \qquad i = 1, 2, 3. \tag{52}$$

These expressions are thus possible polarization states for the Z-particle. It then suffices to evaluate the helicity amplitudes for each $\epsilon_i^*(p)$, $i = 1, 2, 3$, Eqs.(49), and to square the absolute values of the resulting expressions to obtain the cross section.

UNSOLVED PROBLEMS

<u>Collinear Fermions and Gluons</u>

In Ref. 5, we showed how the corrections due to a finite fermion mass could be taken into account at the level of the spin amplitudes. This allowed us to treat the case of photon or gluon emission in directions nearly collinear to a fermion direction to leading order in m/E. Here, m denotes the fermion mass and E is the incoming energy.

The helicity method developed there does not allow us to describe the cases of nearly collinear fermions or nearly collinear gluons, e.g., the process $e^+e^- \to e^+e^-\gamma$ with the outgoing $e^\pm$ near the beam direction.

To treat the case of collinear fermions, Berends, Daverveldt and Kleiss [13] propose to evaluate the spin amplitudes for a given process exactly, i.e., without neglecting the fermion masses. To this end, they first rewrite the massive spinors as

$$\begin{aligned} u_+(p) &= \frac{1}{\sqrt{2(p \cdot k_0)}} (\slashed{p} + m) u_-(k_0), \\ u_-(p) &= \frac{1}{\sqrt{2(p \cdot k_0)}} (\slashed{p} + m) \slashed{k}_1 u_-(k_0), \end{aligned} \tag{53}$$

where $u_-(k_0)$ is a negative helicity massless spinor. The four-vectors k_0 and k_1 must be generally positioned and satisfy the relations

$$k_0^2 = 0, \qquad k_1^2 = -1, \qquad (k_0 \cdot k_1) = 0. \tag{54}$$

The spin amplitudes can then be expressed in terms of the components of the different four-momenta in the usual way.

A similar procedure was followed by Maña and Miquel [14], who developed a Monte Carlo program taking into account the complete fermion mass dependence in the cross section.

Relation to Quaternions

As already stated in the section on the Weyl-Van der Waerden formalism, the helicity method is in fact based on two-dimensional spinor algebra, for which the Pauli matrices $\vec{\sigma}$ are fundamental. It turns out, however, that the matrices $-i\vec{\sigma}$ obey the same multiplication rules as the imaginary units e_i, $i = 1, 2, 3$, of the quaternions

$$e_i e_j = \epsilon_{ijk} e_k - \delta_{ij}, \qquad [\epsilon_{123} = +1] . \qquad (55)$$

Could it be that, for the evaluation of helicity amplitudes, a computer would need less time to perform the associated quaternionic multiplications than for the straightforward two-dimensional matrix multiplications? In that case, it would be very profitable to reformulate the helicity method in terms of quaternions.

Loops in Feynman Amplitudes

The helicity method produced some remarkable simplifications in the calculation of bremsstrahlung processes in the high energy limit. It seems natural to suppose that similar ideas could be fruitful also for the calculation of loop corrections at high energies and large momentum transfers. By explicitly working out a few examples, we found some simplifications leading to a smaller number of Feynman integrals for one-loop processes, but we did not succeed in developing a useful technique based on the helicity method for the Feynman integrals themselves.

CONCLUSION

To avoid all the stress
of the standard mess,
use helicity with success:
it's the road to happiness!

REFERENCES

1. P. De Causmaecker, R. Gastmans, W. Troost and Tai Tsun Wu, Phys. Lett. 105B, 215 (1981); Nucl. Phys. B206, 53 (1982).
2. F.A. Berends, R. Kleiss, P. De Causmaecker, R. Gastmans, W. Troost and Tai Tsun Wu, Nucl. Phys. B206, 61 (1982).
3. P. De Causmaecker, doctoral thesis, University of Leuven (1983).
4. D. Danckaert, P. De Causmaecker, R. Gastmans, W. Troost and Tai Tsun Wu, Phys. Lett. 114B, 203 (1982).
5. CALKUL Collaboration, F.A. Berends, P. De Causmaecker, R. Gastmans, R. Kleiss, W. Troost and Tai Tsun Wu, Nucl. Phys. B239, 382 (1984).
6. R. Gastmans, W. Troost and Tai Tsun Wu, Phys. Lett. 184B, 257 (1987); Nucl. Phys. B291, 731 (1987).

7. S.J. Parke and T.R. Taylor, Phys. Lett. 157B, 81 (1985); Nucl. Phys. B269, 410 (1986).
8. Z. Xu, D.-H. Zhang and L. Chang, Nucl. Phys. B291, 392 (1987).
9. H. Weyl, Gruppentheorie und Quantummechanik (Leipzig, 1928); B.L. van der Waerden, Goettinger Nachrichten (1929), 100.
10. F.A. Berends and W. Giele, Nucl. Phys. B294, 700 (1987).
11. S.-Q. Su, doctoral thesis, University of Leuven (1982).
12. G. Passarino, Nucl. Phys. B237, 249 (1984).
13. F.A. Berends, P.H. Daverveldt and R. Kleiss, Phys. Lett. 148B, 489 (1984); Nucl. Phys. B253, 441 (1985).
14. C. Maña and R. Miquel, proceedings of this workshop.

NEXT-TO-LEADING FACTORIZATION IN QED

O. Nicrosini

Istituto Nazionale di Fisica Nucleare, Sezione di Pavia, and
Dipartimento di Fisica Nucleare e Teorica dell'Università, Pavia, Italy

ABSTRACT

Some possible developments of the structure function formalism are in this paper considered. Transverse momentum dependent structure functions are introduced. The generalization of the evolution at the two-loop level is described. A simple solution to the problem of automatically taking into account box and interference contributions within the formalism is presented for a non resonant cross section.

INTRODUCTION

In the last few years a powerful approach to the calculation of QED radiative corrections in e^+e^- physics, the structure function approach, has been proposed by many authors[1,2]. The method is based on a QCD analogy: the incoming electrons are viewed as dressed by a cloud of soft and collinear photons, and e^+e^- annihilation is considered as the QED analogue of the QCD Drell-Yan process.

The structure function approach developed so far allows one to compute QED radiative corrections to a given process in an inclusive sense, that is summing over all the photonic degrees of freedom but the fraction of energy carried away by the radiation, x. It has been shown that QED radiative corrections can be computed with a high degree of accuracy, namely better than 1 %, as far as inclusive quantities such as the Z^0 line shape[1-3] or the integrated forward-backward asymmetries[4] are concerned. It is however possible to further generalize and improve the formalism: in this paper some feasible developments will be addressed.

A first generalization concerns the possibility of dealing with non inclusive quantities. It is in fact important for some processes, both from the theoretical and experimental point of view, to obtain exclusive expressions, i.e. expressions which allow to reconstruct the entire final state. An example of such a process is the neutrino counting reaction, $e^+e^- \to Z^0 \to \gamma(\nu\bar{\nu})$, where complete information on the final state configuration of the photon is necessary. In order to tackle this problem, it is necessary to use more exclusive structure functions. It is well known in QCD[5] that the ordinary evolution equations and structure functions can be generalized to include, besides the longitudinal degrees of freedom, also the transverse ones. Here we will investigate such a generalization for the QED case, writing down $p_\perp$-dependent evolution equations consistent with the accuracy required in e^+e^- physics at LEP/SLC energies.

A second possible improvement of the formalism is to consider higher order evolution. In the structure function approach as it stands, in fact, a one-loop Altarelli-Parisi vertex is employed. This allows to reproduce the $O(\alpha)$ bremmstrahlung spectrum exactly, but fails in taking into account some of the non leading logarithms starting from $O(\alpha^2)$. Albeit numerically negligible, at least at LEP/SLC energies, these contributions can however be accounted for by implementing a two-loop Altarelli-Parisi vertex[6]. In this paper an analytical expression for the second order bremmstrahlung spectrum in terms of the two-loop contribution to the evolution vertex will be given.

A last development of the structure function formalism concerns the possibility of taking into account initial-final state interference effects. If on one side the structure function approach is a very effective tool for computing radiative corrections to inclusive quantities, allowing a complete calculation of radiative corrections coming from initial and final states[7], one the other side its main limitation is represented so far by the lack of description of real and virtual interference effects between initial and final state radiation, which become important when differential quantities are considered. These effects have been up to now included at hand[7]. In this note we will show a possible solution to the problem. In particular a simple way out is presented for the case of a non resonant cross section.

LONGITUDINAL MOMENTUM DEPENDENT STRUCTURE FUNCTIONS

It is well known that radiative corrections to processes of the tipe $e^+e^- \to X$ can be treated in the formalism of the evolution equations for the initial fermions, in analogy with a Drell-Yan process. In the non-singlet approximation the evolution equation for the electron (positron) takes the form

$$D(x,s) = \delta(1-x) + \frac{\alpha}{2\pi} \int_{m^2}^{s} \frac{dk^2}{k^2} \int_{x}^{1} \frac{dz}{z} P(z) D(\frac{x}{z}, k^2), \qquad (1)$$

where $P(z)$ is the regularized $e \to e\gamma$ vertex, given by

$$P(z) = \frac{1+z^2}{1-z} - \delta(1-z) \int_{0}^{1} dt \frac{1+t^2}{1-t}. \qquad (2)$$

$D(x,s)$ is the probability of finding inside a parent electron, at the scale s, an electron with fraction of longitudinal momentum x.

In order to find solutions of eq. (1), various procedures are possible, according to the approximation level required[1-2]. For example, one can iterate eq. (1) to the desired perturbative order; following this way, the $O(\alpha)$ solution is

$$D(x,s) = \delta(1-x) + \frac{\alpha}{2\pi} \ln \frac{s}{m^2} P(x) \qquad (3)$$

It is however possible to find more accurate solutions. In particular, in ref. [2] a solution which resums to all orders the leading and next to leading singularities has been found.

Corrected cross sections can be computed as convolutions of a bare cross section with the structure functions for the initial fermions:

$$\sigma(s) = \int dx_1\, dx_2\, \sigma_0(x_1 x_2 s)\, D(x_1, s) D(x_2, s), \qquad (4)$$

where

$$s = 4E^2, \qquad E = \text{beam energy},$$

and x_1 and x_2 are the fractions of longitudinal momentum of the electron and positron respectively. The integrand in eq. (4) gives the double differential spectrum of the fractions of energy of the initial fermions:

$$\frac{d\sigma}{dx_1 dx_2} = \sigma_0(x_1 x_2 s) \, D(x_1, s) D(x_2, s).$$

Eq. (4) can be rewritten in the following form[1,2]:

$$\sigma(s) = \int dx \, \sigma_0((1-x)s) \, H(x,s), \tag{5}$$

where now $x = 1 - x_1 x_2$ is the fraction of energy radiated away and the radiator $H(x,s)$ is defined as:

$$H(x,s) = \int_{1-x}^{1} \frac{dz}{z} \, D(z,s) \, D(\frac{1-x}{z}, s). \tag{6}$$

The integrand in eq. (5) gives the differential spectrum of the fraction of energy of the radiated photons:

$$\frac{d\sigma}{dx} = \sigma_0((1-x)s) \, H(x,s),$$

so that the radiator $H(x,s)$ represents the probability of radiating, at a scale s, a fraction x of the centre of mass energy.

The evolution-equation approach can be used as an alternative method which completely agree with the standard Feynman-diagrams tecnique (see [2], [3] and refs. therein) and has also been generalized to both initial and final state corrections in e^+e^- annihilation at the Z^0 resonance[7].

TRANSVERSE MOMENTUM DEPENDENT STRUCTURE FUNCTIONS

Eq. (1) can be generalized to take into account the transverse degrees of freedom too (see ref. [5] and refs. therein). The evolution equation for D in differential form is:

$$\frac{\partial D(x, p_\perp, |k^2|)}{\partial k^2} = \frac{\alpha}{2\pi} \frac{1}{k^2} \int_x^1 \frac{dz}{z} P(z) \qquad (7)$$
$$\cdot \int \frac{d^2 q_\perp}{\pi} \delta(kinematics) D(\frac{x}{z}, p_\perp - \frac{x}{z} q_\perp; |k^2|).$$

$D(x, p_\perp; s)$ represents the probability of finding inside a parent electron, at the scale s, an electron with fraction of longitudinal momentum x and transverse momentum $p_\perp$ with respect to the beam direction. The δ function in eq. (7) insures energy-momentum conservation at the branching vertex; since the conservation law is dependent on the kinematics of the process, it is necessary to distinguish between the space-like and time-like case.

Space-like kinematics

The process in which an on shell fermion decays, emitting a radiation quantum, into an off shell one is described by space-like decay kinematics ($k^2 < 0$). In the frame in which the momentum of the initial fermion is along the z axis, an explicit calculation gives[5]:

$$-\frac{k^2}{z} = \frac{q_\perp^2}{z(1-z)}$$

Time-like kinematics

The time-like decay case ($k^2 > 0$) is, on the contrary, tipical of a process in which an off shell fermion goes towards the energy shell by emitting radiation quanta. In this situation the kinematical constraint is:

$$k^2 = \frac{q_\perp^2}{z(1-z)}$$

Inserting the kinematical relations into eq. (7), one can write down the following integral equation:

$$D(x, p_\perp; s) = \delta(1-x)\delta^{(2)}(p_\perp) + \frac{\alpha}{2\pi} \int_{m^2}^s \frac{dk^2}{k^2} \int_x^1 \frac{dz}{z} P(z) \qquad (8)$$
$$\times \int \frac{d^2 q_\perp}{\pi} \delta(\mathcal{K}(z)k^2 - q_\perp^2) D(\frac{x}{z}, p_\perp - \frac{x}{z} q_\perp; k^2),$$

where $\mathcal{K}(z)$ is a kinematics dependent function given by

$$\mathcal{K}(z) = \begin{cases} 1-z, & \text{space-like case;} \\ z(1-z), & \text{time-like case.} \end{cases} \qquad (9)$$

With one iteration of the source term in eq. (8), one is led to the order α solution

$$D(x, p_\perp; s) = \delta(1-x)\delta^{(2)}(p_\perp) + \frac{\alpha}{2\pi} P(x) \frac{1}{\pi} \frac{1}{p_\perp^2} + O\left(\frac{p_\perp^2}{E_\gamma^2}\right), \qquad (10)$$

which does not depend explicitly on the kinematics. $D(x, p_\perp; s)$ is normalized in such a way that

$$\int d^2 p_\perp D(x, p_\perp; s) = D(x, s)$$

consistently at order α, so that the integrated distribution of eq. (3) is recovered.

Following the same lines as in the x dependent case, one can write the corrected cross section in the following form:

$$\sigma(s) = \int dx_1 dx_2 d^2 p_{\perp 1} d^2 p_{\perp 2} \sigma_0(x_1 x_2 s) D_{e^+}(x_1, p_{\perp 1}, s) D_{e^-}(x_2, p_{\perp 2}, s). \qquad (11)$$

The integrand in eq. (11) defines the differential spectrum $\frac{d\sigma}{dx_1 dx_2 d^2 p_{\perp 1} d^2 p_{\perp 2}}$:

$$\frac{d\sigma}{dx_1 dx_2 d^2 p_{\perp 1} d^2 p_{\perp 2}} = \sigma_0(x_1 x_2 s) D_{e^+}(x_1, p_{\perp 1}, s) D_{e^-}(x_2, p_{\perp 2}, s).$$

Eq. (11) can be rewritten in the form:

$$\sigma(s) = \int dx d^2 p_\perp \sigma_0\left((1-x)s\right) H(x, p_\perp; s), \qquad (12)$$

where the $p_\perp$ dependent radiator $H(x, p_\perp; s)$ is given by the expression:

$$H(x, p_\perp; s) = \int \frac{dz}{z} d^2 p_{\perp 1} d^2 p_{\perp 2} \delta^{(2)}(p_{\perp 1} + p_{\perp 2} - p_\perp) \\ \times D_{e^-}(z, p_{\perp 1}; s) D_{e^+}(\frac{1-x}{z}, p_{\perp 2}; s) \qquad (13)$$

and again $x = 1 - x_1 x_2$ is the fraction of energy radiated away. The integrand in eq. (12) is the energy and transverse momentum spectrum of the emitted photon:

$$\frac{d\sigma}{dx d^2 p_\perp} = \sigma_0 \left((1-x)s\right) H(x, p_\perp; s),$$

so that the $p_\perp$-dependent radiator $H(x, p_\perp; s)$ represents the probability of radiating a photon with fraction of longitudinal momentum x and transverse momentum $p_\perp$. Performing one of the tranverse momentum integrations in eq. (13) one is left with the expression:

$$H(x, p_\perp; s) = \int \frac{dz}{z} d^2 k_\perp D_{e^-}(z, k_\perp; s) D_{e^+}(\frac{1-x}{z}, p_\perp - k_\perp; s),$$

Inserting in the last equation the $O(\alpha)$ iterative solution for $D(x, p_\perp; s)$ eq. (10), one obtains the $O(\alpha)$ expression for $H(x, p_\perp; s)$:

$$H^{(\alpha)}(x, p_\perp; s) = \delta(x)\delta^{(2)}(p_\perp)$$
$$+ \frac{\alpha}{2\pi} P(1-x) \frac{1}{\pi} \left[\frac{1}{p_{\perp_{e^-}}^2} + \frac{1}{p_{\perp_{e^+}}^2}\right] + O\left(\frac{p_\perp^2}{E_\gamma^2}\right). \qquad (14)$$

From the $p_\perp$ dependent radiator one can build an angle dependent radiator $H^{(\alpha)}(x, y; s)$, with $y = \cos\vartheta$, ϑ being the angle between the three-momentum of the emitted photon and the beam axis. The transverse momentum and the cosine of the angle of the emitted photon, measured with respect to the incoming electron, are proportional to one another:

$$p_{\perp_{e^-}}^2 \propto (1-y), \qquad p_{\perp_{e^+}}^2 \propto (1+y).$$

Using the last relation and taking into account the Jacobian of the tranformation from transverse momentum to angular variables one obtains, for the angle dependent radiator, the following expression:

$$H^{(\alpha)}(x, y; s) = \frac{\alpha}{\pi} \frac{1+(1-x)^2}{x} \frac{1}{1-y^2} + O\left(\frac{m^2}{E^2}\right). \qquad (15)$$

Eq. (15) describes the bremsstrahlung spectrum in the extreme relativistic limit.

THE IMPROVED ALTARELLI-PARISI VERTEX

The k^2 and z evolutions in eq. (8) are factorized in the form $P(z)/k^2$. Due to this factorization, eq. (15) does not take into account mass effects which shield the collinear singularity and become important at small angles. This defect can be overcome by defining an effective Altarelli-Parisi vertex in a non factorized form. Let's rewrite eq. (8) as follows:

$$D(x,p_\perp;s) = \delta(1-x)\delta^{(2)}(p_\perp) + \frac{\alpha}{2\pi} \int_{m^2}^{s} dk^2 \int_x^1 \frac{dz}{z} Q(z,k^2)$$
$$\cdot \int \frac{d^2q_\perp}{\pi} \delta(\mathcal{K}(z)k^2 - q_\perp^2) D(\frac{x}{z}, p_\perp - \frac{x}{z}q_\perp; k^2), \quad (16)$$

where $\mathcal{K}(z)$ is the kinematics dependent function given by eq. (9) and $Q(z,k^2)$ is an improved Altarelli-Parisi vertex. The form of $Q(z,k^2)$ can be determined so as to reproduce the bremmstrahlung spectrum to any degree of accuracy. Let's search for a formal solution of eq. (16). By iterating the source term as in the preceding section, one obtains the $O(\alpha)$ solution as follows:

$$D(x,p_\perp;s) = \delta(1-x)\delta^{(2)}(p_\perp) + \frac{\alpha}{2\pi} \frac{1}{\pi} \frac{1}{\mathcal{K}(x)} Q\left(x, \frac{p_\perp^2}{\mathcal{K}(x)}\right). \quad (17)$$

From eq. (17) it is possible to obtain the $p_\perp$-dependent radiator:

$$H^{(\alpha)}(x,p_\perp;s) = \delta(x)\delta^{(2)}(p_\perp) + \frac{\alpha}{2\pi} \frac{1}{\pi\mathcal{K}(1-x)}$$
$$\cdot \left[Q\left(1-x, \frac{p_{\perp e-}^2}{\mathcal{K}(1-x)}\right) + Q\left(1-x, \frac{p_{\perp e+}^2}{\mathcal{K}(1-x)}\right) \right] \quad (18)$$

In order to explicitly compute the improved vertex $Q(z,k^2)$ it is necessary to define the particular kinematical situation one is interested in. Let's focus our attention on the bremmstrahlung process from an incoming electron. The kinematics, of the space-like kind, is defined by:

$$p = (E, E, \mathbf{0})$$
$$q = (xE + \frac{p_\perp^2}{2xE}, xE, \mathbf{p}_\perp) \quad (19)$$

From eq. (19) one can compute the relation between the transverse momentum and the angle of the emitted photon. A straightforward calculation gives:

$$p_{\perp e^-}^2 = 2E^2 x^2 (1-y), \qquad p_{\perp e^+}^2 = 2E^2 x^2 (1+y). \tag{20}$$

Let's now assume for $Q(z, k^2)$ the following non factorized form:

$$Q(z, k^2) = \frac{1}{k^2 + g(z)m^2} \left[\frac{1+z^2}{1-z} + f(z) \frac{k^2}{s} \right], \tag{21}$$

where m is the electron mass and s is the Mandelstam variable. By inserting eqs. (20) and (21) into eq. (18) and taking into account the Jacobian of the tranformation from transverse momentum to angular variables, one is led to the angle dependent radiator

$$H^{(\alpha)}(x, y, s) = \frac{\alpha}{2\pi} \frac{1}{x} \left[2 \frac{1+(1-x)^2}{1 + \frac{m^2}{E^2} - y^2} - x^2 + O\left(\frac{m^2}{E^2}\right) \right], \tag{22}$$

provided that

$$g(z) = 1 - z \quad \text{and} \quad f(z) = -1. \tag{23}$$

The radiator of eq. (22) coincides with the one given in ref. [8] and with the ones given in ref. [9] apart from terms of order m^2/E^2.

So the final form of the improved Altarelli-Parisi vertex is

$$Q(z, k^2) = \frac{1}{k^2 + (1-z)m^2} \left[\frac{1+z^2}{1-z} - \frac{k^2}{s} \right], \tag{24}$$

The vertex of eq. (24) is able to reproduce the bremmstrahlung spectrum with accuracy of order m^2/E^2. This accuracy can be arbitrarily increased by properly modifying the vertex $Q(z, k^2)$.

The $p_\perp$ dependent structure functions method may open a novel approach to the problem of computing electromagnetic radiative corrections, allowing the control of the transverse degrees of freedom of the radiation. In particular it can be applied to the analitic computation of the radiative corrections to the process $e^+e^- \to Z^0 \to \gamma(\nu\bar{\nu})$ (neutrino counting)[8]. It can also be used for the construction of an exclusive algorithm of the Monte Carlo type for QED processes[10].

TWO-LOOP EVOLUTION IN QED

Let us rewrite eq. (1) in the form

$$D(x,s) = \delta(1-x) + \int_{m^2}^{s} \frac{dk^2}{k^2} \int_{x}^{1} \frac{dz}{z} \left[\frac{\alpha}{2\pi} P(z)\right] D(\frac{x}{z}, k^2), \qquad (25)$$

where $P(z)$ is the usual Altarelli-Parisi vertex at the one loop level given by eq. (2). It can be computed as the one loop electron self energy and is responsible for the exponentiation to all orders of perturbation theory of the order α radiation effect.

At the two-loop level one can write

$$\frac{\alpha}{2\pi} P(z) \to \left(\frac{\alpha}{2\pi}\right) P^{(1)}(z) + \left(\frac{\alpha}{2\pi}\right)^2 P^{(2)}(z), \qquad (26)$$

in which a two-loop contribution to the Altarelli-Parisi vertex has been introduced[6]. Inserting eq. (26) into eq. (25) one is left with

$$D(x,s) = \delta(1-x) + \int_{m^2}^{s} \frac{dk^2}{k^2} \int_{x}^{1} \frac{dz}{z} \left[\left(\frac{\alpha}{2\pi}\right) P^{(1)}(z) + \left(\frac{\alpha}{2\pi}\right)^2 P^{(2)}(z)\right] D(\frac{x}{z}, k^2). \qquad (27)$$

By iterating the source term to the second order in α one obtains:

$$\begin{aligned} D(x,s) = &\delta(1-x) + \int_{m^2}^{s} \frac{dk^2}{k^2} \int_{x}^{1} \frac{dz}{z} \left[\left(\frac{\alpha}{2\pi}\right) P^{(1)}(z) \right.\\ &\left. + \left(\frac{\alpha}{2\pi}\right)^2 P^{(2)}(z)\right] \delta\left(1-\frac{x}{z}\right) \\ &+ \int_{m^2}^{s} \frac{dk^2}{k^2} \int_{x}^{1} \frac{dz}{z} \int_{m^2}^{k^2} \frac{dQ^2}{Q^2} \int_{x/z}^{1} \frac{dt}{t} \left[\left(\frac{\alpha}{2\pi}\right) P^{(1)}(t) \right.\\ &\left. + \left(\frac{\alpha}{2\pi}\right)^2 P^{(2)}(t)\right]^2 \delta\left(1-\frac{x}{zt}\right) + O(\alpha^3). \end{aligned} \qquad (28)$$

Expliciting the integrals, D can be written in the form

$$D(x,s) = \delta(1-x) + \left(\frac{\alpha}{2\pi}\right) LP^{(1)}(x)$$
$$+ \left(\frac{\alpha}{2\pi}\right)^2 \left[LP^{(2)}(x) + \frac{1}{2}L^2 \int_x^1 \frac{dz}{z} P^{(1)}(z) P^{(1)}\left(\frac{x}{z}\right)\right], \quad (29)$$

where L is the collinear logarithm given by

$$L = \ln\frac{s}{m^2}.$$

Eq. (29) can be used to evaluate the radiator defined by eq. (6). An explicit calculation gives

$$H(x,s) = \delta(x) + \left(\frac{\alpha}{\pi}\right) LP^{(1)}(1-x)$$
$$+ \frac{1}{2}\left(\frac{\alpha}{\pi}\right)^2 \left[LP^{(2)}(1-x) + L^2 \int_{1-x}^1 \frac{dz}{z} P^{(1)}(z) P^{(1)}\left(\frac{1-x}{z}\right)\right]. \quad (30)$$

From the radiator of eq. (30) the single and double bremmstrahlung spectra can be extracted:

$$\frac{d\sigma^{(1)}}{dx} = \left(\frac{\alpha}{\pi}\right) LP^{(1)}(1-x)\sigma_0\left((1-x)s\right)$$
$$\frac{d\sigma^{(2)}}{dx} = \frac{1}{2}\left(\frac{\alpha}{\pi}\right)^2 \left[L^2 \int_{1-x}^1 \frac{dz}{z} P^{(1)}(z) P^{(1)}\left(\frac{1-x}{z}\right)\right. \quad (31)$$
$$\left. + LP^{(2)}(1-x)\right]\sigma_0\left((1-x)s\right).$$

As is well known[11], the single bremmstrahlung spectrum of eq. (31) is correct only at the leading level. However the substitution

$$L \quad \rightarrow \quad L-1$$

is such that it can be reproduced exactly:

$$\frac{d\sigma^{(1)}}{dx} = \frac{\alpha}{\pi}(L-1)\frac{1+(1-x)^2}{x}\sigma_0\left((1-x)s\right).$$

Operating the same substitution in the double bremmstrahlung spectrum and defining as usual

$$\beta = 2\frac{\alpha}{\pi}(L-1),$$

eq. (31) can be rewritten as

$$\begin{aligned}\frac{d\sigma^{(1)}}{dx} &= \frac{1}{2}\beta P^{(1)}(1-x)\sigma_0\left((1-x)s\right) \\ \frac{d\sigma^{(2)}}{dx} &= \left[\frac{1}{8}\beta^2 \int_{1-x}^{1}\frac{dz}{z}P^{(1)}(z)P^{(1)}\left(\frac{1-x}{z}\right) \right. \\ &\quad \left. + \frac{1}{2}\left(\frac{\alpha}{\pi}\right)^2 (L-1)P^{(2)}(1-x)\right]\sigma_0\left((1-x)s\right).\end{aligned} \qquad (32)$$

Let us now focus our attention on the second order bremmstrahlung spectrum in eq. (32). Two kinds of terms are present. The first one, $\int P^{(1)}P^{(1)}$, comes from order α iteration. It contains the leading and part of the next to leading logarithms. The second one, proportional to $P^{(2)}(1-x)$, is a truly second order effect. It contains terms of the kind $\alpha^2 L$ and $\alpha^2 constant$ and can be computed in such a way as to reproduce the second order bremmstrahlung spectrum exactly[12].

BOX AND INTERFERENCE EFFECTS VIA STRUCTURE FUNCTIONS

In the structure function approach the initial and final state radiatively corrected cross section can be written in the form[7]

$$\sigma_{rad}(s) = \int_0^\varepsilon dx\sigma_0\left((1-x)s\right)H_e(x,s)F_\mu\left(\varepsilon-x,(1-x)s\right), \qquad (33)$$

where $\sigma_0(s)$ is the Born cross section for the process $e^+e^- \to \mu^+\mu^-$ and H and F are given by

$$H(x,s) = \int_{1-x}^{1}\frac{dz}{z}D(z,s)D\left(\frac{1-x}{z},s\right), \qquad (34)$$

$$F(x,s) = \int_0^x dy\, H(y,(1-y)s). \tag{35}$$

The functions D appearing in eq. (34) are the structure functions for the fermions which partecipate to the reaction. They have been computed with different approximations by the authors of refs. [1,2]. The radiator H can be explicitely computed from the definition eq. (34), and it is given by the expression

$$\begin{aligned}H(x,s) = & \Delta(s)\beta x^{\beta-1} - \frac{1}{2}\beta(2-x) \\ & + \frac{1}{8}\beta^2\left\{(2-x)[3\ln(1-x) - 4\ln x] - 4\frac{\ln(1-x)}{x} + x - 6\right\},\end{aligned} \tag{36}$$

where $\beta = \frac{2\alpha}{\pi}(L-1)$, $L = \ln\frac{s}{m^2}$ and

$$\begin{aligned}\Delta(s) = & 1 + \frac{\alpha}{\pi}\left[\frac{3}{2}L + 2(\zeta(2) - 1)\right] + \left(\frac{\alpha}{\pi}\right)^2\left\{\left[\frac{9}{8} - 2\zeta(2)\right]L^2 \right. \\ & + \left[3\zeta(3) + \frac{11}{2}\zeta(2) - \frac{45}{16}\right]L \\ & \left. + \left[-\frac{6}{5}\zeta^2(2) - \frac{9}{2}\zeta(3) - 6\zeta(2)\ln 2 + \frac{3}{8}\zeta(2) + \frac{57}{12}\right]\right\}.\end{aligned} \tag{37}$$

The indices e and μ in eq. (33) refer to the fact that the radiators H_e and F_μ, i.e. the collinear logarithm L, have to be computed at the electron and muon mass respectively. As has already been pointed out[7], eq. (33) takes into account the effects of electromagnetic radiation from both initial and final states, resumming the leading and part of the next to leading logarithms to all orders of perturbation theory. In the radiator of eq. (36) two kinds of terms appear. The one proportional to $\beta x^{\beta-1}$ resums to all orders of perturbation theory the effects of soft photon radiation. The finite order terms take into account hard photon emission up to $O(\beta^2)$. The $O(\alpha^2)$ factor Δ reproduces the correct form factor of the fermion and is such that in the cross section additional finite order terms are included, with an accuracy up to terms of $O(\alpha^2 L)$.

If on one side eq. (33) is a good tool for computing the initial and final state radiation effects, on the other side it is not able to reproduce by itself the contributions coming from real and virtual interference terms, so that in ref. [7] they have been introduced by defining an effective integration kernel

$$\sigma_k(s) = \sigma_0(s) + \sigma_{box}(s) + \sigma_{int}(s),$$

where σ_{box} and σ_{int} are the contributions to the total cross section coming from box and inteference diagrams respectively.

Let us now discuss the possibility of taking automatically into account these kinds of contributions. They come from those diagrams in which one or more virtual photon lines connect an initial electron with a final muon and the interference between initial and final real photon radiation. If the Born cross section is a non resonant one, they give a contribution of the following form[13]

$$d\sigma_{(i)}(\vartheta, s) \simeq d\sigma_0(\vartheta, s)\varepsilon^{2\beta_{int}},$$

where $\beta_{int} = 4\frac{\alpha}{\pi}\ln\tan\frac{\vartheta}{2}$.

A simple way of handling these contributions in the structure function formalism is to carefully take into consideration the energy scales involved in the process.

As far as initial or final state radiative corrections are concerned, the energy scale governing photon radiation is the invariant mass squared of incoming or outgoing fermions: in fact from the point of view of photon radiation initial and final states are completely independent from one another. This is the reason why the energy scales which initial and final radiators H_e and F_μ in eq. (33) are computed at are s and $(1-x)s$ respectively. When considering box and interference contributions, on the contrary, initial and final states are not independent any more. For example a virtual photon connecting the initial electron and the final muon scattered at small or large angles "sees" an energy scale of the order

$-t$ or $-u$ respectively, where t and u are the Mandelstam variables

$$t = -2E^2(1-c), \qquad u = -2E^2(1+c),$$

with $c = \cos\vartheta$, ϑ being the angle between the incoming electron and the outgoing muon. Moreover it is well known from finite order calculations[14] that the contributions coming from box and interference diagrams are of the form $\ln\frac{t}{u}$. Since the energy scale appears in the radiators only through the collinear logarithm L, with a proper rescaling one can absorb automatically this kind of logarithms. The energy scale which allows one to take into account box and interference diagrams is given by

$$\tilde{s} = s\frac{t}{u},$$

By inserting the scale $\tilde{s}$ in the resummed terms of the radiators of eq. (33) one has

$$d\sigma(\vartheta, s) = \int_0^\varepsilon dx \, d\sigma_0\left(\vartheta, (1-x)s\right)$$
$$\times H_e\left(x, s\frac{t}{u}\right) F_\mu\left(\varepsilon - x, (1-x)s\frac{t}{u}\right).$$

Perfoming the calculation in the soft approximation one obtains:

$$d\sigma(\vartheta, s) \simeq d\sigma_0(\vartheta, s) F_\mu\left(\varepsilon, s\frac{t}{u}\right) \int_0^\varepsilon dx \, H_e\left(x, s\frac{t}{u}\right) \quad (38)$$
$$= d\sigma_0(\vartheta, s)\Delta_e(s)\Delta_\mu(s)\varepsilon^{\beta_e\left(s\frac{t}{u}\right)+\beta_\mu\left(s\frac{t}{u}\right)}.$$

β depends on the scale only through the collinear logarithm L, so that

$$\beta_e\left(s\frac{t}{u}\right) + \beta_\mu\left(s\frac{t}{u}\right) = \beta_e(s) + \beta_\mu(s) + 2\beta_{int},$$

since

$$\beta\left(s\frac{t}{u}\right) = \beta(s) + 2\frac{\alpha}{\pi}\ln\left(\frac{t}{u}\right).$$

Inserting the last expression into eq. (38) one obtains

$$d\sigma(\vartheta,s) \simeq d\sigma_0(\vartheta,s)\Delta_e(s)\Delta_\mu(s)\varepsilon^{\beta_e(s)+\beta_\mu(s)+2\beta_{int}}, \qquad (39)$$

which coincides with the cross section obtained in ref. [13]. Eq. (39) takes into account and resums to all orders the leading and part of the next to leading logarithms coming from initial and final state electromagnetic radiation ($\varepsilon^{\beta_e(s)}$ and $\varepsilon^{\beta_\mu(s)}$) and the leading logarithms coming from box and interference effects ($\varepsilon^{2\beta_{int}}$). The $O(\alpha)$ box and interference contribution contained in eq. (39) is given by the form

$$d\sigma_{(i)}(\vartheta,s) = d\sigma_0(\vartheta,s) 2\beta_{int} \ln\varepsilon = 8\frac{\alpha}{\pi} d\sigma_0(\vartheta,s) \ln\tan\frac{\vartheta}{2} \ln\varepsilon,$$

which coincides, at the leading level, with the $O(\alpha)$ expression given in ref. [14]. The comparison with this finite order result allows to compute a ϑ-dependent K factor containing additional non leading finite order contributions. The $O(\alpha)$ K factor for the pure QED case has the form

$$K^{QED}(\vartheta) = 1 + 2\frac{\alpha}{\pi}\frac{-2}{1-c^2}\left[c\left(\ln^2\sin\frac{\vartheta}{2} + \ln^2\cos\frac{\vartheta}{2}\right)\right]$$
$$\left[-\cos^2\frac{\vartheta}{2}\ln\sin\frac{\vartheta}{2} + \sin^2\frac{\vartheta}{2}\ln\cos\frac{\vartheta}{2}\right] + 2\ln^2\sin\frac{\vartheta}{2} \qquad (40)$$
$$\left[-2\ln^2\cos\frac{\vartheta}{2} - \text{Li}_2\left(\sin^2\frac{\vartheta}{2}\right) + \text{Li}_2\left(\cos^2\frac{\vartheta}{2}\right)\right].$$

Combining eq. (40) with eq. (33) taken at the scale $\tilde{s}$ one has

$$d\sigma^{QED}(\vartheta,s) = K^{QED}(\vartheta) \int_0^\varepsilon dx \, d\sigma_0^{QED}(\vartheta,(1-x)s)$$
$$\times H_e\left(x, s\frac{t}{u}\right) F_\mu\left(\varepsilon - x, (1-x)s\frac{t}{u}\right). \qquad (41)$$

Eq. (41) resummes to all orders radiation, box and interference contributions, and reproduces exactly the $O(\alpha)$ result[14].

If the Born cross section is a resonant one, the simple rescaling argument previously described does not work. The reason is probably that a resonance has an intrinsic energy scale given by the width Γ. Physically this means that there is a delay $\tau = 1/\Gamma$ between resonance formation and decay which inhibits initial and final states from interfering. This delay effect should be of course taken into account to be consistent. On this problem work is in progress.

CONCLUSIONS

In this paper some possible developments of the structure function formalism formalism have been reviewed.

Transverse momentum dependent structure functions have been introduced, showing the possibility of taking the transverse degrees of freedom of the radiation under control. In particular an improved Altarelli-Parisi vertex has been defined and computed with the required accuracy.

The generalization of the evolution at the two-loop level has been described. An explicit expression for the $O(\alpha^2)$ bremmstrahlung spectrum has been given.

A simple solution to the problem of automatically taking into account box and interference contributions within the formalism has been presented for a non resonant cross section. The K-factor for the pure QED case has been explicitly computed.

ACKNOWLEDGMENTS

It is a pleasure to thank G. Bonvicini, S. Catani, M. Greco, M. Mangano, G. Passarino and L. Trentadue for useful comments and discussions at various stages of this work.

REFERENCES

[1] E.A. Kuraev and V.S. Fadin, Yad. Fiz. **41** (1985) 753 [Sov. J. Nucl. Phys. **41** (1985) 466]; G. Altarelli and G. Martinelli in "Physics at LEP", CERN-Yellow Report 86-02, eds. J.Ellis and R.Peccei (Geneva, February 1986).

[2] O. Nicrosini and L. Trentadue, Phys. Lett. **196 B** (1987) 551.

[3] J.P. Alexander, G. Bonvicini, P.S. Drell and R. Frey, Phys. Rev. **D**, vol. 37 n. 1, 56-70 (1988); L. Trentadue, proceedings of this Conference.

[4] J.E. Campagne and R. Zitoun, LPNHEP-88-06; G. Montagna, O. Nicrosini and L. Trentadue, CERN preprint TH.5445/89, submitted to Phys. Lett **B**.

[5] A. Bassetto, M. Ciafaloni and G. Marchesini, Nucl. Phys. **B163** (1980) 477.

[6] G. Curci, W. Furmanski and R. Petronzio, Nucl. Phys. **B175** (1980) 27-92; J. Kalinowski, K. Konishi, P.N. Scharbach and T.R. Taylor, Nucl. Phys. **B181** (1981) 253-276; E.G. Floratos, R. Lakaze and C. Kounnas, Phys. Lett. **98B** (1981) 89, 285.

[7] O. Nicrosini and L. Trentadue, Z. Phys. **C39** (1988) 479-486.

[8] O. Nicrosini and L. Trentadue, Nucl. Phys. **B318** (1989) 1.

[9] G. Bonneau and F. Martin, Nucl. Phys. **B27** (1971) 381; F.A. Berends and R. Kleiss, Nucl. Phys. **B260** (1985) 32.

[10] G. Bonvicini and L. Trentadue, Nucl. Phys. **B323** (1989) 253.

[11] E.A. Kuraev and V.S. Fadin, ref. [1].

[12] F.A. Berends, G.J.H. Burgers and W.L. van Neerven, Nucl. Phys. **B297** (1988) 429-478.

[13] D.R. Yennie, S.C. Frautschi and H. Suura, Ann. Phys. (N.Y.) **13**, 379 (1961); M. Greco, G. Pancheri and Y. Srivastava, Nucl. Phys. **B101** (1975) 11, **B171** (1980) 118.

[14] F.A. Berends, R. Kleiss and S. Jadach, Nucl. Phys. **B202** (1982)63.

THE STRUCTURE OF GLUON RADIATION IN QCD [1]

Stephen PARKE

Fermi National Accelerator Laboratory [2]
P.O. Box 500, Batavia, IL 60510.

and

Michelangelo MANGANO

Istituto Nazionale di Fisica Nucleare
S. Piero a Grado, Pisa, ITALY

Abstract

For massless QCD the hard scattering amplitudes are naturally written in terms of the dual color expansion. Here I present this expansion for purely gluonic processes and processes involving quark-antiquark pairs and gluons. The properties of the sub-amplitudes as well as explicit algebraic expressions are given for a number of these processes. Finally, I demonstrate how to recover massless QED amplitudes from the dual expansion of massless QCD.

1 Introduction

In perturbative QCD the calculation of multi-gluon scattering amplitudes, even at tree level, is very challenging. Part of the reason for the difficulty is that up to recently there has been no systematic way to efficiently identify the appropriate gauge invariant subsets of the full amplitude. Here I summarize what has been discovered on how to make this division[1]. By insuring that the gauge invariant subsets are invariant under cyclic permutations of the external gluons tremendous cancellations occurr at the amplitude level and the sub-amplitudes so defined have remarkable factorization properties. The generalization to QCD processes involving one or more quark-antiquark pairs is also given.

---

[1] Presented by SP.

[2] Fermilab is operated by the Universities Research Association Inc. under contract with the United States Department of Energy.

© 1990 American Institute of Physics

2 Duality and Gauge Invariance

Consider an $SU(N)$ Yang-Mills theory, then at *tree level* in perturbation theory, any vector particle scattering amplitude, with colors $a_1, a_2 \ldots a_n$, external momenta $p_1, p_2 \ldots p_n$ and helicities $\epsilon_1, \epsilon_2 \ldots \epsilon_n$, can be written as

$$\mathcal{M}_{ng} = \sum_{perm'} tr\left(\lambda^{a_1}\lambda^{a_2}\ldots\lambda^{a_n}\right) m(p_1, \epsilon_1; p_2, \epsilon_2; \cdots; p_n, \epsilon_n), \quad (2.1)$$

where the sum, *perm'*, is over all $(n-1)!$ *non-cyclic* permutations of $1, 2, \ldots, n$ and the λ's are the matrices of the symmetry group in the fundamental representation. This expansion is known as the dual expansion because of the invariance of the sub-amplitudes under cyclic permutations[1].

The proof that one can always make this expansion is very simple using the identities $[\lambda^a, \lambda^b] = i\sqrt{2}f_{abc}\lambda^c$ and $tr(\lambda^a\lambda^b) = \delta^{ab}$. In any tree level Feynman diagram, replace the color structure function at some vertex using

$$f_{abc} = -(i/\sqrt{2})\, tr(\lambda^a\lambda^b\lambda^c - \lambda^c\lambda^b\lambda^a). \quad (2.2)$$

Now each leg attached to this vertex has a λ matrix associated with it. At the other end of each of these legs there is either another vertex or this is an external leg. If there is another vertex, use the λ associated with this internal leg to write the structure function of this vertex $f_{cde}\,\lambda^c$ as $-i\,[\lambda^d, \lambda^e]/\sqrt{2}$. Continue this processes until all vertices have been treated in this manner. Then this Feynman diagram has been placed in the form of eqn(2.1). Repeating this procedure for all Feynman diagrams for a given process completes the proof.

The sub-amplitudes $m(1, 2, \ldots, n) \equiv m(p_1, \epsilon_1; p_2, \epsilon_2; \ldots p_n, \epsilon_n)$ of eqn(2.1) satisfy a number of important properties and relationships.

(1) $m(1, 2, \ldots, n)$ is gauge invariant.
(2) $m(1, 2, \ldots, n)$ is invariant under cyclic permutations of $1, 2, \ldots, n$
(3) $m(n, n-1, \ldots, 1) = (-1)^n m(1, 2, \ldots, n)$
(4) The Dual Ward Identity:

$$m(1,2,3,\ldots,n) + m(2,1,3,\ldots,n) + m(2,3,1,\ldots,n) \quad (2.3)$$
$$+ \cdots + m(2,3,\ldots,1,n) = 0$$

(5) Factorization of $m(1, 2, \cdots, n)$ in the soft, collinear and multi-gluon pole limits.
(6) Incoherence to leading order in number of colors:

$$\sum_{colors} |\mathcal{M}_{ng}|^2 = N^{n-2}(N^2-1) \sum_{perm'} \left\{|m(1,2,\cdots,n)|^2 + \mathcal{O}(N^{-2})\right\}. \quad (2.4)$$

This set of properties for the sub-amplitudes, we will refer to as duality and the expansion in terms of these dual sub-amplitudes the dual expansion. Properties (1) and (2) can be seen directly from the properties of linear independence, for arbitrary N, and invariance under cyclic permutations of $tr\,(\lambda^1\lambda^2\ldots\lambda^n)$. Whereas (3) and (4) follow by studying the sum of Feynman diagrams which contribute to each sub-amplitude. The sum of Feynman diagrams which make the Dual Ward Identity is such that each diagram is paired with another with opposite sign so that the combination contained in eqn(2.3) trivially vanishes. Property (5) will be discussed in great detail in section IV and the incoherence to leading order in the number of colors (6) follows from the color algebra of the $SU(N)$ gauge group.

To the string theorist this expansion and the duality properties (1) to (6), see [2], are quite familar since the string amplitude, in the zero slope limit, reproduces the Yang-Mills amplitude on mass shell [3]. Each sub-amplitude is then represented by the zero slope limit of a string diagram, and the sub-amplitude could be obtained by using the usual Koba-Nielsen formula [4]. The traces of λ matrices are just the Chan-Paton factors[5]. For the string amplitude the properties (1) through (6) are satisfied even before the zero slope limit is taken. Also from the string diagrams it is simple to see which Feynman diagrams contribute to a given sub-amplitude, e.g. Fig. 1. The coefficients for the contributing diagrams are obtained by the procedure developed earlier in this section for rewriting the color factors. The relationship between the string diagram and our dual sub-amplitudes suggests that a Yang-Mills amplitude expressed in terms of these dual sub-amplitudes will assume a particularly simple form.

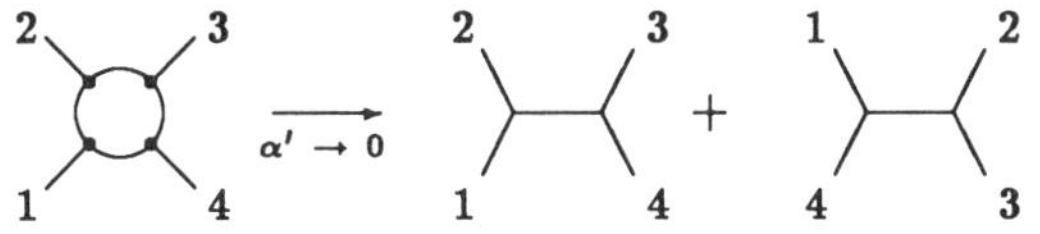

Figure 1: The zero-slope limit of the four gluon string diagram in terms of Feynman diagrams (tri-gluon couplings only).

The gauge invariance and properties under cyclic and reverse permutations allows the calculation of far fewer than the $(n-1)!$ sub-amplitudes that appear in the dual expansion. In fact the number of sub-amplitudes that are needed is just the number of different orderings of positive and negative helicities around a circle. Of course some of the sub-amplitudes vanish because of the partial helicity conservation of tree level Yang-Mills and others are simply related to one another through the properties (2) through (4).

3 Spinor Products

To evaluate the sub-amplitudes we have used the helicity basis for the polarization vectors which was introduced by Xu, Zhang and Chang[6]. This technique requires the introduction of the concept of a spinor product of two light-like momentum vectors. We define the following symbols for the chiral spinors associated with the light-like momenta, p_i, and their spinor products:

$$|i\pm\rangle \equiv \frac{1}{2}(1\pm\gamma_5)u(p_i) , \qquad \langle i\pm| \equiv \bar{u}(p_i)\frac{1}{2}(1\mp\gamma_5) \tag{3.1}$$

$$\langle ij\rangle = \langle i-|j+\rangle , \qquad [ij] = \langle i+|j-\rangle = sign(p_i^0 p_j^0)\langle ji\rangle^*. \tag{3.2}$$

The important properties of these spinor products that will be needed in this paper are that both $\langle ij\rangle$ and $[ij]$ are odd under interchange of i and j and are complex square roots of the Lorentz invariant $S_{ij} \equiv (p_i + p_j)^2$;

$$\langle ij\rangle \equiv \sqrt{|S_{ij}|} \exp(i\phi_{ij}), \tag{3.3}$$

$$[ij] \equiv \sqrt{|S_{ij}|} \exp(i\tilde{\phi}_{ij}) \tag{3.4}$$

If both momenta having positive energy, the phase factor ϕ_{ij} is defined, in a popular representation of the gamma matrices, by

$$\cos\phi_{ij} = \frac{(p_i^1 p_j^+ - p_j^1 p_i^+)}{\sqrt{p_i^+ p_j^+}}$$

$$\sin\phi_{ij} = \frac{(p_i^2 p_j^+ - p_j^2 p_i^+)}{\sqrt{p_i^+ p_j^+}}. \tag{3.5}$$

Where $p^\pm = (p^0 \pm p^3)$ and since all $p_i^2 = 0$ the spinor product for this representation of gamma matrices is undefined for a momentum vector in the minus 3 direction. If one or more of the momenta in $\langle ij\rangle$ have negative energy, ϕ_{ij} is calculated with minus the momenta with negative energy and then $n\pi/2$ is added to ϕ_{ij} where n is the number of negative momenta in the spinor product. The associated phase factor, $\tilde{\phi}_{ij}$, for $[ij]$ can be found using equation (3.2) or calculated from S_{ij} using the identity $S_{ij} \equiv \langle ij\rangle [ji]$.

4 Factorization Properties

The most important and remarkable properties of the Yang-Mills dual sub-amplitudes are their factorization properties, whose origin can be traced back to the string picture. In this section we give the factorization properties of the gluon sub-amplitudes in

(1) the soft gluon limit,
(2) when two gluons become collinear and
(3) when three gluons add to form an on mass-shell gluon
i.e. on the three gluon pole.

For arbitrary n-gluon scattering these factorization properties of the sub-amplitudes will extend up to factorization on the $[n/2]$-gluon poles.

First, we consider the soft gluon limit. Consider the sub-amplitudes when gluon 1 has an energy which is small compared to all the other energies in the process. Then the gluon sub-amplitudes must satisfy

$$m(1^+, 2\ldots, n) \overset{1^+ \ soft}{\longrightarrow} \left\{ \frac{g \langle n\ 2 \rangle}{\langle n\ 1 \rangle \langle 1\ 2 \rangle} \right\} m(2, 3\ldots, n) \qquad (4.1)$$

$$m(1^-, 2\ldots, n) \overset{1^- \ soft}{\longrightarrow} \left\{ \frac{g\ [n\ 2]}{[n\ 1][1\ 2]} \right\} m(2, 3\ldots, n). \qquad (4.2)$$

The factors in braces are square roots of the eikonal factor

$$\frac{g^2\ (p_n \cdot p_2)}{(p_n \cdot p_1)(p_1 \cdot p_2)}.$$

This soft gluon factorization and the incoherence of these sub-amplitudes to leading order in the number of colors, N, leads to the soft gluon factorization of the full matrix element squared as proposed by Bassetto, Ciafaloni and Marchesini [7],

$$\sum_{colors} |\mathcal{M}_{ng}|^2 \overset{1\ soft}{\longrightarrow} \sum_{ij} \left(\frac{g^2\ (p_i \cdot p_j)}{(p_i \cdot p_1)(p_1 \cdot p_j)} \right) |A_{ij}(2, \cdots, n)|^2. \qquad (4.3)$$

In the limit when two gluons become collinear, Altarelli and Parisi [8] demonstrated that the double poles associated with this collinear pair do not appear in the full amplitude squared i.e. there is a cancellation of one power of the propagator of the sum of the two collinear gluons. This cancellation occurs at the amplitude level rather than the square of the amplitude in this dual formulation. Therefore the squared sub-amplitudes diverge no more rapidly than a single power of the propagator for the collinear gluons, this is the Altarelli and Parisi observation. The origin of this behaviour of the dual sub-amplitudes stems from the factorization properties of string amplitudes.

To demonstrate this square root divergence of the sub-amplitudes in the collinear limit, consider the case when the momenta of particles 1 and 2 become parallel. Let $1 \to z\,P$ and $2 \to (1-z)\,P$ with $P^2 = 0$, and z is the momentum fraction of particle 1. Then the sub-amplitudes become

$$m(1^+,2^+,3,\ldots) \;\overset{1^+\|\,2^+}{\longrightarrow}\; \left\{\frac{ig\,[12]}{\sqrt{z(1-z)}}\right\} \frac{-i}{S_{12}}\, m(P^+,3,\ldots) \qquad (4.4)$$

$$m(1^+,2^-,3,\ldots) \;\overset{1^+\|\,2^-}{\longrightarrow}\; \left\{\frac{ig\,z^2\langle 12\rangle}{\sqrt{z(1-z)}}\right\} \frac{-i}{S_{12}}\, m(P^+,3,\ldots) \qquad (4.5)$$

$$+ \left\{\frac{ig\,(1-z)^2\,[12]}{\sqrt{z(1-z)}}\right\} \frac{-i}{S_{12}}\, m(P^-,3,\ldots)$$

$$m(1^-,2^-,3,\ldots) \;\overset{1^-\|\,2^-}{\longrightarrow}\; \left\{\frac{ig\,\langle 12\rangle}{\sqrt{z(1-z)}}\right\} \frac{-i}{S_{12}}\, m(P^-,3,\ldots). \qquad (4.6)$$

Note that either $\langle 12\rangle$ or $[12]$ appears in the numerator of each term. Also, it is useful to interpret the factor in braces as the "three gluon sub-amplitude" in the limit when two gluons become collinear. This three gluon sub-amplitude has the square root suppression of the pole as well as having the square root of the appropriate Altarelli-Parisi gluon-fusion function. From this result and the incoherence of the sub-amplitudes in the square of the matrix element the standard results of Altarelli and Parisi are obtained in a simple manner.

The sub-amplitudes also factorize in the three particle channel; here let $P = 1+2+3$, then as $P^2 \to 0$ it is easy to see that

$$m(1,2,3,4,5,6) \;\to\; m(1,2,3,-P)\,\frac{-i}{P^2}\,m(P,4,5,6) \qquad (4.7)$$

for the helicity structure three positive and three negative. Since helicity is conserved in the four gluon process, the helicity of the intermediate gluon is determined for this helicity structure and the four positive - two negative helicity sub-amplitude has no three particle poles.

Of course the full matrix element must also factorize. This is trivial in Feynman diagram language but here it is not so obvious because of the way we have added diagrams together. The color factors almost factorizes for an $SU(N)$ gauge group,

$$tr\,(\lambda^1\lambda^2\ldots\lambda^n) \;=\; \sum_x tr\,(\lambda^1\ldots\lambda^m\lambda^x)\,tr\,(\lambda^x\lambda^{m+1}\ldots\lambda^n) \qquad (4.8)$$

$$+\,\frac{1}{N}tr\,(\lambda^1\ldots\lambda^m)\,tr\,(\lambda^{m+1}\ldots\lambda^n).$$

This "factorization" property of the traces follows from the identity

$$\sum_a \lambda^a_{ij} \lambda^a_{kl} = (\delta_{il}\delta_{jk} - \frac{1}{N}\delta_{ij}\delta_{kl}). \tag{4.9}$$

The $1/N$ term could destroy the full factorization, but it does not. Terms proportional to $1/N$ vanish at the pole because of the Dual Ward Identity for the sub-amplitudes. Therefore, all the gluon amplitudes discussed in this paper satisfy, as expected, the factorization property

$$\mathcal{M}_{n+n'} \rightarrow \sum \mathcal{M}_{n+1} \frac{-i}{P^2} \mathcal{M}_{n'+1} \tag{4.10}$$

as $P^2 \rightarrow 0$ for $n, n' \geq 2$. The sum is over the color and helicity of the intermediate state.

5 Pure Gluon Amplitudes

For four gluon scattering only the helicity conserving amplitudes are non zero. Using the convention that all particles are labelled with their helicities and momenta as if they were outgoing, i.e. the incoming particles have negative energies, the helicity conserving sub-amplitudes are given by

$$\begin{aligned} m_{2+2-}(1,2,3,4) &= -ig^2 \frac{\langle IJ\rangle^2 [KL]^2}{S_{12} S_{23}} \\ &= ig^2 \frac{\langle IJ\rangle^4}{\langle 12\rangle\langle 23\rangle\langle 34\rangle\langle 41\rangle}. \end{aligned} \tag{5.1}$$

The momenta I and J (K and L) in the numerator are the momenta of the negative (positive) helicity gluons independent of their ordering in the sub-amplitude, whereas the order of the spinor products in the denominator is only determined by the order of the momenta in the sub-amplitude. Using the properties of the spinor product is simple to demonstrate that eqn(5.1) satisfies the four particle Dual Ward Identity (2.3).

In squaring the four gluon amplitude and summing over colors the $\mathcal{O}(N^{-2})$ terms in eqn(2.4) can be shown to vanish by using only the general properties, especially the Dual Ward Identity, of the sub-amplitudes. Therefore,

$$\sum_{colors} |\mathcal{M}_{4g}|^2 = N^2(N^2-1) \sum_{perm'} |m(1,2,3,4)|^2, \tag{5.2}$$

and the square of each sub-amplitude is very simple because the spinor product is the square root of twice the dot product. The final result is the standard four gluon matrix element squared.

$$\sum_{\text{hel.}}\sum_{\text{colors}} |\mathcal{M}_{4g}|^2 = N^2(N^2-1) g^4 \left(\sum_{i>j} S_{ij}^4\right) \sum_{\text{perm}'} \frac{1}{S_{12}S_{23}S_{34}S_{41}}. \quad (5.3)$$

Here we have not averaged over incoming helicities or colors.

For five gluon scattering only those Feynman diagrams, or part there of, with color structure the same as the diagrams of Fig. 2 contribute to the $m(1,2,3,4,5)$ sub-amplitude.

Again, it is a straight forward, simple calculation [1] to show that the only nonzero sub-amplitudes have either two or three negative helicity gluons and that the three positive - two negative helicity sub-amplitude is given by

$$m_{3+2-}(1,2,3,4,5) = ig^3 \frac{\langle IJ \rangle^4}{\langle 12 \rangle \langle 23 \rangle \langle 34 \rangle \langle 45 \rangle \langle 51 \rangle}. \quad (5.4)$$

Where I and J are again the momenta of the negative helicity gluons and the denominator ordering is determined by the order of the momenta in the sub-amplitude. The two positive - three negative helicity amplitude is obtained from this last equation by complex conjugation. By using the Fierz properties of the spinor product it is easy to demonstrate that eqn(5.4) satisfies the five particle Dual Ward Identity, eqn(2.3).

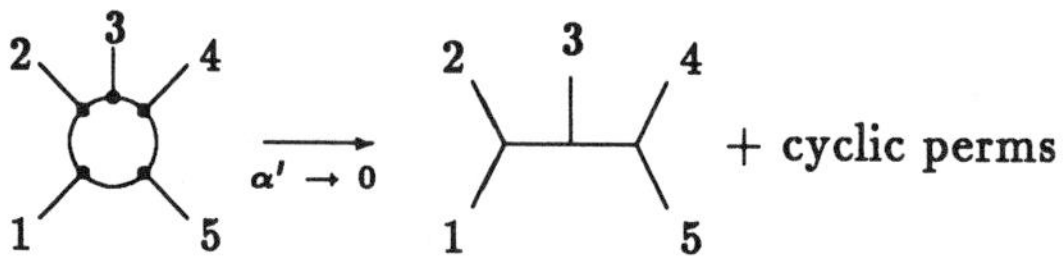

Figure 2: The zero-slope limit of the five gluon string diagram in terms of Feynman diagrams (tri-gluon couplings only).

Again, the general properties of the sub-amplitude can be used to show that the $\mathcal{O}(N^{-2})$ terms in eqn(2.4) vanish for the five gluon process giving the following standard result [9] that

$$\sum_{\text{hel.}}\sum_{\text{colors}} |\mathcal{M}_{5g}|^2 = 2 N^3(N^2-1) g^6 \left(\sum_{i>j} S_{ij}^4\right) \sum_{\text{perm}'} \frac{1}{S_{12}S_{23}S_{34}S_{45}S_{51}}. \quad (5.5)$$

Here we have not averaged over incoming helicities or colors.

For the six gluon process only those Feynman diagrams, or part there of, with the same color structure as the diagrams of Fig. 3 contribute to the $m(1,2,3,4,5,6)$ sub-amplitude. Then, by using the appropriate reference momenta for the polarization vectors it is easy to see that the only non-zero sub-amplitudes are those with four positive - two negative, two positive - four negative and three positive - three negative helicities. After a lengthy calculation we have obtained the following expressions for the six gluon sub-amplitudes.

The sub-amplitudes for the four positive - two negative helicity processes are a straight forward generalization of the four and five-gluon sub-amplitudes;

$$m_{4+2-}(1,2,3,4,5,6) = ig^4 \frac{\langle IJ\rangle^4}{\langle 12\rangle\langle 23\rangle\langle 34\rangle\langle 45\rangle\langle 56\rangle\langle 61\rangle}. \tag{5.6}$$

Again, I and J represent the momenta of the negative helicity gluons. Different permutations can be obtained as before by keeping fixed the numerator and permuting the momenta in the denominator. The two positive - four negative helicity sub-amplitude is obtained from eqn(5.6) by complex conjugation.

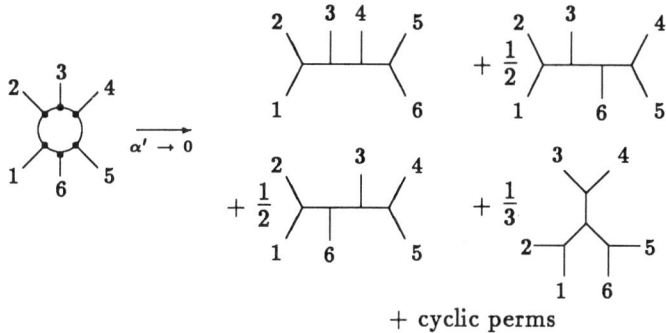

Figure 3: The zero-slope limit of the six gluon string diagram in terms of Feynman diagrams (tri-gluon couplings only).

The three positive - three negative helicity sub-amplitudes are not as simple. To exhibit the factorization on the three particle channels these sub-amplitudes are

$$m_{3+3-}(1,2,3,4,5,6) = ig^4 \left[\frac{\alpha^2}{t_{123}S_{12}S_{23}S_{45}S_{56}} + \frac{\beta^2}{t_{234}S_{23}S_{34}S_{56}S_{61}} \right. \tag{5.7}$$
$$\left. + \frac{\gamma^2}{t_{345}S_{34}S_{45}S_{61}S_{12}} + \frac{t_{123}\beta\gamma + t_{234}\gamma\alpha + t_{345}\alpha\beta}{S_{12}S_{23}S_{34}S_{45}S_{56}S_{61}} \right]$$

where the $t_{ijk} \equiv (p_i + p_j + p_k)^2 = S_{ij} + S_{jk} + S_{ki}$. The coefficients α, β and γ for the three distinct orderings of the helicities are given in Table I. With this representation it is a simple exercise to show that these sub-amplitudes factorize on the three particle pole into a product of two four particle sub-amplitudes, eqn(5.1), times the three particle propagator.

<u>Table I</u>
Coefficients for the m_{3+3-} Sub-amplitudes:
where $\langle I|K|J\rangle \equiv \langle I+|K\cdot\gamma|J+\rangle$, which is linear in K
and if $K^2 = 0$ is given by $[IK]\langle KJ\rangle$.

	$1^+2^+3^+4^-5^-6^-$ $X = 1 + 2 + 3$	$1^+2^+3^-4^+5^-6^-$ $Y = 1 + 2 + 4$	$1^+2^-3^+4^-5^+6^-$ $Z = 1 + 3 + 5$
α	0	$-[12]\langle 56\rangle\langle 4|Y|3\rangle$	$[13]\langle 46\rangle\langle 5|Z|2\rangle$
β	$[23]\langle 56\rangle\langle 1|X|4\rangle$	$[24]\langle 56\rangle\langle 1|Y|3\rangle$	$[51]\langle 24\rangle\langle 3|Z|6\rangle$
γ	$[12]\langle 45\rangle\langle 3|X|6\rangle$	$[12]\langle 35\rangle\langle 4|Y|6\rangle$	$[35]\langle 62\rangle\langle 1|Z|4\rangle$

The six gluon sub-amplitudes satisfy the three distinct Dual Ward Identities obtained from the following equation

$$m(1,2,3,4,5,6) + m(2,1,3,4,5,6) + m(2,3,1,4,5,6) \\ + m(2,3,4,1,5,6) + m(2,3,4,5,1,6) = 0 \quad (5.8)$$

using the helicity ordering of the first term as either $m(1+,2+,3+,4+,5-,6-)$, $m(1+,2+,3+,4-,5-,6-)$ or $m(1+,2-,3+,4-,5+,6-)$. These three Identities are extremely powerful and relate sub-amplitudes with different orderings of the helicities.

Given the simplicity of the sub-amplitudes with two negative helicities and all the others positive, equations (5.1), (5.4) and (5.6), it is obvious that the generalization to arbitrary n is

$$m_{(n-2)+2-}(1,2,\ldots,n) = ig^{n-2} \frac{\langle IJ\rangle^4}{\langle 12\rangle\langle 23\rangle\cdots\langle n1\rangle} \quad (5.9)$$

where once again I and J are the momenta of the negative helicity gluons. Apart from this being the natural square root of the expression given by Parke and Taylor [10], [7], it also *satisfies* the Dual Ward Identity for arbitrary n.

The complete square of the six-gluon amplitude, including the non-leading color terms is

$$\sum_{colors} |\mathcal{M}_{6g}|^2 =$$

$$N^4(N^2-1) \sum_{perm'} |m(1,2,3,4,5,6)|^2 \qquad (5.10)$$
$$+ \frac{1}{N^2} \Big(m^*(1,2,3,4,5,6)\big[m(1,3,5,2,6,4)$$
$$+ m(1,3,6,4,2,5) + m(1,4,2,6,3,5)\big] + c.c\Big).$$

Note that the sub-amplitudes add incoherently to leading order in the number of colors and the simplicity of the non-leading color terms is achieved by the properties of the sub-amplitudes, especially the Dual Ward Identity equation (2.3). This result together with the expressions for the sub-amplitudes, eqn(5.6) and (5.7), can be used to calculate the matrix element squared by evaluating the sub-amplitudes as complex numbers. Owing to the simplicity of the sub-amplitudes and the simplicity of the leading and non-leading terms in the number of colors this method of calculation is appreciable faster than previous numerical algorithms [12], [13].

The ordering of the gluons in the non-leading color terms is of particular import. These terms are the only possible ones which have no two or three particle propagators in common with the original ordering $(1,2,3,4,5,6)$ and as such are less singular in the collinear limit than the leading part in N. In fact the non-leading color terms are finite in the collinear limit so that in this limit they are completely irrelevant compared to the leading color terms. Also by comparing numerically the leading to non-leading pieces for N= 3, the non-leading terms contribute in general only a few percent to the total cross-section. This result is even true in the soft gluon limit. Therefore the non-leading terms can be ignored given that this calculation is only to tree level, and the other uncertainties in any Monte Carlo application are much larger than this uncertain. The smallness of the non-leading color terms and the fact that the leading color terms are just the squares of the simple sub-amplitudes implies that the square of this matrix element is easy to obtain.

6 The Addition of Quarks

The dual basis is modified by the addition of a quark-antiquark pair in the scattering amplitude in the following way. Consider a quark and an antiquark with colors α and $\bar{\alpha}$ respectively then we write the amplitude as

$$\mathcal{M}_{\bar{q}qng} = \sum_{perm} (\lambda^{a_1}\lambda^{a_2}\ldots\lambda^{a_n})_{\alpha\bar{\alpha}}\, m_{\bar{q}q}(\,p_{\bar{q}},\epsilon_{\bar{q}};\,p_q,\epsilon_q;\,p_1,\epsilon_1;\cdots;\,p_n,\epsilon_n) \qquad (6.1)$$

where the sum, *perm*, is over all n! permutation of the gluons. This expansion of the quark amplitude in terms of this color basis is well known and in particular was used by Kunszt in Reference [13]. We will call the color basis in Equation (6.1) the *quark dual basis*[14],[15].

For the amplitude squared, we have an expression very similar to Equation (2.4):

$$\sum_{colors} |\mathcal{M}_{\bar{q}qng}|^2 = N^{n-3}(N^2-1) \sum_{\{1,\ldots,n\}} \{|m_{\bar{q}q}(\bar{q},q,1,\cdots,n)|^2 + \mathcal{O}(N^{-2})\}. \quad (6.2)$$

Notice however the change in the exponent of the leading power of N. The explicit form of the sub-leading terms for $n = 2, 3, 4$ is given in reference [15].

As in the pure gluon case the amplitudes with all particles or all but one particle having the same helicity vanishes at tree level. Also the quark and the antiquark must have opposite helicity or otherwise the amplitude vanishes from chirality conservation in massless QCD. The amplitudes with one gluon the same helicity as the quark or antiquark and all other gluons having the opposite helicity have simple expressions:

$$m_{\bar{q}q}(\bar{q}^+, q^-, g_1^+, \ldots, g_I^-, \ldots, g_n^+) = ig^n \frac{\langle \bar{q}I\rangle\langle qI\rangle^3}{\langle \bar{q}q\rangle\langle q1\rangle\langle 12\rangle\cdots\langle n\bar{q}\rangle}.$$

$$m_{\bar{q}q}(\bar{q}^-, q^+, g_1^+, \ldots, g_I^-, \ldots, g_n^+) = ig^n \frac{\langle \bar{q}I\rangle^3\langle qI\rangle}{\langle \bar{q}q\rangle\langle q1\rangle\langle 12\rangle\cdots\langle n\bar{q}\rangle}. \quad (6.3)$$

Other amplitudes are more complex, e.g. the general form of the quark-antiquark four gluon amplitude has the following pole structure as dictated by duality[2]:

$$m_{\bar{q}q}(\bar{q}, q, g_1, g_2, g_3, g_4) = ig^4 \left[\frac{P_1}{t_{\bar{q}q1}S_{\bar{q}q}S_{q1}S_{23}S_{34}} + \frac{P_2}{t_{q12}S_{q1}S_{12}S_{34}S_{4\bar{q}}} \right.$$

$$\left. + \frac{P_3}{t_{123}S_{12}S_{23}S_{4\bar{q}}S_{\bar{q}q}} + \frac{P_s}{S_{\bar{q}q}S_{q1}S_{12}S_{23}S_{34}S_{4\bar{q}}} \right]. (6.4)$$

The numerators P_i are complicated and I refer you to reference [15] for explicit expressions for these quantities. The sub-amplitudes defined by equation (6.1) have similar properties to the purely gluonic ones in the soft, collinear and multi-particle pole limits.

To construct the QED results from the non-abelian amplitudes all that is needed is to replace $\lambda_{\alpha\bar{\alpha}}$ in equation (6.1) by $\delta_{\alpha\bar{\alpha}}$, for details see reference [16]. For example one of the helicity amplitudes for the amplitudes involving an electron, positron and n photons can be written as

$$\mathcal{M}(\bar{e}^+, e^-; \gamma_1^+, \ldots, \gamma_I^-, \ldots, \gamma_n^+) = ie_0^n \frac{\langle \bar{e}I\rangle\langle eI\rangle^3}{\langle \bar{e}e\rangle} \sum_{\{1,\ldots,n\}} \frac{1}{\langle e1\rangle\langle 12\rangle\cdots\langle n\bar{e}\rangle}$$

$$= ie_0^n \frac{\langle \bar{e}I\rangle\langle eI\rangle^3}{\langle \bar{e}e\rangle^2} \prod_j \frac{\langle \bar{e}e\rangle}{\langle ej\rangle\langle j\bar{e}\rangle}. \quad (6.5)$$

In this example one can see that for the abelian amplitude the photons are emitted independently of each other whereas for the non-abelian amplitude there is a correlation between the emitted gluons.

To extend this quark dual basis to more than one quark antiquark pair I refer you to the paper by Mangano [16], and here I briefly sketch the color basis. For two quark-antiquark pairs of different flavors with colors α, α' and $\bar\alpha$, $\bar\alpha'$ the dual color expansion is

$$\mathcal{M}_{\bar{q}q\bar{q}'q'ng} = \sum_{\{\sigma,\tau\}} (\prod_\sigma \lambda)_{\alpha\bar{\alpha}'}(\prod_\tau \lambda)_{\alpha'\bar{\alpha}}\, m^x_{\bar{q}q\bar{q}'q'}(\sigma,\tau)$$

$$+ \frac{1}{N} (\prod_\sigma \lambda)_{\alpha\bar{\alpha}}(\prod_\tau \lambda)_{\alpha'\bar{\alpha}'}\, m^0_{\bar{q}q\bar{q}'q'}(\sigma,\tau) \qquad (6.6)$$

where the sum is over all partitions, $\{\sigma,\tau\}$, of all permutations of the n gluons. The first term is the contribution in which the color flow connects α to $\bar\alpha'$ and α' to $\bar\alpha$ whereas the second term comes from the color flow connecting α to $\bar\alpha$ and α' to $\bar\alpha'$. For two quark pairs of the same flavor one must add $\mathcal{M}_{\bar{q}q'\bar{q}'qng}$ to $\mathcal{M}_{\bar{q}q\bar{q}'q'ng}$. Similar factorization properties to $m_{\bar{q}q}$ also hold for $m^x_{\bar{q}q\bar{q}'q'}$ and $m^0_{\bar{q}q\bar{q}'q'}$.

Finally, the simplicity of the helicity amplitudes with two partons having opposite helicity from all the others, eqn(5.9) and (6.3) has start an industry of approximation methods for these multi-parton processes, see reference [17] for details.

7 Conclusion

Here we have demonstrated that the *dual color basis* is the natural one for QCD amplitudes. The sub-amplitudes defined by this basis have remarkable factorization properties and are quite simple given the complexity of processes of QCD. From purely gluonic processes to processes involving quarks and other vector particles, this basis displays the underlying physics of the process.

References

[1] M. Mangano, S. Parke and Z. Xu, *in* Proc. of "Les Rencontres de Physique de la Vallee d'Aoste", La Thuile, Italy, (1987), ed. M. Greco, Editions Frontières, p.513;
M. Mangano, S. Parke and Z. Xu, Nucl. Phys. B298 (1988) 653;
F. A. Berends and W. Giele, Nucl. Phys. B294 (1987) 700.

[2] See for example M. Jacob and S. Mandelstam's lectures contained in 'Lectures on Elementary Particles and Quantum Field Theory', Brandeis 1970, S.

Deser, M. Grisaru and H. Pendleton eds., MIT Press and reference contained in these lectures.

[3] For a comprehensive review see: J. Schwarz, Phys. Rep. 89 (1982), 224.

[4] Z. Koba and H. B. Nielsen, Nucl. Phys. B10 (1969), 633; B12 (1969), 517: Evaluation of the integrals and taking the zero slope limit is extremely tedious for more than five external particles so the sub-amplitudes were evaluated using Feynman perturbation theory.

[5] J. Paton and Chan Hong-Mo, Nucl. Phys. B10 (1969) 519.

[6] Z. Xu, Da-Hua Zhang and L. Chang, Tsinghua University Preprints, Beijing, The People's Republic of China, TUTP-84/4, TUTP-84/5, TUTP-84/6 and Nucl. Phys. B291 (1987) 392.

[7] A. Bassetto, M. Ciafaloni and P. Marchesini, Phys. Rep. 100 (1983), 201.

[8] G. Altarelli and G. Parisi, Nucl. Phys. B126 (1977), 298.

[9] T. Gottschalk and D. Sivers, Phys. Rev. D21 (1980), 102;
F. A. Berends, R. Kleiss, P. De Causmaeker, R. Gastmans and T.T. Wu, Phys. Lett. 103B (1981), 124.

[10] S. Parke and T. Taylor, Phys. Rev. Lett. 56 (1986) 2459.

[11] S. Parke and T. Taylor, Nucl. Phys. B269 (1986), 410.

[12] S. Parke and T. Taylor, Nucl. Phys. B269 (1986), 410;
J. Gunion and J. Kalinowski, Phys. Rev. D34 (1986), 2119.

[13] Z. Kunszt, Nucl. Phys. B271 (1986), 333.

[14] M. Mangano and S. Parke, Proc. of the 'International Europhysics Conference on High Energy Physics', Uppsala, Sweden, (1987), ed. O. Botner, p.201.

[15] M. Mangano and S. Parke, Nucl. Phys. B299 (1988) 673.

[16] M. Mangano, Nucl. Phys. B315 (1989) 391.

[17] Z. Kunszt and W. J. Stirling, Phys. Rev. D30 (1988),2439;
C. J. Maxwell, Phys. Lett. 192B (1987), 190;
M. Mangano and S. Parke, Phys. Rev. D39 (1989), 758;
C. J. Maxwell, Nucl. Phys. B316, (1989), 312.

RENORMALIZATION QROUP IMPROVED CALCULATION OF RADIATIVE CORRECTIONS IN ANNIHILATION AND SCATTERING CHANNELS

E.A.Kuraev

Inclusive cross-sections of hard processes with radiative corrections (RC) in all order of perturbation theory (Pt) taken into account may be written in form of cross-section of Drell-Yan process. Structure functions as well as a fragmentation functions entering Drell-Yan picture obey the renormqroup Alta-relly-Parizi-Lipatov eguations. They describe the r.c. in leading logarathmic appsoximation (lla)

$$\frac{\alpha}{\pi} L = \frac{\alpha}{\pi} \ln Q^2/m^2 \sim 1 \quad , \quad \frac{\alpha}{\pi} \ll 1 , \tag{1}$$

Q - large momenta of problem. To go beyond lla one may introduce the K - factor $K = 1 + \frac{\alpha}{\pi} a$: this quarantee the correct evaluation of next-to leading terms of form $(\frac{\alpha}{\pi})(\frac{\alpha}{\pi} L)^n$, and correspond to experimental re-quirement to calculation of r.c. with the accuracy of order $\sim 0.1\%$.

To find out K-factor one must compare the explicit calculation of cross section up to two loop level with the results of renorm.qroup analysis and Yennie-Frautchy-Suura infrared approximation. One may cinsider the case of soft additional phitons and pair emission.

1. Lowest orderd calculation in soft limit.

Consider first the single-proton annihilation e^+e^-. The r.c. associated with the initial state are the universal component of corrections to cross sections of any process with arbitrary final state.

Types of Feynman dyagrams discribing cross-section with the lowest order rc are: virtnal corrections to Dirac form factor

Sinqle proton emission

emission of two photons and lepton pair

Neqlecting terms of order m_e^2/s compared to unity, $S = 4E^2$, E - c.m. beam energy, we see that only Dirac form factor contribute

$$F_1(s) = 1 + \frac{\alpha}{\pi} F_1^{(2)}(s) + \left(\frac{\alpha}{\pi}\right)^2 F_1^{(4)}(s) + \cdots$$

$$F_1^{(2)}(s) = B(s)\ln\frac{\lambda}{m_e} - \frac{1}{4}L^2 + \frac{3}{4}L + \frac{\pi^2}{3} - 1 - i\pi\left(\frac{1}{2}L - \frac{3}{4}\right), \; B(s) = L - 1 - i\pi, \; L = \ln s/m_e^2,$$

$$\operatorname{Re} F_1^{(4)}(s) = -\frac{1}{2}\operatorname{Re} B^2(s)\ln^2\left(\frac{\lambda}{m_e}\right) + \operatorname{Re}\left(B(s) F_1^{(2)}(s)\right)\ln\left(\frac{\lambda}{m_e}\right) + \frac{1}{32}L^4 - \frac{31}{144}L^3 +$$

$$+ \left(\frac{229}{288} - \frac{5\pi^2}{24}\right)L^2 + \left(-\frac{1627}{864} + \frac{5}{9}\pi^2 + \frac{3}{2}\zeta(3)\right)L + \text{Const}, \; \zeta(3) = \sum_1^\infty \frac{1}{n^3} = 1,202. \quad (2)$$

λ - fictitions "photon mass".

Cross section of single photon annihilation with emission of one solt photon with the energy not exceeding $\Delta\varepsilon \ll \varepsilon$

$\Delta\varepsilon$ - energy resolution:

$$\sigma^{(\delta)} = \tilde{\sigma}(s) \cdot \frac{\alpha}{\pi}\left[1 + \frac{2\alpha}{\pi}\operatorname{Re} F_1^{(2)}(s)\right]\left[2\operatorname{Re} B(s) \cdot \ln\frac{m_e}{\lambda} + \frac{1}{2}L^2 - \frac{\pi^2}{3} - 2(L-1)\ell\right] \quad (3)$$

where

$$\tilde{\sigma}(s) = \sigma_B(s) \big/ |1 - \Pi(s)|^2, \quad \ell = \ln\frac{E}{\Delta E}, \quad (4)$$

$\tilde{\sigma}_B(s)$ — cross section in Born approximation, $\Pi(s)$ — vacuum polarization. Emission of two soft photons with total c.m. energy not exceeding ΔE, $\omega_1 + \omega_2 < \Delta E$ give:

$$\sigma^{\delta\delta} = \tilde{\sigma}(s)\left(\frac{\alpha}{\pi}\right)^2 \frac{1}{2}\left\{\left[2\operatorname{Re}B(s)\ln\frac{m_e}{\lambda} + \frac{1}{2}L^2 - \frac{\pi^2}{3} - 2(L-1)\ell\right]^2 - \frac{2}{3}\pi^2(L-1)^2\right\}.$$

Finally the cross section of an additional solf e^+e^- pair production $E_+ + E_- < \Delta E$, $2m_e \ll \Delta E \ll E$ have the form (logarithmic accuracy):

$$\sigma^{e^+e^-} = \tilde{\sigma}(s)\left(\frac{\alpha}{\pi}\right)^2\left[\frac{1}{18}(L-2\ell)^3 - \frac{5}{18}(L-2\ell)^2 + \frac{1}{9}\left(\frac{28}{3} - \pi^2\right)(L-2\ell)\right], \quad (6)$$

we neglect in $\sigma^{e^+e^-}/\tilde{\sigma}$ the terms of order 10^{-4} since their contribution exceed the accuracy requirement to r.c. $\sim 10^{-1}\%$. The same reason is why we don't consider emission of soft heavy pairs such as $\mu^+\mu^-$, $\tau^+\tau^-$. The Simple evaluation shows that one may neglect the $\tau^+\tau^-$ vacuum polarization contribution.

We assume in derivation of (4,5,6) that the cross section $\tilde{\sigma}(s)$ is a smooth function of S: the change in argument due to emission of solf photons and pairs may be neglected.

Sum of (4,5,6) and taking into consideration the virtnal corrections (2) one obtains for cross section of single-photon e^+e^- pair annihilation in which allowed the emission of additional photons and e^+e^- pairs with the total energy don't exceed $\Delta E \ll E$:

$$\sigma(s) = \tilde{\sigma}(s)\left\{1 + \frac{\alpha}{\pi}\left[-2\ell(L-1) + \frac{3}{2}L + \frac{\pi^2}{2} - 2\right] + \left(\frac{\alpha}{\pi}\right)^2\left[\frac{1}{2}(-2\ell(L-1))^2 + \left(\frac{3}{2}L + \frac{\pi^2}{3} - 2\right)(-2\ell)(L-1) + L^2\left(-\frac{1}{3}\ell + \frac{11}{8} - \frac{\pi^2}{3}\right) + L\left(\frac{2}{3}\ell^2 + \frac{10}{9}\ell - \frac{203}{48} + \frac{11}{12}\pi^2 + 3\zeta(3)\right) - \frac{4}{9}\ell^3 - \frac{10}{9}\ell^2 - \frac{2}{3}\left(\frac{28}{3} - \pi^2\right)\ell\right]\right\} \quad (7)$$

Note that all the dependence on "photon mass" λ disappeared in sum (7).

2. Drell-Yan picture, renormalization group, K-factor.

One may consider the single-photon annihilation processas a Drell-Yan ones and apply to it's investigation the methods developed in QCD, based on renormalization group ideas and factorization of mass singularities. In frames of 11a (1) the cross section may be represended in form:

$$\sigma(s) = \int_{x_1+x_2 > 2 - \frac{\Delta E}{E}}^{1} dx_1 \int^1 dx_2 \sum_{A = e^+, e^-} D_{e^-}^A(x_1, s) D_{e^+}^{\bar A}(x_2, s) \tilde\sigma_{A\bar A}(x_1, x_2, s); \quad (8)$$

right hand side of (8) may be drawn in graphical form

$$\begin{array}{c}\text{(9)}\end{array}$$

D_e^A (x, s) is the distribution function - density of prolability to find out parton of sort A with energy fraction x and momentum square up to s. The restriction on the reqion of inteqration with respect to x follows from the requirement that the excluded energy carried away by the undetected particles must not exceed Δ E.

D_e^A (x, s) is free from mass sinqularisities. It obey the Lipatov equations (known in QCD as an equation Altarelly-Parizi).

With the accuracy requirement $\sim 0,1\%$ one may retain in sum (8) only term $A_1 = e-$. addendum $A = e^+$ qives contribution of order $\sim (\alpha L/\pi)^4 < 0,1\%$. We put now the quantity $\mathcal{D}_e^\ell = \mathcal{D}$ in form $\mathcal{D} = \mathcal{D}^{NS} + \mathcal{D}^S$. Sinqlet part $\mathcal{D}^S$ arise from such a mechanism of pair production

It has the form:

$$\mathcal{D}^S(x,s) = \frac{1}{8x}\left(\frac{\alpha L}{\pi}\right)^2 \left[\frac{4}{3}(1-x^3) + x(1-x) + 2x(1+x)\ln x\right]. \quad (10)$$

It's contribution don't exceed 0.5%.

Nonsinglet part $\mathcal{D}^{NS}$ satisfy the equation:

$$\mathcal{D}^{NS}(x,Q^2) = \delta(x-1) + \int_{m^2}^{Q^2} \frac{\alpha(Q'^2)}{2\pi} \frac{dQ'^2}{Q'^2} \int_x^1 \frac{dz}{z} P(z)\mathcal{D}^{NS}\left(\frac{x}{z}, Q'^2\right), \quad (11)$$

where $\alpha(Q^2)$ is a "running coupling constant":

$$\alpha(Q^2) = \alpha\Big/\left(1 - \frac{\alpha}{3\pi}\ln\frac{Q^2}{m^2}\right); \quad P(z) = \frac{1+z^2}{1-z} - \delta(1-z)\int_0^1 \frac{dx(1+x^2)}{1-x}. \quad (12)$$

We put here for completeness the set of Lipatov equations:

$$\mathcal{D}_e^e(x,s) = \delta(x-1) + \int_{m^2}^s \frac{dt\,\alpha(t)}{2\pi t}\left[\int_x^1 \frac{dy}{y}\mathcal{D}_e^e(y,t)P_e^e\left(\frac{x}{y}\right) + \int_x^1 \frac{dy}{y}\mathcal{D}_e^\gamma(y,t)P_\gamma^e\left(\frac{x}{y}\right)\right];$$

$$\mathcal{D}_e^{\bar{e}}(x,s) = \int_{m^2}^s \frac{dt\,\alpha(t)}{2\pi t}\left[\int_x^1 \frac{dy}{y}\mathcal{D}_e^{\bar{e}}(y,t)P_e^{\bar{e}}\left(\frac{x}{y}\right) + \int_x^1 \frac{dy}{y}\mathcal{D}_e^\gamma(y,t)P_\gamma^{\bar{e}}\left(\frac{x}{y}\right)\right];$$

$$\mathcal{D}_e^\gamma(x,s) = -\frac{2}{3}\int_{m^2}^s \frac{dt\cdot\alpha(t)}{2\pi t}\cdot\mathcal{D}_e^\gamma(x,t) + \int_{m^2}^s \frac{dt\,\alpha(t)}{2\pi t}\left[\int_x^1 \frac{dy}{y}\mathcal{D}_e^e(y,t)P_e^\gamma\left(\frac{x}{y}\right)\right.$$

$$\left. + \int_x^1 \frac{dy}{y}\mathcal{D}_e^{\bar{e}}(y,t)P_{\bar{e}}^\gamma\left(\frac{x}{y}\right)\right],$$

where $\alpha(t)$, $P_e^e(z) = P(z)$ are given in (12),

$$P_\gamma^e(z) = P_\gamma^{\bar{e}}(z) = z^2 + (1-z)^2, \quad P_e^\gamma(z) = P_{\bar{e}}^\gamma(z) = \frac{1}{z}(1+(1-z)^2).$$

Note that the divergences in (12) disappear in (11) and (8). This cancellation of divergences is the well known cancellation of infrared divergences.

The equation (10) in differential form

$$\frac{\partial \mathcal{D}_e^i(x,s)}{\partial \ln(s/m^2)} = -W_i \mathcal{D}_e^i(x,s) + \sum_j \int_x^1 dx' W_{j \to i}(x',x) \mathcal{D}_e^i(x',s), \quad (11\text{ a})$$

$$W_i = \sum \int_0^1 dx' W_{i \to j}(x,x').$$

have a simple interpretation: change of numler of partons of sort i into electron $\mathcal{D}_e^i$ when it's momentum square changed by $d \ln s/m^2$ goes from decay to another partons and creation when partons of different sorts decay. I.e. the evolution equation (11, 11 a) are the statistical balance equation.

Representation (8) is valid for arbitrary values $\Delta E/E \sim 1$, but the analytical formulas may be obtained in the case $\Delta E/E \ll 1$. In this case the condition $x_1 + x_2 > 2 - \frac{\Delta E}{E}$ is equivalent to $x_1 x_2 > 1 - \frac{\Delta E}{E}$. One may omit $\mathcal{D}^s$ and for smooth $\tilde{\sigma}(s)$ the equation (8) accept the form

$$\sigma(s) = \tilde{\sigma}(s) \cdot R\left(1 - \frac{\Delta E}{E}, s\right), \quad (13)$$

$$R(x,s) = \int_x^1 dx_1 \int_{x/x_1}^1 dx_2 \, \omega^{Ns}(x_1,s) \mathcal{D}^{Ns}(x_2,s). \quad (14)$$

The equation for $R(x,s)$ follows from (11):

$$R(x,s) = 1 + \int_{m^2}^{s} \frac{\alpha(t) dt}{\pi t} \int_x^1 d\tau \, P(\tau) R\left(\frac{x}{\tau}, t\right). \quad (15)$$

In the case $1 - x \ll 1$ equation (15) may be solved by means of Mellin transform. Result is:

$$R(x,s) = \frac{(1-x)^b}{\Gamma(1+b)} \exp\left\{\frac{1}{2} b \left(\frac{3}{2} - 2c\right)\right\} \quad (16)$$

C = 0,577 is the Enler constant.

Expansion of (16) in powers :

$$\sigma(s) = \tilde{\sigma}(s) \left\{ 1 + \frac{\alpha l}{\pi} \left(\frac{3}{2} - 2l\right) + \left(\frac{\alpha l}{\pi}\right)^2 \left(2l^2 - \frac{10}{3} l + \frac{11}{8} - \frac{\pi^2}{3}\right) \right\} \quad (17)$$

agrees with direct calculation (7).

For small resolution energy $\Delta E \ll E$ the expansion parameter (17) $\frac{\alpha}{\pi} \ln \frac{s}{m^2} \ln \frac{E}{\Delta E}$ may be of order of one. It happens, say, in production of heavy narrow resonances with quantum numbers of photon. Width or resonance is small compared to it's mass $\Delta E \sim \Gamma \ll M$. In this case the accuracy of (17) is insufficient. Besides the cross-section is a sharp function of energy in resonance region. So the general expression (8) is valid. It's accuracy may be increased by means K-factor

$$K = 1 + \frac{\alpha}{\pi}\left(\frac{\pi^2}{3} - \frac{1}{2}\right) \qquad (18)$$

in right hand part (8). The form of K-factor can be deduced from nonleading terms in (7).

We modify now the expression for $\mathcal{D}$ in (8) for two different organization of experiment: include or exclude events with e^+e^- pair production in taking on statistic.

To obtain the quantity $\mathcal{D}^{NS} = \mathcal{D}^{\gamma}$ for the case when pairs are excluded we first may put $\alpha(Q^2) = \alpha = 1/137$ in integral equation to take into account the virtual pair production we add to solution appropriate terms of order $(\alpha/\pi)^2 L^3$ and $(\alpha/\pi)^2 L^2$. Then we replace $L \to L-1$ in 11a solutions to agree with the soht photon's limit investigated by Yennie Franchi and Suura. We obtain:

$$\mathcal{D}^{\gamma}(x,s) = \frac{1}{2}\beta(1-x)^{\frac{\beta}{2}-1}\left(1 + \frac{3}{8}\beta - \frac{\beta^2}{48}\left(\frac{1}{3}L + \pi^2 - \frac{47}{8}\right)\right) - \frac{1}{4}\beta(1+x) +$$
$$+ \frac{\beta^2}{32}\left(4(1+x)\ln\frac{1}{1-x} + \frac{1+3x^2}{1-x}\ln\frac{1}{x} - 5 - x\right), \beta = \frac{2\alpha}{\pi}(L-1), L = \ln s/m_e^2. \qquad (19)$$

In the case of allowed pairs we are to to modify $\mathcal{D}^S$ and part of $\mathcal{D}^{NS}$ in such a way to take into account soft photon emission:

$$\mathcal{D}^{e^+e^-}(x,s) = \theta\left(1-x-\frac{2m}{E}\right)\left(\frac{\alpha}{\pi}\right)^2 \left\{\frac{(1-x-\frac{2m}{E})^{\beta/2}}{12(1-x)}\left(x-\frac{5}{3}\right)^2\left(1+x^2+\right.\right.$$
$$\left.\left.+\frac{1}{6}\beta\left(x-\frac{5}{3}\right)\right)+\frac{1}{4}L^2\left(\frac{2}{3}\frac{(1-x^3)}{x}+\frac{1-x}{2}+(1+x)\ln x\right)\right\}, \tag{20}$$

where $\theta(x) = \begin{cases} 1, & x>0 \\ 0, & x<0 \end{cases}$ - function reflect the threshold behavior $\mathcal{L} = L + 2\ln(1-x)$

So, finally:
$$\sigma(s) = \int_{\xi_1}^{1} dx_1 \int_{\xi_2}^{1} dx_2 \, \mathcal{D}(x_1,s)\mathcal{D}(x_2,s)\,\tilde{\sigma}(x_1 x_2 s)\cdot K, \tag{21}$$

$$\mathcal{D} = \begin{cases} \mathcal{D}^\gamma + \mathcal{D}^{e^+e^-}, & \text{Pairs allowed} \\ \mathcal{D}^\gamma & \text{Pairs excluded} \end{cases} \tag{22}$$

We give here also a convolution formula (pairs allowed). Changing order of integration in (21) we obtain:

$$\sigma(s) = \int_0^{\varepsilon} dx \, \tilde{\sigma}(s(1-x)) F(x,s),$$

$$F(x,s) = \beta x^{\beta-1}\left(1+\frac{\alpha}{\pi}\left(\frac{\pi^2}{3}-\frac{1}{2}\right)+\frac{3}{4}\beta-\frac{1}{24}\beta^2\left(\frac{1}{3}L+2\pi^2-\frac{3\pi}{4}\right)\right)-\beta\left(1-\frac{1}{2}x\right)+$$

$$+\frac{1}{8}\beta^2\left(4(2-x)\ln\frac{1}{x}-\frac{1}{x}(1+3(1-x)^2)\ln(1-x)-6+x\right)+\left(\frac{\alpha}{\pi}\right)^2\left\{\frac{1}{6x}\left(x-\frac{2m}{E}\right)^{\beta/2}\cdot\left(L+\right.\right.$$

$$\left.+2\ln x-\frac{5}{3}\right)^2\left(2-2x+x^2+\frac{1}{3}\beta(L+2\ln x-\frac{5}{3})\right)+\frac{1}{2}L^2\left(\frac{2(1-(1-x)^3)}{3(1-x)}+\right.$$

$$\left.\left.+(2-x)\ln(1-x)+\frac{1}{2}x\right)\right\}\theta\left(x-\frac{2m}{E}\right) \tag{22}$$

The quantities ε, ε_1, ε_2 in (21) (22) are to be choised according to experimental set-up.

II Scattering channel.

Scattering of electrons on nucleons and nucleys is the main instrument for testing the hadronic structure of target. Now accent is made on deep inelastic scattering of high electrons (and muons). Extraction of information about strong interactions imples the knowlebge of r.c. of electromagnetic nature. A lot of attention was paid to this subject from theoretical as well as experimental point of view.

For the case of soft photons the r.c. in higher orders of p.t. are evaluated by means of well known method of exponentiation of lowest order p.t. result. But the situation for hard photons is much more complicated.

Consider now the process of deep inelastic scattering (dis) of high-energy electron. It's cross-section with r.c. taken into account will be described of such a dyagram

and have a form

$$\frac{\varepsilon_2 d^3\sigma(P_1,P_2)}{d^3 P_2} = \int dz_1 \int dz_2 \, z_2^{-2} \sum_{AA'} \mathcal{D}_e^A(z_1,Q^2)\bar{\mathcal{D}}_{A'}^e(z_2,Q^2) \frac{\tilde{\varepsilon}_2 d^3\sigma_{AA'}^{hard}(z_1P_1,\frac{P_2}{z_2})}{d^3\tilde{P}_2}\bigg|_{\tilde{P}_2=\frac{P_2}{z_2}} \quad (1)$$

The structure functions $\mathcal{D}_e^A$, $\bar{\mathcal{D}}_A^e$ are related by Gribov-Lipatov equation

$$\mathcal{D}_e^A(z,Q^2) = \bar{\mathcal{D}}_A^e(z,Q^2).$$

Its convenient to use the variables used commonly in dis description

$$q = P_1 - P_2, \quad \nu = 2P P_1 = 2M\varepsilon_1, \quad x = \frac{-q^2}{2Pq}, \quad y = \frac{2Pq}{\nu}, \quad \tau = \frac{M^2}{\nu}, \quad (2)$$
$$Q^2 = -q^2 = \nu x y.$$

Using variables (2) and relabeling $\mathcal{D}_e^\ell \equiv \mathcal{D}$ the equation (1) takes a form:

$$\frac{d\sigma(x,y)}{dxdy} = \int_0^1 \int_0^1 \frac{dz_1 dz_2}{z_1 z_2^2} \mathcal{D}(z_1, Q^2) \mathcal{D}(z_2, Q^2) \frac{y \, d\sigma^{hard}(\tilde{x},\tilde{y})}{\tilde{y} \, d\tilde{x} \, d\tilde{y}} \quad (3)$$

where

$$\tilde{x} = \frac{z_1 y x}{z_1 z_2 + y - 1}, \quad \tilde{y} = \frac{z_1 z_2 + y - 1}{z_1 z_2}, \quad \tilde{Q}^2 = \frac{z_1}{z_2} Q^2.$$

The quantity $\mathcal{D}$ depends on set up of the experiment: $\mathcal{D} = \mathcal{D}^\delta$ (pair excuded), or $\mathcal{D} = \mathcal{D}^\delta + \mathcal{D}^{e^+e^-}$ (pair allowed). Lower limits of z_1, z_2 integrations in (3) depends on experimental conditions (see below). Quantities $\mathcal{D}^\delta, \mathcal{D}^{e^+e^-}$ are given above, $L = \ell_n Q^2/m_e^2$. Hard cross-section entering (3) in 11a coincides with the cross-section in lowest order of p.t. on coupling $\alpha(Q^2)$:

$$\frac{d\sigma^{hard}(x,y)}{dxdy}\bigg|_{ela} = \frac{d\sigma^{(0)}(x,y,Q^2)}{dxdy},$$

$$\frac{d\sigma^{(0)}(\tilde{x},\tilde{y},\tilde{Q}^2)}{\tilde{y} d\tilde{x} d\tilde{y}} = \frac{z_2}{z_1} \cdot \frac{4\pi\alpha\left(\frac{z_1 Q^2}{z_2}\right)}{Q^2 x y^2} \bigg[\left(1-y-\frac{M^2}{Q^2}x^2 y^2\right) \cdot \frac{Q^2}{2Mx} W_2\left(\frac{xyz_1}{z_1 z_2 + y - 1}\right), \quad (4)$$

$$\frac{z_1 Q^2}{z_2}\bigg) + xy^2/M \, W_1\left(\frac{xyz_1}{z_1 z_2 + y - 1}, \frac{z_1 Q^2}{z_2}\right)\bigg], \quad \alpha(Q^2) = \alpha/|1 - \Pi(q^2)|,$$

$W_{1,2}(x, Q^2)$ — structure functions of target. (Don't confuse them with structure functions $\mathcal{D}(x,s)$ of partons).

Essentially inelastic contribution to (3) correspond to condition

$$(p+\tilde{q})^2 = M_{th}^2, \quad \tilde{q} = z_1 p_1 - \frac{1}{z_2} p_2, \quad \tilde{q}^2 = -\tilde{Q}^2, \quad (5)$$

where $M_{th} = M + m_\Lambda$ — inelastic threshold of target. In terms of (2) we have the region of integration (shaded area in Figure).

$$z_1 z_2 + y - 1 - xy z_1 > z_1 \delta, \quad \delta = (m_{th}^2 - m^2)/v. \tag{6}$$

Elastic peak and it's radiative tail (TEP) correspond to condition

$$(p+\tilde{q})^2 = M^2, \quad z_1 z_2 + y - 1 - xy z_1 = 0 \tag{7}$$

TEP contribution orginated from curve (7) don't equal to zero due to δ - character of structure functions $W_{1,2}$ in elastic limit:

$$M W_1 (x, Q^2) = \tfrac{1}{2} \delta(x-1) (F_1(q^2) + F_2(q^2))^2,$$
$$\tfrac{1}{M} W_2 (x, Q^2) = \tfrac{2}{Q^2} \delta(x-1) \left(F_1^2(q^2) + \tfrac{Q^2}{4M^2} F_2^2(q^2) \right), \tag{8}$$

where $F_{1,2}$ - Pauli form-factors.

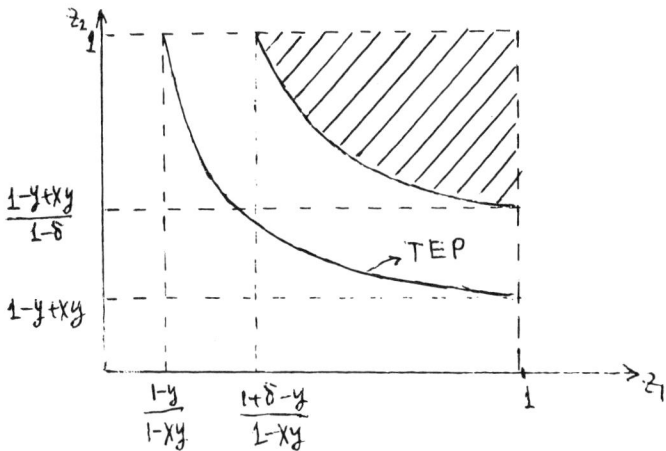

Extension beyond the lla

Problem of accounting of the terms $\alpha (\alpha L)^n$ in QCD is solved. Structure functions $\mathcal{D}(x, Q^2)$ as well as $d\sigma^{hard}$ are to be modified. In QED parameter $\tfrac{\alpha}{\pi} \ln Q^2/m_e^2 \leq 0,1$ is really small so it's sufficient to use (3) with function $\mathcal{D}(x,Q^2)$ given above and to modify $d\sigma^{hard}/dx dy$:

$$\frac{d\sigma^{hard}}{dxdy} = \frac{d\sigma^{(0)}(x,y)}{dxdy} + \frac{d\sigma^{(1)}(x,y)}{dxdy} \quad (9)$$

where $d\sigma^{(1)}$ is the hard correction to $d\sigma^{(0)}$ of relative order α. To find it out it's necessary to calculate $d\sigma/dxdy$ in α^3 order and then substract from this expression the right hand part of (3) in the same order with $d\sigma^{hard} = d\sigma^{(0)}$. The result is

$$\frac{d\sigma^{hard}(x,y)}{dxdy} = \frac{d\sigma^{(0)}(x,y)}{dxdy}\left\{1 - \frac{\alpha}{\pi}\left(\frac{1}{2}\ell_m^2\left(\frac{1-xy}{1-z_+}\right) + \frac{1}{2} + \frac{\pi^2}{6} + f\left(\frac{1-y-zxy}{(1-xy)(1-z_+)}\right)\right) + $$

$$+ \frac{\alpha}{\sqrt{x}}\int_0^{z_+}dz\int_{z_-}^{z_+}d\tau\left\{\frac{xy(\tau+z_+)}{\sqrt{y^2+4\tau xy}}\left[\frac{G(z,z_2)}{(1-z)^2} + \frac{G(z,z_1)}{(1-y+z)^2} + \frac{2G(z,z)}{xyz^2(\tau+z_+)}\right]+$$

$$+ \frac{\mathcal{P}}{1-z}\left[\frac{1}{(1-xy)|z-z_1|}\left(\left(1+\frac{1}{z^2}\right)G(z,z) - \left(1+\frac{1}{z_1^2}\right)G(z,z_1)\right) - \frac{1}{(1-z_+)|z-z_2|}\left((1+$$

$$+ \frac{1}{z^2})G(z,z) - (1+\frac{1}{z_2^2})G(z,z_2))\right] + \frac{\alpha(Q^2)MW_2(x',Q'^2)}{z_2(1-z)xy\tau}\left[-\frac{2(1-z)(1-y)}{\sqrt{y^2+4\tau xy}} + \right.$$

$$\left. + \frac{(1-y-z)(z_2-z)}{|z_2-z|} - \frac{(1-z(1+y))(z-z)}{|z-z|}\right]\right\},$$

where

$$f(x) = \int_0^x\frac{dt}{t}\ell_n(1-t), \quad z_1 = \frac{1-y+z}{1-xy}, \quad z_2 = \frac{1-z}{1-z_+}, \quad z_+ = y(1-x), \quad x' = \frac{xyz}{xyz+z},$$

$$Q'^2 = Q^2 z, \quad z_\pm(z) = [2xy(\tau+z_+)]^{-1}[2xy(\tau+z)+(z_+-z)(y\pm\sqrt{y^2+4\tau xy})].$$

Symbol $\mathcal{P}$ means the integration in meaning of main value. Meaning of integration variable

$$z = (M_x^2 - M^2)/V, \quad z = Q'^2/Q^2$$

M_x - invariant mass of adrons created after collizion.

Note that $d\sigma^{hard}$ don't contain any sinqularities: collinear at $z = z_{1,2}$, infrared at $z = 1$ and non physical at $z = z_+$.

Region of integration in plan (z_1, z) is shown in picture

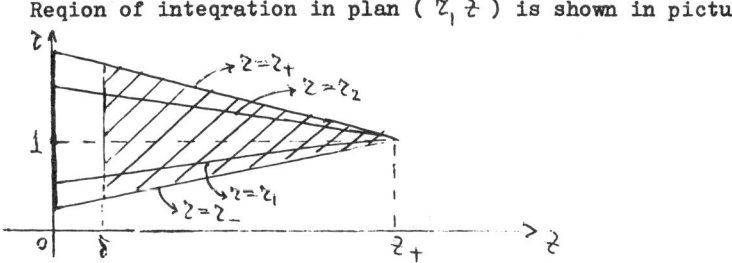

Shaded area correspond to really inelastic scattering Solid line $z = 0$ correspond to radiative tail of elastic scattering. In conclusion it's necessary to say that the advantage of peak approximation (photon emitting angle to initial and final electron momenta is small) in 11a ($d\sigma \sim d\sigma_i \cdot \frac{\alpha L}{\pi}$) may fail if target structure functions $W_{1,2}$ not smooth. More preferable may be the noncollinear kinematic of photon emission. The lost of quantity L compensated by the advantage in parton cross section magnitude. So the nonlending terms in will be important.

In our work [3] we calculate the corrections of such type. For reliable structure functions $W_{1,2}$ the calculation was made by D.Bardeen et al. The contributions from two noncollinear real hard photons as well as e^+e^- pair was done by N.P.Merenkov.

References

1   Yad.Fiz. 41 (1985), p.733. E.A.Kuraev, F.S.Fadin.

2   Yad.Fiz. 47 (1988), p.1593. E.A.Kuraev, N.P.Merenkov, V.S.Fadin.

3   Yad.Fiz. 45(1987), p.782. E.A.Kuraev, N.P.Merenkov, V.S.Fadin.

APPLICATION OF STRUCTURE FUNCTION METHOD TO POLARIZED PARTICLES

V.S.Fadin

Institute of Nuclear Physics, Novosibirsk, USSR

ABSTRACT

In this paper the structure function method is developed for radiative correction calculations for polarized particles. Structure functions for longitudinally polarized particles are calculated. Suitable formalism for arbitrary polarized particles is presented.

1. INTRODUCTION

As it is well known, in the Standard Model of the electroweak interactions the most important terms in radiative corrections to various processes at high energy and large momentum transfer are those, which contain large logarithms of the type $\ln Q^2/m^2$, where Q^2 stands for the squared momentum transfer, m denotes electron mass. Mass singularity of these terms is originated from collinear divergences, i.e. the divergences of integrals at zero angles between the momenta of interacting massless particles where the denominators of corresponding propagators turn out to be zero. In the QED case these propagators are photon and electron ones. Pole approximations suitable for the calculation of the main logarithmic contributions are known as equivalent photon [1,2] and quasireal electron [3,4] methods.

In the cases when high accuracy of calculations is needed, the presence of the large logarithms enforces one to sum up perturbation series in spite of the small value of the coupling constant α. The method of such summation was developed in the QCD where the coupling constant α_s is much larger than α (see, for example, review papers [5,6]). The method is based on the factorization theorems permitting to divide the contributions of large and small distances. As a result, the cross-sections of the so-called hard processes, i.e. the processes with large momentum transfers, are presented in the form of convolution structure functions of initial particles and fragmentation functions of final particles with a hard part of the cross-sections. The whole contribution of large distances is shown to be included in the structure and fragmentation functions. The hard part of the cross-section is determined by distances of the order of $1/Q$. Therefore, being expressed via the coupling constant at these distances, $\alpha(Q^2)$, it contains no large logarithms and can be calculated in terms of perturbation theory. The structure and fragmentation functions obey the equations of renormalization-group type [7-9] which are known in the QCD as Altarelle-Parisi equations.

The use of the structure function method developed in the QCD for radiative corrections (RC) calculations in the QED has been initiated in Ref.[10], where RC to one photon e^+e^- annihilation cross-section were calculated. In an analogous case, this method was applied in the Z^0-boson energy region in Ref.[11]. Afterwards the method has been used by several groups of authors (see, for example, Refs.[12-22]) for the calculation of total and differential cross-section with different degree of accuracy.

It is known that experiments with polarized e^+e^- beams can provide the most stringent testing of the Standard Model of the strong and electroweak interactions and be the most sensitive to possible manifestations of new Physics (see, for example, Refs.[23-25]). Production of polarized e^+e^- beams is planned at SLC in the nearest future. The question on polarized beams at the LEP collider is also discussed. Thus it seems urgent to develop the structure function method for RC calculations in the case of polarized particles. The corresponding formalism is presented below. For definetness we deal with the total annihilation cross-section of polarized e^+e^- pairs. Beyond doubt the applicability region of the structure function expressions derived is not restricted by this case. A wide range of questions can be considered in a simular way.

2. Structure Functions for Longitudinal Polarized Particles

Let us begin with the case of longitudinal polarizations. This case can be considered in a simple manner one the basis of Refs. [9,10]. The cross-section $\sigma_{e_\lambda \bar{e}_{\bar{\lambda}}}(s)$ of e^+e^- annihilation (λ ($\bar{\lambda}$) is $e^-(e^+)$ helicity and $\sqrt{s}$ is total energy in the c.m.s.) is presented as follows (cf. [10]):

$$\sigma_{e_\lambda \bar{e}_{\bar{\lambda}}}(s) = \sum_{A,z;B,\bar{z}\ \varepsilon_1\ \varepsilon_2} \int_{\varepsilon_1}^{1}\int_{\varepsilon_2}^{1} dx_1 dx_2\, \mathcal{D}_{e_\lambda}^{Az}(x_1,s) \quad (1)$$

$$\cdot \mathcal{D}_{\bar{e}_{\bar{\lambda}}}^{B\bar{z}}(x_2,s)\, \sigma_{Az B\bar{z}}^{(hard)}(x_1 x_2 s).$$

Here $\mathcal{D}_{e_\lambda}^{Az}(x_1,s)$ and $\mathcal{D}_{\bar{e}_{\bar{\lambda}}}^{B\bar{z}}(x_2,s)$ are the so-called structure functions having clear physical meaning [8]: $\mathcal{D}_{A\lambda}^{Bz}(x,Q^2)$ is the density of the probability to find in a physical particle A with helicity λ a bare particle B with helicity z, longitudinal momentum share X and the squared transverse momentum up to Q^2. $\sigma_{A\lambda B\bar{z}}^{(hard)}(Q^2)$ is the so-called hard cross-section which is free from electron mass singularities being expressed via the "running" coupling

$$\alpha(Q^2) = \frac{\alpha}{1-\mathcal{P}(Q^2)} \quad (2)$$

where $\mathcal{P}(Q^2)$ is the real part of the photon vacuum polarization operator.

The limits of integration $\varepsilon_1, \varepsilon_2$ are determined by the experimental conditions.

The structure functions $\mathcal{D}(x, Q^2)$ are universal, that is independent of process under consideration. They obey the equations of renormalization group type [7-10]:

$$\mathcal{D}_{A\lambda}^{BZ}(x,s) = \delta_A^B \delta_\lambda^Z \delta(1-x) + \int_{m^2}^{s} \frac{dQ^2}{Q^2} \frac{d(Q^2)}{2\pi}$$

$$\times \sum_{C,\gamma} \int_x^1 \frac{dy}{y} \mathcal{D}_{A\lambda}^{C\gamma}(y, Q^2) P_{C\gamma}^{BZ}\left(\frac{x}{y}\right), \quad (3)$$

where $P_{A\lambda}^{BZ}(x)$ are the so-called Altarelli-Parisi Kernels. In Ref.[9] they are presented for the QCD case. It is easy to transform them into the QED case interesting to us:

$$P_{e\lambda}^{eZ}(x) = P_{\bar{e}\lambda}^{\bar{e}Z}(x) = \delta_\lambda^Z P_e^e(x), \quad P_{\gamma\lambda}^{\gamma Z}(x) = -\frac{2}{3}\delta_\lambda^Z \delta(1-x), \quad (4)$$

$$P_e^e(x) = \frac{1+x^2}{1-x} - \delta(1-x)\int_0^1 \frac{dz(1+z^2)}{1-z};$$

$$P_{e\lambda}^{\gamma Z}(x) = P_{\bar{e}\lambda}^{\gamma Z}(x) = \frac{1}{2}P_e^\gamma(x) + \frac{\lambda z}{2|\lambda z|} \Delta P_e^\gamma(x),$$

(5)

$$P_e^\gamma(x) = \frac{1+(1-x)^2}{x}, \quad \Delta P_e^\gamma(x) = \frac{1-(1-x)^2}{x} = 2-x;$$

$$P_{\gamma\lambda}^{eZ}(x) = P_{\gamma\lambda}^{\bar{e}Z}(x) = \frac{1}{2}P_\gamma^e(x) + \frac{\lambda z}{2|\lambda z|} \Delta P_\gamma^e(x),$$

(6)

$$P_\gamma^e(x) = x^2 + (1-x)^2, \quad \Delta P_\gamma^e(x) = x^2 - (1-x)^2 = 2x-1$$

The singularities at the point $X = 1$ involved in $P_e^e(X)$ (4) are cancelled under integration $P_e^e(X)$ with any smooth function of X. This cancellation is known as the infrared singularity cancellation with taking real photon emission (the first term in P_e^e) and virtual corrections (the second term in P_e^e) into account.

With the use of Kernels (4)-(6) equations (3) can be solved perturbatively. In the QED case one can restrict oneself to several first iterative terms, since the effective expansion parameter $\alpha \ln Q^2/m^2$ is much smaller than 1 at any reasonable values of energy.

The small $1-X$ region, where the value of $\ln \frac{1}{1-X}$ becomes large just as $\ln Q^2/m^2$ needs particular treatment. This region is important for the structure functions $\mathcal{D}_{e\lambda}^{ez} = \mathcal{D}_{\bar{e}\lambda}^{\bar{e}z}$ which are singular here. In the case of fine energy resolution (due to either the experimental resolution or the resonance behaviour of the cross-section under consideration) this region determines the cross-section of a process. Since the e^+e^- annihilation cross-section we are interested in is expressed via the structure functions $\mathcal{D}_{e\lambda}^{ez} = \mathcal{D}_{\bar{e}\lambda}^{\bar{e}z}$ let us consider these functions in more detail.

It is convenient to divide $\mathcal{D}_{e\lambda}^{ez}$ into the singlet (S) and nonsinglet (NS) parts:

$$\mathcal{D}_{e\lambda}^{ez} = \left(\mathcal{D}^{NS}\right)_\lambda^z + \left(\mathcal{D}^S\right)_\lambda^z ,$$

$$\left(\mathcal{D}^S\right)_\lambda^z = \mathcal{D}_{e\lambda}^{\bar{e}z} , \quad \left(\mathcal{D}^{NS}\right)_\lambda^z = \mathcal{D}_{e\lambda}^{ez} - \mathcal{D}_{e\lambda}^{\bar{e}z}$$

(7)

The singlet contribution $\left(\mathcal{D}^S\right)_\lambda^z = \mathcal{D}_{e\lambda}^{\bar{e}z}$ appears only from the α^2-order. At this order one can easily obtain from (3)

$$\mathcal{D}_{e\lambda}^{\bar{e}\,z}(x,Q^2) = \left(\frac{\alpha}{2\pi}\right)^2 \frac{1}{2}\left(\ln\frac{Q^2}{m^2}\right)^2 \int_x^1 \frac{dy}{y}\sum_{z'} P_{\gamma z'}^{\bar{e}\,z}\left(\frac{x}{y}\right)$$

$$\times \int_y^1 \frac{dz}{z} P_{e\lambda}^{\gamma z'}\left(\frac{y}{z}\right)\delta(1-z) = \left(\frac{\alpha}{4\pi}\ln\frac{Q^2}{m^2}\right)^2 \quad (8)$$

$$\times\left[\frac{1}{3x}(1-x)(4+7x+4x^2)-2(1+x)\ln\frac{1}{x}\right.$$

$$\left. -\frac{\lambda z}{|\lambda z|}\left(2(1+x)\ln\frac{1}{x}-5(1-x)\right)\right].$$

This contribution is not singular at $x \to 1$.

The nonsinglet part satisfies the equation of type (3) with the only term containing $P_{e\lambda}^{ez}$ in the right hand side. It is seen from formulae (4) that this Kernel is diagonal with respect to the helicity indexes in accordance with the helicity conservation. Therefore, the nonsinglet part of $\mathcal{D}_{e\lambda}^{ez}$ is proportional to δ_λ^z i.e.

$$\left(\mathcal{D}^{NS}\right)_\lambda^z = \delta_\lambda^z \mathcal{D}^{NS} \quad (9)$$

where $\mathcal{D}^{NS}$ is the nonsinglet part of structure function $\mathcal{D}_e^e$ summed over polarisation states of the final electron and averaged over those of the initial electron. This function was considered in detail in the paper [10].

It is convenient to separate explicitly from the $\mathcal{D}^{NS}$ the contribution of real e^+e^- pair production which should be or should not be taken into account depending on the experimental conditions. The part $\mathcal{D}^{(\gamma)}(x,Q^2)$ of the structure function $\mathcal{D}^{NS}$ obtained by omitting the contribution of the real e^+e^- pairs can be presented in the form

$$\mathcal{D}^{(\gamma)}(x,s) = \frac{\beta}{2}(1-x)^{\beta/2-1}\left(1+\frac{3\beta}{8}-\frac{\beta^2}{48}\left(\frac{L}{3}+\pi^2\right.\right.$$
$$\left.\left.-\frac{47}{8}\right)\right)-\frac{\beta}{4}(1+x)+\frac{\beta^2}{32}\left(4(1+x)\ln\frac{1}{1-x}+\frac{1+3x^2}{1-x}\ln\frac{1}{x}-5-x\right), \quad (10)$$

where

$$\beta = \frac{2\alpha}{\pi}(L-1), \quad L = \ln\frac{s}{m^2}. \quad (11)$$

The function $\mathcal{D}^{(\gamma)}(x,s)$ incorporates the effects of soft bremmstrahlung in all orders of perturbation theory. Hard photon contributions are taken into account up to the α^2-order.

The contribution of the real e^+e^- pair emission $\mathcal{D}^{(e)}(x,s)$ to $\mathcal{D}^{NS}$ in the same approximation as the $\mathcal{D}^{(\gamma)}(x,s)$ has the form

$$\mathcal{D}^{(e)} = \left(\frac{\alpha}{\pi}\right)^2\left[1-x-\frac{4m}{\sqrt{s}}\right]^{\beta/2}\frac{1}{12(1-x)}\left(\mathcal{L}-\frac{5}{3}\right)^2 \quad (12)$$
$$\times\left(1+x^2+\frac{\beta}{6}\left(\mathcal{L}-\frac{5}{3}\right)\right)\theta\left(1-x-\frac{4m}{\sqrt{s}}\right),$$

where $\theta(x)$ is the theta-function,

$$\mathcal{L} = \ln\left(\frac{s(1-x)^2}{m^2}\right). \quad (13)$$

Simularly to Ref.[10], we have neglected the production of additional $\mu^+\mu^-$, $\tau^+\tau^-$ pairs and hadrons, as well as their contributions to the structure functions via vacuum polarization.

Thus, the expression for $\mathcal{D}_{e\lambda}^{e2}$ has the form

$$\mathcal{D}_{e\lambda}^{e2} = \delta_\lambda^2 \left(\mathcal{D}^{(\gamma)} + \mathcal{D}^{(e)} \right) + \mathcal{D}_{e\lambda}^{\bar{e}2} \qquad (14)$$

where $\mathcal{D}^{(\gamma)}, \mathcal{D}^{(e)}$ and $\mathcal{D}_{e\lambda}^{\bar{e}2}$ are given by (10), (12) and (8), respectively.

3. Generalization to arbitrary polarizations

In the general case, for the description of polarization of particle A the probabilities w_λ^A to find helicity λ are not sufficient, and we need to introduce a polarization density matrix $\rho_{\lambda\lambda'}^A$. The probabilities w_λ^A are given by the diagonal terms of this matrix: $w_\lambda^A = \rho_{\lambda\lambda}^A$.

Similarly, in the general case it is impossible to confine oneself to functions $\mathcal{D}_{A\lambda}^{B\tau}$ and we need to define a certain structure density matrix. The way of definition of such a matrix is indicated by the physical meaning of $\mathcal{D}_{A\lambda}^{B\tau}(x,Q^2)$ which was already discussed: it is the probability density to find in a physical particle A with helicity λ a bare particle B with helicity τ, longitudinal momentum share X and the squared transverse momentum up to Q^2, i.e.

$$\mathcal{D}_{A\lambda}^{B\tau}(x,Q^2) = \sum_N \int \frac{d^3 p_B}{(2\pi)^3 \, 2E_B} \delta\left(X - \frac{p_{B\parallel}}{p_A}\right) \quad (15)$$

$$\times \left| \langle B,\tau; N | U(0,-\infty) | A,\lambda \rangle \right|^2$$

Here $U(t,t_o)$ is the time translation operator in the interaction representation, summation is done over all states containing the particle B. The matrix element (15) must be calculated in the infinite momentum frame of the particle A ; the squared transverse to P_A momenta must be bounded to Q^2. The corresponding structure density matrix can be defined as

$$\mathcal{D}^{B\zeta\zeta'}_{A\lambda\lambda'}(x, Q^2) = \sum_N \int \frac{d^3 P_B}{(2\pi)^3 2 E_B} \delta\left(x - \frac{P_{B\parallel}}{P_A}\right) \quad (16)$$

$$\times \langle B, \zeta; N | U(0, -\infty) | A, \lambda \rangle \langle B, \zeta'; N | U(0, -\infty) | A, \lambda' \rangle^*.$$

With this definition the cross-section of annihilation of electron and positron with polarization density matrixes $\rho^e_{\lambda\lambda'}$ and $\rho^{\bar{e}}_{\bar{\lambda}\bar{\lambda}'}$ has the form (cf. (1))

$$\sigma(s) = \sum_{\lambda,\lambda',\bar{\lambda},\bar{\lambda}'} \rho^e_{\lambda\lambda'} \rho^{\bar{e}}_{\bar{\lambda}\bar{\lambda}'} \sum_{\substack{A, \zeta, \zeta' \\ B, \bar{\zeta}, \bar{\zeta}'}} \int_{\varepsilon_1}^{1} \int_{\varepsilon_2}^{1} dx_1 dx_2 \quad (17)$$

$$\times \mathcal{D}^{A\zeta\zeta'}_{e\lambda\lambda'}(x_1, s) \mathcal{D}^{B\bar{\zeta}\bar{\zeta}'}_{\bar{e}\bar{\lambda}\bar{\lambda}'}(x_2, s) \sigma^{(hard)}_{A\zeta\zeta' B\bar{\zeta}\bar{\zeta}'}(x_1 x_2 s).$$

Here $\sigma^{(hard)}_{A\zeta\zeta' B\bar{\zeta}\bar{\zeta}'}$ stands for the quantity which gives hard cross-section in convolution with corresponding polarization density matrixes $\rho^A_{\zeta\zeta'}$ and $\rho^B_{\bar{\zeta}\bar{\zeta}'}$. It is seen from (17) that $\sum_{\lambda\lambda'} \mathcal{D}^{B\zeta\zeta'}_{A\lambda\lambda'} \cdot \rho^A_{\lambda\lambda'}$ is the unnormalized polarization matrix density of the particle B.

The structure matrixes $\mathcal{D}^{B\zeta\zeta'}_{A\lambda\lambda'}$ obey the equations (cf. (3)):

$$\mathcal{D}^{B\zeta\zeta'}_{A\lambda\lambda'}(x, s) = \delta^B_A \delta^\zeta_\lambda \delta^{\zeta'}_{\lambda'} \delta(1-x)$$

$$+ \int_{m^2}^{s} \frac{dQ^2}{Q^2} \frac{d(Q^2)}{2\pi} \sum_{C,\sigma,\sigma'} \int_{x}^{1} \frac{dy}{y} \mathcal{D}_{A\lambda\lambda'}^{C\sigma\sigma'}(y,Q^2) P_{C\sigma\sigma'}^{B22'}\left(\frac{x}{y}\right), \quad (18)$$

where $P_{A\lambda\lambda'}^{B22'}$ are the generalised Altarelli-Parisi Kernels. Using definition (16) it is easy to obtain (cf.[9]):

$$P_{A\lambda\lambda'}^{B22'}(x) = \frac{x(1-x)}{2 p_\perp^2} \sum_{c,\sigma} V_{A\lambda}^{B2C\sigma} V_{A\lambda'}^{B2'C\sigma *} \quad (19)$$

where $V_{A\lambda}^{B2C\sigma}$ is $A \to BC$ vertex, x is part of momentum of the particle A carried away by the particle B, $p_\perp$ denotes the transverse to p_A component of the particle B momentum.

Straightforward calculations give [26]

$$P_{e\lambda\lambda'}^{e22'}(x) = P_{\bar{e}\lambda\lambda'}^{\bar{e}22'}(x) = \delta_\lambda^2 \delta_{\lambda'}^{2'}\left[P_e^e(x) + (\delta_{\lambda\lambda'}-1)\Delta P_e^e(x)\right], \quad (20)$$

$$\Delta P_e^e(x) = 1-x;$$

$$P_{e\lambda\lambda'}^{\gamma 22'}(x) = \delta_{\lambda\lambda'} \delta^{22'} P_{e\lambda}^{\gamma 2}(x); \quad (21)$$

$$P_{\gamma\lambda\lambda'}^{e22'}(x) = \delta_{\lambda\lambda'} \delta^{22'} P_{\gamma\lambda}^{e2}(x). \quad (22)$$

The kernels $P_e^e, P_{e\lambda}^{\gamma 2}$ and $P_{\gamma\lambda}^{e2}$ are given by Eqs. (4)-(6).

From (20)-(22) it is seen that only Kernels $P_{e\lambda\lambda'}^{e22'} = P_{\bar{e}\lambda\lambda'}^{\bar{e}22'}$ are nondiagonal with respect to the same particle helicities. Therefore, it follows from Eq. (18) that only the structure matrixes $\mathcal{D}_{e\lambda\lambda'}^{e22'} = \mathcal{D}_{\bar{e}\lambda\lambda'}^{\bar{e}22'}$

can have such nondiagonal terms different from Born ones. Let us divide these matrices into singlet and nonsinglet parts, as it was done for $\mathcal{D}_{e\lambda}^{ez}$ (7). Then the singlet part will be diagonal:

$$(\mathcal{D}^{(S)})_{\lambda\lambda'}^{zz'} = \mathcal{D}_{e\,\lambda\lambda'}^{\bar{e}\,zz'} = \delta_{\lambda\lambda'}\delta^{zz'}\mathcal{D}_{e\lambda}^{\bar{e}z}, \qquad (23)$$

where $\mathcal{D}_{e\lambda}^{\bar{e}z}$ is given by Eq.(8).

The nonsinglet part of $\mathcal{D}_{e\lambda\lambda'}^{ezz'}$ has the following form (cf. (9)) owing to chirality conservation:

$$(\mathcal{D}^{NS})_{\lambda\lambda'}^{zz'} = \delta_{\lambda}^{z}\delta_{\lambda'}^{z'}\mathcal{D}_{\lambda\lambda'}^{NS}. \qquad (24)$$

Since

$$\mathcal{D}_{++}^{NS} = \mathcal{D}_{--}^{NS} = \mathcal{D}^{NS} = \mathcal{D}^{(\delta)} + \mathcal{D}^{(e)}, \qquad (25)$$

where $\mathcal{D}^{(\delta)}$ and $\mathcal{D}^{(e)}$ are given by Eqs.(10)-(13) it is necessary to calculate only $\mathcal{D}_{+-}^{NS}$ and $\mathcal{D}_{-+}^{NS}$ which are equal due to $\mathcal{P}$-invariance. It is convenient to define

$$\Delta\mathcal{D}^{NS} = \mathcal{D}_{++}^{NS} - \mathcal{D}_{+-}^{NS} \qquad (26)$$

This function obeys the equation

$$\Delta\mathcal{D}^{NS}(x,s) = \int_{m^2}^{s}\frac{dQ^2}{Q^2}\frac{\alpha(Q^2)}{2\pi}\int_{x}^{1}\frac{dy}{y}\left[\mathcal{D}^{NS}(y,Q^2)\right.$$

$$\times \Delta P_e^e\left(\frac{x}{y}\right) + \Delta \mathcal{D}^{NS}(x,Q^2)\left(P_e^e\left(\frac{x}{y}\right) - \Delta P_e^e\left(\frac{x}{y}\right)\right)]. \tag{27}$$

Here P_e^e and ΔP_e^e are given by (4) and (20), respectively. Using the relation $\mathcal{D}^{NS} = \mathcal{D}^{(\gamma)} + \mathcal{D}^{(e)}$ and expressions (10)-(13) one can solve Eq.(27) perturbatively. It is convenient to represent $\Delta \mathcal{D}^{NS}$ as the sum of the contribution of additional e^+e^- pair production and the pure photonic contribution, as it was done for $\mathcal{D}^{NS}$:

$$\Delta \mathcal{D}^{NS} = \Delta \mathcal{D}^{(\gamma)} + \Delta \mathcal{D}^{(e)} \tag{28}$$

For $\Delta \mathcal{D}^{(\gamma)}$ calculation one needs to put $\alpha(Q^2) = \alpha$ and $\mathcal{D}^{NS} = \mathcal{D}^{(\gamma)}$ in Eq.(27). As the result, we obtain with the accuracy adopted

$$\Delta \mathcal{D}^{(\gamma)}(x,Q^2) = \frac{\alpha}{2\pi}\left(\ln\frac{Q^2}{m^2}\right)(1-x) + \frac{\left(\alpha \ln\frac{Q^2}{m^2}\right)^2}{8\pi^2}$$
$$\times\left[(1-3x)\ln\frac{1}{x} + (1-x)\left(-4\ln\frac{1}{1-x} + 1\right)\right]. \tag{29}$$

After all, for $\Delta \mathcal{D}^{(e)}$ calculation it is sufficient to take into account in Eq.(27) the difference $\alpha(Q^2)$ from α; with the same accuracy we get:

$$\Delta \mathcal{D}^{(e)}(x,s) = \int_{m^2}^{s}\frac{dQ^2}{Q^2}\frac{\alpha^2}{6\pi^2}\ln\frac{Q^2}{m^2}\int_{x}^{1}\frac{dy}{y}\delta(1-y)\left(1-\frac{x}{y}\right)$$
$$= \frac{\alpha^2\left(\ln\frac{s}{m^2}\right)^2}{12\pi^2}(1-x). \tag{30}$$

4. SUMMARY

Here we have applied the structure function method for radiative correction calculations to the case of polarized particles. For longitudinal polarization radiative corrected cross-sections are expressed via structure functions $\mathcal{D}_{A\lambda}^{B\,2}(x,Q^2)$ having simple probabilistic meaning. These functions are calculated with the approximation of Ref.[10]: single logarithmic terms are Kept up to $(\alpha \ln Q^2/m^2)^2$ and the effects of soft photon bremmstrahlung (double logarithmic terms) are included in all orders of perturbation theory.

In the general case it is impossible to confine oneself to functions $\mathcal{D}_{A\lambda}^{B\,2}$. The generalization of these functions - structure density matrix $\mathcal{D}_{A\lambda\lambda'}^{B\,2\,2'}$ - is introduced. The equation for this matrix is obtained, which is the generalization of Altarelli-Parisi equations, and the corresponding Kernels are calculated. Equations for relevant matrix elements are solved with the approximation discussed above.

REFERENCES

1. E.J.Williams, Proc. Roy. Soc. A139, 163 (1933); Phys. Rev. 45, 729 (1934); Mat.-Pys. Meddalingen 13, 4 (1935).
2. C.F.Weizsäcker, Zeits. f. Phys. 88, 612 (1934).
3. P.Kessler, Nuovo Cimento 17, 809 (1960).
4. V.N.Baier, V.S.Fadin, V.A.Khose, Nucl. Phys. B65, 381 (1973).
5. G.Altarelli, Phys. Rep. 81, 1 (1982).
6. J.C.Collins, D.E.Soper, G.Sterman, Preprint ITP-SB-89--31 (1989).
7. V.N.Gribov, L.N.Lipatov, Yad. Fiz. 15, 781, 1218 (1972) Sov. J. Nucl. Phys. 15, 938, 675 (1972).
8. L.N.Lipatov, Yad. Fiz. 20, 181 (1974).
9. G.Altarelli, G.Parisi, Nucl. Phys. B126, 298 (1977).

10. E.A.Kuraev, V.S.Fadin, Yad. Fiz. 41, 753 (1985)
    Sov. J. Nucl. Phys. 41, 466 (1985).
11. G.Altarelli, G.Martinelli, in: Physics at LEP,
    CERN-Yellow Report 8602, eds. J.Ellis and R.Peccei
    (Geneva, February 1986) vol.1, 47.
12. E.A.Kuraev, N.P.Merenkov, V.S.Fadin, Yad. Fiz. 47,
    1593 (1988).
13. V.S.Fadin, V.A.Khoze, Yad. Fiz. 47, 1693 (1988).
14. O.Nicrosini, L.Trentadue, Phys. Lett. 169B, 551
    (1987).
15. F.A.Berends, G.J.H.Burgers, W.L. van Neerven, Nucl.
    Phys. B297, 429 (1988); E Nucl. Phys. B304, 921
    (1988).
16. R.N.Cahn, Phys. Rev. D36, 2666 (1987).
17. G.Bonvicini, L.Trentadue, Preprint UM-HE-88-36 (Submitted to Nucl. Phys.B)
18. O.Nicrosini, L.Trentadue, Preprint FNT/T-88/12,
    UPRF-88-75, Nucl. Phys. B318, 1 (1989).
19. J.P.Alexander, G.Bonvicini, P.S.Drell, R.Frey, Phys.
    Rev. D37, 56 (1988).
20. O.Nicrosini, L.Trentadue, Z. Phys. C39, 479 (1988).
21. F.Aversa, M.Greco, Preprint LNF-89/025 (PT) (Submitted to Physics Letters).
22. F.A.Berends, G.J.H.Burgers, W.Hollik, W.L. van Neerven, Phys. Lett. B203, 177 (1988).
23. B.W.Lynn, M.E.Peskin, R.G.Stuart, Physics at LEP,
    CERN 86-02, v.1, 90 (1986).
24. M.Consoli, A.Sirlin, Physics at LEP, CERN 86-02,
    v.1, 63 (1986).
25. S.Jadach, J.H.Kühn, R.G.Stuart, MPI-PAE/PTh71/87
    (1987).
26. S.A.Lutsenko Diploma work, Novosibirsk State University, 1988.

RENORMALIZATION IN THE STANDARD MODEL OF THE ELECTROWEAK INTERACTIONS

Giampiero PASSARINO

Dipartimento di Fisica Teorica, Università di Torino, Torino, Italy
INFN, Sezione di Torino, Italy

Abstract

The problem of predicting the W mass in the standard model is considered as an example of how to include radiative corrections. Renormalization is performed by fitting the bare parameters of the Lagrangian in terms of three data points, namely α, G_F and M_Z. The ρ–parameter is a computable quantity and no ad hoc definition of $\sin^2 \theta_W$ is introduced. Dyson resummation of radiative corrections is briefly discussed.

Electroweak interactions are now entering in a new era of precision experiments. In a systematic investigation of the validity of the standard model we must therefore consider the radiative corrections as an indispensable part of the verification of the theory. The model contains three free parameters, namely the $SU(2)$ coupling constant g, the sine of the weak mixing angle $s_\theta = \sin\theta_W$ and M, the zero'th order W mass. The relation of these parameters to measurable quantities is the content of a renormalization scheme.

A first example of the extension from QED to the standard model of a renormalization scheme is perhaps given by the calculation of Ross and Taylor [1]. However the first complete calculation of $e^+e^- \to \mu^+\mu^-$ in the standard electroweak theory and away from the Z^0 peak was performed in ref. 2, where renormalization was implemented in a way that has become known as *on shell* scheme. At that time this procedure was understood as a sort of compromise because admittedly some very interesting radiative corrections to vector boson masses were missing. Renormalization from low energy data was thereafter explained in full details by Green and Veltman [3], leading finally to the vector boson mass shift discovered independently by Veltman [4] and Antonelli, Consoli and Corbò [4]. Since then the on shell scheme has become very popular and therefore the definition of $\sin^2\theta_w = 1 - M_W^2/M_Z^2$, proposed by Sirlin [5], is more or less commonly accepted. Another fundamental ingredient of the standard model is the ρ-parameter, introduced by Ross and Veltman [6]. The previous definition of $\sin^2\theta_w$ makes ρ equal to one by construction, while the importance of the ρ-parameter for confrontation with experimental data and its dependence on isospin breaking has been pointed out and repeatedly reviewed by Veltman [7].

Here we want to emphasize the importance of a renormalization scheme were, given the data, we subtract radiative corrections and determine g, θ_W and M. In this way no additional constraint is superimposed and hopefully some informations on the Higgs sector will be available in a near future from the comparison with experimental data. Once three measurements are obtained the rest must follow because of gauge invariance. As soon as a fourth measurement will be available, for example the W mass, then we can compute the ρ-parameter. Our scheme [8,9] is therefore nothing more than a set of fitting equations as introduced by Veltman in a prophetic paper [10] on mass differences in the standard model.

One quantity we can predict from the knowledge of α, G_F and M_Z is the W mass. This is of course done in the on shell scheme by introducing the factor Δr. Here we give our solution for M_W in a scheme where ρ is some computable quantity, not one by construction.

An unwanted feature of one loop calculations in the standard model is that the $Z^0 - \gamma$ transition at $q^2 = 0$ is not zero. A solution has been given in different approaches by Kennedy and Lynn [11], Hollik [12], and the author [9]. The way we perform renormalization avoids the explicit introduction of renormalization constants. Three quantities are computed, electric charge, Fermi coupling constant and position of the Z^0 peak. By comparison g, s_θ and M are fixed.

$$g = f_g(\alpha, G_F, M_Z),$$
$$s_\theta = f_s(\alpha, G_F, M_Z),$$
$$M = f_M(\alpha, G_F, M_Z)$$

These quantities are by themselves infinite and scale dependent, but infinities and the arbitrary mass scale cancel out of all physical results, including ρ as it should be. Consider the electric charge defined to be the residue of the pole at zero momentum transfer in $e\mu$ scattering. In a one loop renormalization what we get is a replacement of the type [3]

$$4\pi\alpha = g^2 s_\theta^2 (1 + \delta\alpha)$$

$\delta\alpha$ denotes the contributions of wave-function factors, two vertex diagrams, γ propagator insertions and $Z^0 - \gamma$ transition. It is a well known fact that purely e.m. contributions, exclusive of photon propagator diagrams, cancel. Vector bosons however modify this situation. First we compute infinities (and finite parts as well) containing m_μ and m_e. They cancel as expected and therefore we only consider the massless limit. Consider

$$\delta\alpha = \delta^W + \delta^V + \delta^\gamma + \delta^{Z\gamma}$$

δ^W is given in ref. 3 in terms of B form factors [2]. The two vertex diagrams give δ^V which can be cast into the form

$$\delta^V = \frac{g^2}{32\pi^2} \sum_{f=e,\mu} \left[C_f(0) - \frac{v_\theta + 1}{4 c_\theta^2} C_{ff}(0) \right]$$

where $v_\theta = 4 s_\theta^2 - 1$ and using the C form factors [2]

$$C_{ff}(s) = -2 C_{24} + (C_{11} + C_{23}) s + 1$$
$$C_f(s) = 6 C_{24} - (C_0 + C_{11} + C_{23}) s - 1$$

The non e.m. contributions add up to a non zero result

$$\delta^W + \delta^V = \frac{g^2}{16\pi^2} (\Delta - \ln M^2)$$

where $\Delta = -2/(n-4) + \gamma - \ln \pi$. Also vector bosons give a new contribution to the $Z^0 - \gamma$ transition at zero momentum

$$\delta^{Z\gamma} = -v_\theta \frac{c_\theta}{s_\theta} \frac{S_{Z\gamma}(0)}{M^2} \neq 0$$

with

$$S_{Z\gamma}(0) = -\frac{g^2}{8\pi^2} \frac{s_\theta}{c_\theta} M^2 (\Delta - \ln M^2)$$

Being $\delta^W + \delta^V + \delta^{Z\gamma}$ different from zero we apparently cannot express the electric charge renormalization via the summation of photon self-energy diagrams alone. A possible way out is the following. Start with a triplet of vector bosons B^i_μ and a singlet B^0_μ. The relevant terms in the interaction Lagrangian are

$$\mathcal{L} = \frac{i}{4} (\bar{\nu}, \bar{e}) \gamma^\mu \begin{pmatrix} g B^3_\mu + g'' B^0_\mu & \sqrt{2} g W^+_\mu \\ \sqrt{2} g W^-_\mu & -g B^3_\mu + g'' B^0_\mu \end{pmatrix} (1 + \gamma^5) \begin{pmatrix} \nu \\ e \end{pmatrix}$$
$$+ \frac{i}{4} g''' B^0_\mu \bar{e} \gamma^\mu (1 - \gamma^5) e - \frac{1}{2} g^2 F^2 W^+_\mu W^-_\mu - \frac{1}{4} F^2 (g B^3_\mu + g' B^0_\mu)^2$$

where for simplicity we considered only one fermion doublet. Moreover F is the Higgs vacuum expectation value. Before diagonalizing the mass matrix we introduce a new coupling constant $\bar{g}$, related to g by

$$\bar{g} = g (1 + g^2 \Gamma)$$

Where Γ is for the moment a free constant coefficient. Next we define fields A_μ and Z^0_μ by

$$Z^0_\mu = \frac{\bar{g} B^3_\mu + g' B^0_\mu}{(\bar{g}^2 + g'^2)^{1/2}}, \qquad A_\mu = \frac{-g' B^3_\mu + \bar{g} B^0_\mu}{(\bar{g}^2 + g'^2)^{1/2}}$$

The final result follows by using

$$g' = g'' = -\frac{s_\theta}{c_\theta} \bar{g}, \qquad g''' = g' + g'', \qquad F = \sqrt{2} \frac{M}{\bar{g}}$$

For the quadratic terms we get

$$\mathcal{L}_2 \approx -M^2 W_\mu^+ W_\mu^- - \frac{1}{2}\frac{M^2}{c_\theta^2} Z_\mu^0 Z_\mu^0 + \bar{g}^2 M^2 \Gamma \left(Z_\mu^0 Z_\mu^0 - \frac{s_\theta}{c_\theta} Z_\mu^0 A_\mu + 2 W_\mu^+ W_\mu^- \right)$$

Charged and neutral currents receive extra contributions

$$\mathcal{L}_{CC}^{extra} = -\frac{i\bar{g}^3}{2\sqrt{2}} \Gamma \left[W_\mu^+ \bar{\nu}\gamma^\mu (1+\gamma^5) e + W_\mu^- \bar{e}\gamma^\mu (1+\gamma^5) \nu \right]$$

$$\mathcal{L}_{NC}^{extra} = \frac{i}{4}\bar{g}^3 \Gamma \left(s_\theta A_\mu + c_\theta Z_\mu^0 \right) \left[\bar{e}\gamma^\mu (1+\gamma^5) e - \bar{\nu}\gamma^\mu (1+\gamma^5) \nu \right]$$

At this point we fix Γ by requiring

$$\Gamma = \frac{1}{8\pi^2} B_0(0, M, M)$$

where B_0 is the scalar two-point function[2] at zero momentum and with two equal masses. With this choice the vector boson self-energies are modified. Dropping from now on the bar we get

$$S_{\gamma\gamma} = \frac{g^2 s_\theta^2}{16\pi^2} p^2\, \Pi_{\gamma\gamma}$$

$$S_{ZZ} = \frac{g^2}{16\pi^2 c_\theta^2} \left[(S_{ZZ}^0 - 32\pi^2 M^2 \Gamma) - 2s_\theta^2 (S_{Z\gamma}^0 + 16\pi^2 M^2 \Gamma) + s_\theta^4 p^2\, \Pi_{\gamma\gamma} \right]$$

$$S_{Z\gamma} = \frac{g^2}{16\pi^2}\frac{s_\theta}{c_\theta} \left[(S_{Z\gamma}^0 + 16\pi^2 M^2 \Gamma) - s_\theta^2 p^2\, \Pi_{\gamma\gamma} \right]$$

$$S_{WW} = \frac{g^2}{16\pi^2} \left[(\Pi_{WW}^0 + s_\theta^2 \Pi_{WW}^1) p^2 + (\Sigma_{WW}^0 - 32\pi^2 M^2 \Gamma) + s_\theta^2 \Sigma_{WW}^1 \right]$$

The explicit expressions will be given in appendix. The important fact to be noted here is $S_{Z\gamma}(0) = 0$ which makes $\delta^{Z\gamma} = 0$ in the corrections to the electric charge. Moreover the extra $ee\gamma$ ($\mu\mu\gamma$) vertex gives rise to a contribution

$$\delta^{extra} = -\frac{g^2}{16\pi^2} (\Delta - \ln M^2)$$

Thus

$$\delta^{Z\gamma} = 0, \qquad \delta^W + \delta^V + \delta^{extra} = 0$$

and photon propagator diagrams alone contribute to the shift in e^2. At the same time we consider the eeZ^0 coupling. The infinite parts coming from wave-function factors and vertices add up to

$$\mathcal{L}_{eeZ}|_{I.P.} = -(2\pi)^4 \, i \frac{ig^3}{32\pi^2} c_\theta \Delta \bar{v}\gamma^\mu (1+\gamma^5) u$$

which is exactly cancelled by the extra eeZ^0 vertex proportional to Γ. By means of the same mechanism all the required cancellations take place. There is no parity violation in low energy for the e.m. current, no ultraviolet divergence contained in the $\bar{\nu}\nu\gamma$ coupling and moreover the neutrino charge remains zero.

The second experimental quantity we need is given by G_F or rather by the muon lifetime

$$\frac{1}{\tau_\mu} = \frac{m_\mu^5}{192\,\pi^3} \frac{g^4}{32\,M^4} \left(1 + \delta_\mu^{e.m.} + \delta_\mu^{weak}\right)$$

where

$$\delta_\mu^{e.m.} = \frac{\alpha}{2\pi}\left(\frac{25}{4} - \pi^2\right)$$

If we compute δ_μ^{weak} from wave-function factors, four vertex diagrams and five box diagrams [3], we obtain

$$\delta^W + \delta^V + \delta^B = \frac{g^2}{\pi^2}\left[\tfrac{3}{4} + \tfrac{1}{2}(\Delta - \ln M^2) + \tfrac{1}{4}\frac{7 - 4s_\theta^2}{4s_\theta^2}\ln c_\theta^2\right]$$

The extra contribution from $\mathcal{L}_{CC}^{extra}$ gives

$$\delta_\mu^{extra} = -4g^2\Gamma = -\frac{g^2}{2\pi^2}(\Delta - \ln M^2)$$

Therefore we have an infrared and ultraviolet finite correction beside W self-energy diagrams

$$\delta_\mu = \frac{g^2}{4\pi^2}\left[3 + \frac{7-4s_\theta^2}{4s_\theta^2}\ln c_\theta^2\right] + \delta_\mu^{e.m.}$$

We are now in a position to write down the one loop dressed propagators.

$$\Delta_{\mu\nu}^{\gamma\gamma} = \frac{1}{(2\pi)^4 i}\frac{\delta_{\mu\nu}}{p^2}\left[1 - \frac{g^2 s_\theta^2}{16\pi^2}\Pi_{\gamma\gamma}\right]^{-1}$$

$$\Delta_{\mu\nu}^{ZZ} = \frac{1}{(2\pi)^4 i}\delta_{\mu\nu}\left\{p^2 + \frac{M^2}{c_\theta^2} - \frac{g^2}{16\pi^2 c_\theta^2}\left[\overline{S}_{ZZ}^0 - 2s_\theta^2\overline{S}_{Z\gamma}^0 + s_\theta^4 p^2\,\Pi_{\gamma\gamma}\right]\right\}^{-1}$$

$$\Delta_{\mu\nu}^{Z\gamma} = \frac{1}{(2\pi)^4 i}\delta_{\mu\nu}\frac{g^2}{16\pi^2}\frac{s_\theta}{c_\theta}\left(\frac{\overline{S}_{Z\gamma}^0}{p^2} - s_\theta^2\Pi_{\gamma\gamma}\right)$$
$$\times \left\{p^2 + \frac{M^2}{c_\theta^2} - \frac{g^2}{16\pi^2 c_\theta^2}\left[\overline{S}_{ZZ}^0 - 2s_\theta^2\overline{S}_{Z\gamma}^0 + s_\theta^4 p^2\,\Pi_{\gamma\gamma}\right]\right\}^{-1}$$

where with $\overline{S}$ we denote the corresponding S opportunely subtracted. Before deriving an expression for M_W we discuss in some details how renormalization actually works for $e^+e^- \to \mu^+\mu^-$. With the outlined procedure we only need to worry about self-energy diagrams. In actual calculations the propagators must be calculated with a certain accuracy in the coupling constant. In the present case we use a simple recipe, consider diagrams with one closed loop (no self-energy loops) and bare propagators, and tree diagrams with dressed propagators where the self-energy is computed by considering one loop diagrams only. Order g^4 terms which can arise in deriving propagators for the neutral sector are consistently neglected because they belong to the two loop renormalization. However when needed these terms can be easily accommodated in the procedure. In fitting the parameters the combination $gs_\theta = e$ is easily fixed in terms of α from $e\mu$ scattering

$$\frac{1}{g^2 s_\theta^2} = \frac{1}{4\pi\alpha} + \frac{1}{16\pi^2}\Pi_{\gamma\gamma}(0)$$

Thus photon exchange in $e^+e^- \to \mu^+\mu^-$ gives an amplitude proportional to

$$-(2\pi)^4 i\frac{4\pi\alpha}{p^2}\left\{1 - \frac{\alpha}{4\pi}\left[\Pi_{\gamma\gamma}(p^2) - \Pi_{\gamma\gamma}(0)\right]\right\}^{-1}$$

where the combination of self-energies is ultraviolet finite. For the Z^0 propagator the situation is slightly more complicated. Given

$$\Delta_Z^{-1} = p^2 + \frac{M^2}{c_\theta^2} - S_{ZZ}(p^2)$$

we first obtain

$$\Delta_Z = \frac{c_\theta^2}{g^2}\left[\frac{c_\theta^2}{g^2}p^2 + \frac{M^2}{g^2} - \frac{c_\theta^2}{g^2}S_{ZZ}(p^2)\right]^{-1}$$

Using the second fitting equation, obtained from $\mu-$ decay, the combination M^2/g^2 is fixed

$$8\frac{M^2}{g^2} = \frac{1}{G'_F} + \frac{1}{2\pi^2}\overline{\Sigma}_{WW}(0) = \frac{1}{G_F} + \frac{1}{2\pi^2}\left[\overline{\Sigma}_{WW}(0) + \frac{2\pi^2}{G_F}\delta_G\right]$$

where

$$\delta_G = 1/2\,\delta_\mu, \qquad G'_F = \frac{G_F}{1+\delta_G}$$

We have verified that the two different choices, namely the use of a modified Fermi constant G'_F or the separate inclusion of the finite corrections δ_G, give the same numerical answers. Therefore we get

$$\Delta_Z = 8\,G'_F\frac{c_\theta^2}{g^2}\left\{8\,G'_F\frac{c_\theta^2}{g^2}p^2 + 1 + \delta_G + \frac{G'_F}{2\pi^2}\overline{\Sigma}_{WW}(0)\right.$$
$$\left. - \frac{G'_F}{2\pi^2}\left[\overline{S}^0_{ZZ}(p^2) - 2s_\theta^2\overline{S}^0_{Z\gamma}(p^2) + s_\theta^4 p^2\Pi_{\gamma\gamma}(p^2)\right]\right\}^{-1}$$

The third fitting equation, namely the position of the Z^0 pole, gives

$$\frac{M^2}{c_\theta^2} = M_Z^2 + \frac{g^2}{16\pi^2 c_\theta^2}\,\mathrm{Re}\left[\overline{S}^0_{ZZ}(-M_Z^2) - 2s_\theta^2\overline{S}^0_{Z\gamma}(-M_Z^2) - s_\theta^4 M_Z^2\Pi_{\gamma\gamma}(-M_Z^2)\right]$$
$$\equiv M_Z^2 + \frac{g^2}{16\pi^2 c_\theta^2}\,\mathrm{Re}\,f_Z\left(-M_Z^2\right)$$

The previous equation defines $f_Z(p^2)$. By combining the second and the third equation we easily obtain a solution for c_θ^2/g^2

$$8 G'_F M_Z^2 \frac{c_\theta^2}{g^2} = 1 + \frac{G'_F}{2\pi^2} \text{Re} \left[\overline{\Sigma}_{WW}(0) - f_Z(-M_Z^2) \right]$$

Finally Δ_Z may be cast into the following form

$$\Delta_Z = 8 G'_F M_Z^2 \frac{c_\theta^2}{g^2} \left\{ \left[1 + \delta_G + \frac{G'_F}{2\pi^2} \overline{\Sigma}_{WW}(0) \right] (p^2 + M_Z^2) \right.$$
$$\left. - \frac{G'_F}{2\pi^2} \text{Re} \left[M_Z^2 f_Z(p^2) + p^2 f_Z(-M_Z^2) \right] - i \frac{G'_F M_Z^2}{2\pi^2} \text{Im} f_Z(p^2) \right\}^{-1}$$
$$= 8 G'_F M_Z^2 \frac{c_\theta^2}{g^2} P_Z^{-1}$$

The quantity in braket is ultraviolet finite. Inside the function f_Z we used the zero'th order value for s_θ^2, which follows from the fitting equations

$$\bar{s}_\theta^2 = \tfrac{1}{2} \left[1 - \sqrt{1 - \frac{2\pi\alpha}{G'_F M_Z^2}} \right]$$

Using Δ_Z and $\Delta_{\mu\nu}^{Z\gamma}$ we obtain that the $Z^0 - Z^0$ propagator diagram together with the two $Z^0 - \gamma$ transition diagrams give a finite answer.

To determine M_W, for instance from the total cross section of $e^+e^- \to W^+W^-$, what we need is

$$g^2 \Delta_W = \frac{g^2}{p^2 + M^2 - S_{WW}(p^2)}$$

The pole of the propagator, $p^2 = -M_W^2$, gives M_W. First we introduce

$$f_W(p^2) = \frac{1}{g^2} S_{WW}(p^2)$$

Since

$$g^2 \Delta_W = \left[\frac{p^2}{g^2} + \frac{M^2}{g^2} - \frac{1}{16\pi^2} f_W(p^2)\right]^{-1}$$

we use the second fitting equation to obtain

$$g^2 \Delta_W = 8 G'_F \left\{\frac{8 G'_F}{g^2} p^2 + 1 + \frac{G'_F}{2\pi^2} \left[\overline{\Sigma}_{WW}(0) - f_W(p^2)\right]\right\}^{-1}$$

Finally we need G_F/g^2, which can be derived from the remaining equations. Actually a solution for s_θ^2 must be provided and up to first order we get

$$s_\theta^2 = \overline{s}_\theta^2 \left(1 + \frac{\alpha}{4\pi} s_1\right)$$

$$s_1 = \frac{1}{\overline{c}_\theta^2 - \overline{s}_\theta^2} \left(\overline{c}_\theta^2 \Pi_F + \frac{1}{M_Z^2 \overline{s}_\theta^2} \Sigma_F\right) - \Pi_{\gamma\gamma}(-M_Z^2) - \frac{1}{M_Z^2 \overline{s}_\theta^2} \overline{S}^0_{Z\gamma}(-M_Z^2)$$

where we have introduced the ultraviolet finite combinations

$$\Pi_F = \Pi_{\gamma\gamma}(-M_Z^2) - \Pi_{\gamma\gamma}(0)$$
$$\Sigma_F = \overline{\Sigma}^0_{WW}(0) + \overline{s}_\theta^2 \Sigma^1_{WW}(0) - \overline{S}^0_{ZZ}(-M_Z^2) + \overline{S}^0_{Z\gamma}(-M_Z^2)$$

Once more the corrections to s_θ^2 are not finite. This is in principle all what we need in our scheme where at this order of approximation only one loop self-energy diagrams are considered and (self-energy)^2 contributions are neglected. In this way we obtain an $O(\alpha)$ resummation and the remaining terms should be left for the two loop renormalization together with two loop irreducible diagrams. However the presence of large leading logarithms due to light fermion insertions in the self-energies suggests a possible improvement, already at this level. First by including terms $O(g^4)$ in the Z^0 propagator we modify the third fitting equation

$$\frac{M^2}{c_\theta^2} = M_Z^2 + \frac{g^2}{16\pi^2 c_\theta^2} \operatorname{Re} f_Z(-M_Z^2)$$

$$- \frac{g^4 s_\theta^2}{256\pi^4 c_\theta^2} \frac{1}{M_Z^2} \frac{\operatorname{Re}\left[\overline{S}^0_{Z\gamma}(-M_Z^2) + s_\theta^2 M_Z^2 \Pi_{\gamma\gamma}(-M_Z^2)\right]^2}{1 - \frac{g^2 s_\theta^2}{16\pi^2} \Pi_{\gamma\gamma}(0)}$$

Next we expand s_θ^2 up to second order everywhere but in the W propagator where we use the lowest order value $\overline{s}_\theta^2$. Thus

$$s_\theta^2 = \bar{s}_\theta^2 \left(1 + \frac{\alpha}{4\pi} s_1 + \frac{\alpha^2}{16\pi^2} s_2\right)$$

Consequently we get

$$8 \frac{G'_F}{g^2} = \frac{1}{M_Z^2 \bar{c}_\theta^2} \left(1 + \frac{\alpha}{4\pi} g_1 + \frac{\alpha^2}{16\pi^2} g_2\right)$$

After some algebra the coefficients g_1 and g_2 can be cast into the following form

$$g_1 = \frac{1}{\bar{c}_\theta^2 - \bar{s}_\theta^2} \left(\bar{s}_\theta^2 \Pi_F + \frac{1}{M_Z^2 \bar{s}_\theta^2} \Sigma_F\right) - \frac{1}{M_Z^2 \bar{s}_\theta^2} \overline{S}_{Z\gamma}^0 (-M_Z^2)$$

$$g_2 = \frac{\bar{s}_\theta^2}{\left(\bar{c}_\theta^2 - \bar{s}_\theta^2\right)^3} \left(\bar{c}_\theta^2 \Pi_F + \frac{1}{M_Z^2 \bar{s}_\theta^2} \Sigma_F\right)^2$$

The coefficient g_2 turns out to be ultraviolet finite. This fact is obvious if we remember that only irreducible diagrams are considered and therefore one loop renormalization suffices to render everything finite. The last formula for g^2 corresponds to a $O(\alpha^2)$ resummation. The W propagators becomes

$$g^2 \Delta_W = 8 G'_F \bar{c}_\theta^2 F_W^{-1}(x)$$

$$F_W(x) = -\left(1 + \frac{\alpha}{4\pi} P_W^1 + \frac{\alpha^2}{16\pi^2} P_W^2\right) x + \bar{c}_\theta^2 + Q_W$$

$$P_W^1 = \frac{1}{\bar{c}_\theta^2 - \bar{s}_\theta^2} \left(\bar{s}_\theta^2 \Pi_F + \frac{1}{M_Z^2 \bar{s}_\theta^2} \Sigma_F\right) - \frac{1}{M_Z^2 \bar{s}_\theta^2} \overline{S}_{Z\gamma}^0 (-M_Z^2)$$

$$\quad - \frac{1}{\bar{s}_\theta^2} \Pi_{WW}^0 (-xM_Z^2) - \Pi_{WW}^1 (-xM_Z^2)$$

$$P_W^2 = \frac{\bar{s}_\theta^2}{\left(\bar{c}_\theta^2 - \bar{s}_\theta^2\right)^3} \left(\bar{c}_\theta^2 \Pi_F + \frac{1}{M_Z^2 \bar{s}_\theta^2} \Sigma_F\right)^2$$

$$Q_W = \frac{1}{M_Z^2 \bar{s}_\theta^2} \left\{ \overline{\Sigma}_{WW}^0(0) - \overline{\Sigma}_{WW}^0(-xM_Z^2) \right.$$

$$\left. + \bar{s}_\theta^2 \left[\Sigma_{WW}^1(0) - \Sigma_{WW}^1(-xM_Z^2)\right] \right\}$$

where $x = -p^2/M_Z^2$. With the above result we have shown that the method is capable of dealing with resummation of terms at any given order and all coefficients $g_n, n \geq 2$ are ultraviolet finite. However we mention that resummation is in any case dictated by the consistent use of dressed propagators and it is not a matter of taste. For instance if m_t is large then two loop irreducible diagrams must be included. As a first step we use the explicit expressions for the self-energies and prove that all ultraviolet divergent factors cancel out in the above formula. In this way the W propagator is finite and scale independent. The term $-x + \bar{c}_\theta^2$, is nothing but the lowest order pole. For the numerical evaluation of M_W we split Π_F into hadronic part plus rest [12]. The hadronic contribution is computed perturbatively with effective quark masses, derived from a fit to a dispersion integral [12], $m_u = m_d = 0.041\,GeV, m_c = 1.5\,GeV, m_s = 0.15\,GeV$ and $m_b = 4.5\,GeV$.

The top quark mass is left as a free parameter. As far as a quadratic dependence on m_t is concerned only the combination Σ_F is affected. For large m_t we use the asymptotic expressions for the B form factors [2] to obtain

$$\Sigma_F \sim -\tfrac{3}{4} m_t^2$$

$$F_W(x) \sim -x + \bar{c}_\theta^2 + \tfrac{3}{8} \frac{G_F m_t^2}{\pi^2} \frac{\bar{c}_\theta^2}{\bar{c}_\theta^2 - \bar{s}_\theta^2} x$$

$$= -x + \bar{c}_\theta^2 + \tfrac{3}{4} \frac{\alpha}{4\pi} \frac{m_t^2}{M_Z^2} \frac{x}{\bar{s}_\theta^2 (\bar{c}_\theta^2 - \bar{s}_\theta^2)^2}$$

Thus M_W clearly increases with m_t, a situation which has an obvious correspondence with what we get from low energy data. There M_Z is not input parameter and $\delta M_W^2 \sim const$, $\delta M_Z^2 \sim -m_t^2/M_Z^2$ as $m_t \to \infty$. In general, for very large m_t, the contribution from Σ_F tends to cancel the large leading logarithms from Π_F while a very large m_H has the opposite effect. A simplified formula for M_W which only takes into account the large contributions is the following

$$\frac{M_Z^2 \bar{c}_\theta^2}{M_W^2} = 1 + \frac{\alpha}{4\pi} \frac{1}{\bar{c}_\theta^2 - \bar{s}_\theta^2} (\bar{s}_\theta^2 \Pi_F + \sigma_F) + \frac{\alpha^2}{16\pi^2} \frac{\bar{s}_\theta^2}{(\bar{c}_\theta^2 - \bar{s}_\theta^2)^3} (\bar{c}_\theta^2 \Pi_F + \sigma_F)^2$$

$$\sigma_F = -\tfrac{3}{4} \frac{m_t^2}{M_Z^2 \bar{s}_\theta^2}$$

Data points are now $\alpha = 1/137.0359895$ and $\sqrt{2} G_F = 1.16637 \times 10^{-5}\,GeV^{-2}$. As a result for $M_Z = 91\,GeV$ we find M_W to be

	W mass (all masses in GeV)		
m_t/m_H	10	100	1000
50	79.53	79.45	79.30
100	79.81	79.73	79.57
150	80.11	80.02	79.87
200	80.47	80.39	80.23

As a general remark we stress that for $m_t \approx 200\,\text{GeV}$ the neglected higher order effects, including irreducible diagrams, become more and more important as perturbation theory tends to be less effective. To give an indication of the phenomenon we consider $M_Z = 91\,\text{GeV}$, $m_t = 200\,\text{GeV}$, $m_H = 100\,\text{GeV}$ and compute the coefficients P_W^i up to third order, corresponding to $O(\alpha^3)$ resummation in g^2. We find $P_W^1 = 6.7 \times 10^{-3}$, $P_W^2 = 1.6 \times 10^{-3}$ and $P_W^3 = 1.7 \times 10^{-4}$. We have made several checks for our numerical calculation. The B form factors are computed using the FORTRAN code QFORMF [13], written in REAL*16 precision for the VAX/8600. Cancellation of ultraviolet factors is guarantee since the results are invariant against variations in Δ, the quantity substituted for $1/n - 4$ in the various divergent functions. Finally we have used α, G_F, M_W and M_Z as input in order to compute ρ according to the results of ref. 9. By requiring $\rho = 1$ we obtain a value for m_t for each pair M_W, M_Z, for example $M_Z = 91.9\,GeV$ and $M_W = 80.9\,GeV$ correspond to $m_t = 94.11\,GeV$ (when $m_H = 100\,GeV$). Reversing the procedure we use the same value of M_Z and the derived m_t to compute M_W. We found agreement, within few MeV, between the two values of M_W.

Appendix

The various ingredients occurring in the vector boson self-energies and transitions can be related to the two point scalar functions. In the following we give the complete list and refer to [2] for the explicit expression of the functions $B_0 \ldots B_{21}$.

$$\Pi_{\gamma\gamma} = \tfrac{2}{3} - 12\,B_{21}(M,M) + 7\,B_0(M,M)$$
$$+ 4 \sum_f \left[B_f(m_e, m_e) + \tfrac{4}{3} B_f(m_u, m_u) + \tfrac{1}{3} B_f(m_d, m_d) \right]$$

$$S^0_{Z\gamma} = p^2 \{ \tfrac{2}{3} - 10\,B_{21}(M,M) + \tfrac{13}{2} B_0(M,M)$$
$$+ \sum_f [B_f(m_e, m_e) + 2\,B_f(m_u, m_u) + B_f(m_d, m_d)] \} - 2\,M^2 B_0(M,M)$$

$$S^0_{ZZ} = p^2\, \Pi^0_{ZZ} + \Sigma^0_{ZZ}$$

$$\Pi^0_{ZZ} = \tfrac{2}{3} - 9\,B_{21}(M,M) + \tfrac{25}{4} B_0(M,M) - B_{21}(M_0, m_H)$$
$$- B_1(M_0, m_H) - \tfrac{1}{4} B_0(M_0, m_H)$$
$$+ \tfrac{1}{2} \sum_f [B_f(m_e, m_e) + B_f(m_\nu, m_\nu) + 3\,B_f(m_u, m_u) + 3\,B_f(m_d, m_d)]$$

$$\Sigma^0_{ZZ} = -2\,M^2 B_0(M,M) + \tfrac{1}{2} M_0^2 B_1(M_0, m_H) + \tfrac{5}{4} M_0^2 B_0(M_0, m_H)$$
$$- \tfrac{1}{2} m_H^2 B_1(M_0, m_H) - \tfrac{1}{4} m_H^2 B_0(M_0, m_H)$$
$$- \tfrac{1}{2} \sum_f [m_\nu^2 B_0(m_\nu, m_\nu) + m_e^2 B_0(m_e, m_e)$$
$$+ 3\,m_u^2 B_0(m_u, m_u) + 3\,m_d^2 B_0(m_d, m_d)]$$

$$\Sigma^0_{WW} = \tfrac{9}{2}(M_0^2 - M^2)B_1(M_0, M) + \tfrac{1}{4}(13M_0^2 - 21M^2)B_0(M_0, M)$$
$$+ \tfrac{1}{2}(M^2 - m_H^2)B_1(M, m_H) + \tfrac{1}{4}(5M^2 - m_H^2)B_0(M, m_H)$$
$$+ \sum_f \left[(m_e^2 - m_\nu^2)B_1(m_\nu, m_e) - m_\nu^2 B_0(m_\nu, m_e)\right.$$
$$\left. + 3(m_d^2 - m_u^2)B_1(m_u, m_d) - 3 m_u^2 B_0(m_u, m_d)\right]$$

$$\Sigma^1_{WW} = 2(M^2 - M_0^2)\left[2B_1(M_0, M) + B_0(M_0, M)\right]$$
$$- 2M^2\left[2B_1(\lambda, M) + B_0(\lambda, M)\right]$$

$$\Pi^0_{WW} = \tfrac{2}{3} - 9B_{21}(M_0, M) - 9B_1(M_0, M) + \tfrac{7}{4}B_0(M_0, M)$$
$$- B_{21}(M, m_H) - B_1(M, m_H) - \tfrac{1}{4}B_0(M, m_H)$$
$$+ 2\sum_f \left[B_{21}(m_\nu, m_e) + B_1(m_\nu, m_e) + 3B_{21}(m_u, m_d) + 3B_1(m_u, m_d)\right]$$

$$\Pi^1_{WW} = 8B_{21}(M_0, M) - 2B_0(M_0, M) + 8B_1(M_0, M) - 8B_{21}(\lambda, M)$$
$$- 8B_1(\lambda, M) + 2B_0(\lambda, M)$$

where $B_f = 2B_{21} - B_0$ and λ is the photon mass.

References

[1] D.A. Ross and J.C. Taylor, Nucl. Phys. B51,25 (1973).

[2] G. Passarino and M. Veltman, Nucl. Phys. B160,151 (1979).

[3] M. Green and M. Veltman, Nucl. Phys. B169,137 (1980).

[4] F. Antonelli, M. Consoli and G. Corbò, Phys. Lett. 91B,90 (1980); M. Veltman, Phys. Lett. 91B,95 (1980).

[5] A. Sirlin, Phys. Rev. D22,971 (1980).

[6] D. Ross and M. Veltman, Nucl. Phys. B95,135 (1975).

[7] M. Veltman, Vector Bosons and Higgs System, TASI Summer School 1984.

[8] G. Passarino, Torino preprint DFTT/G-88-3.

[9] G. Passarino, Talk given at the Ringbergworkshop "Electroweak Radiative Corrections", April 3-7 1989, and Torino preprint DFTT/G-89-1

[10] M. Veltman, Nucl. Phys. B123,89 (1977).

[11] D.C. Kennedy and B.W. Lynn, SLAC-PUB-4039(1988).

[12] W. Hollik, DESY preprint, DESY 88-188, December 1988; F. Jegerlehner, Z. Phys. C32,195 (1986).

[13] G. Passarino, QFORMF, a program for the computation of one-loop form factors.

Radiative Corrections to WW Scattering in the Standard Model

R. BOUAMRANE
Randall Laboratory of Physics
University of Michigan
Ann Arbor, MI 48109, USA

1 Introduction

The standard model of electroweak interactions has been consistent with all known experimental data to this date. However, there remain some important components of this model as yet unverified experimentally, namely the vector boson self–interactions and the Higgs sector. Both are needed to insure renormalizability without spoiling gauge invariance.

If the Higgs exists and is heavy but still light enough for perturbation theory to remain valid ($m_{Higgs} < 1\ TeV$), this calculation might provide a basic guideline for measuring deviations from the minimal Standard Model in future high energy WW scattering experiments. On the other hand, if the Higgs mass is much higher than $1\ TeV$, this calculation allows us to *refine* the unitarity bounds by taking into account the one–loop radiative corrections to the J=0 partial–wave amplitude. (Of course, we do not know what higher order corrections will do.)

2 The model

The model we use is the minimal $SU(2) \times U(1)$ standard model. The Higgs sector is restricted to a single complex doublet. The free parameters of the model are:

g	weak coupling constant
s_θ, c_θ	the sine and cosine of the weak mixing angle
M	mass of the charged vector boson
m	mass of the Higgs boson
$m_t^\alpha, m_b^\alpha, m_e^\alpha$	masses of quarks and leptons (α is the generation index).

We assume massless neutrinos.

All our calculations are done in the limit $m^2 \gg s, t, u \gg M^2, m_t^2$ where s, t, and u are the usual Mandelstam variables. Only longitudinally polarized vector bosons are considered. Furthermore, we consider the following three independent amplitudes:

$$\begin{aligned}
\mathcal{A}_a &= \mathcal{A}(W_L^+ W_L^- \to W_L^+ W_L^-) \\
\mathcal{A}_b &= \mathcal{A}(W_L^0 W_L^0 \to W_L^+ W_L^-) \\
\mathcal{A}_c &= \mathcal{A}(W_L^0 W_L^0 \to W_L^0 W_L^0)
\end{aligned} \quad (1)$$

3 Tree level calculations

The tree level amplitudes for $W_L W_L$ scattering have been extensively studied [1,2]. At energies very high with respect to the W mass, we have the following:

$$\begin{aligned}
\mathcal{A}_a &= \frac{g^2 m^2}{4M^2}\left(\frac{s}{-s+m^2} + \frac{t}{-t+m^2}\right) \\
\mathcal{A}_b &= \frac{g^2 m^2}{4M^2}\left(\frac{s}{-s+m^2}\right) \\
\mathcal{A}_c &= \frac{g^2 m^2}{4M^2}\left(\frac{s}{-s+m^2} + \frac{t}{-t+m^2} + \frac{u}{-u+m^2}\right)
\end{aligned}$$

Taking $m^2 \gg s,t,u$ we have for the J=0 partial-wave amplitude:

$$\begin{aligned}
a_a^0 &= \frac{g^2 s}{128\pi M^2} \\
a_b^0 &= \frac{g^2 s}{64\pi M^2} \\
a_c^0 &= 0
\end{aligned} \quad (2)$$

The most important feature of the J=0 partial-wave tree level amplitude for WW scattering is the cancellation at high energy ($s \gg m^2, M^2$) of the s^2 and s terms from the W exchange diagrams and similar terms from the four-W vertex and the Higgs exchange diagrams. This cancellation makes sure that tree level unitarity is respected at very high energies. This is true provided that the Higgs mass m is not too large (less than 1 TeV [1]). In the case $m^2 \gg s \gg M^2$, the J=0 partial-amplitude stays linear as a function of s. (It is constant for $W^0 W^0 \to W^0 W^0$). Tree level unitarity is violated at a critical energy $\sqrt{s^*} \simeq 1.7\, TeV$ [2]. This bound is obtained by considering the requirements of partial-wave unitarity on the two-channel system consisting of $W_L^+ W_L^-$, and $\frac{1}{\sqrt{2}} W_L^0 W_L^0$, with amplitudes given by (2). This is achieved by calculating the largest eigenvalue of the following 2 × 2 matrix:

$$\begin{pmatrix} a_a^0 & \frac{1}{\sqrt{2}} a_b^0 \\ \frac{1}{\sqrt{2}} a_b^0 & \frac{1}{2} a_c^0 \end{pmatrix} \quad (3)$$

4 Renormalization

All renormalization schemes involve some sort of redefinition of the parameters and the fields in the Lagrangian. For example [3]:

$$\begin{aligned}
g &\to g(1+\delta_g) \\
M &\to M(1+\delta_M) \\
c_\theta &\to c_\theta(1+\delta_{c_\theta}) \\
m &\to m(1+\delta_m) \\
W_\mu^\pm &\to W_\mu^\pm(1+\delta_c) \\
W_\mu^0 &\to W_\mu^0(1+\delta_0) + \delta_{0A} A_\mu \\
A_\mu &\to A_\mu(1+\delta_A) + \delta_{A0} W_\mu^0 \\
\phi^\pm &\to \phi^\pm(1+\delta_H) \\
\phi^0 &\to \phi^0(1+\delta_H) \\
H &\to H(1+\delta_H) + \frac{M}{g}\delta_t
\end{aligned} \qquad (4)$$

Depending on the scheme prefered, the quantities δ_g, δ_M, etc. are chosen to compensate part of the one-loop corrections of the processes used as data input to fix g, M, etc. So, in principle, the one-loop corrections to these processes must be computed before any prediction on the four-W amplitude is made. This of course would be the case when doing an exact calculation. In the limit in which we are interested, namely $m^2 \gg s,t,u \gg M^2$, the quantities δ_g, δ_M, etc. can be chosen as:

$$\delta = a_3 m^2 \log(m^2) + a_2 m^2 + a_1 \log(m^2) + a_0 , \qquad (5)$$

where the a_i's are constants to be determined.

The radiative corrections to the four-W amplitude are obtained by adding the contributions from all the diagrams shown in figs. 1 and 2. Diagrams with crosses involve the quantities δ_g, etc.. Their associated Feynman rules were derived from the extra terms generated in the Lagrangian after the redefinitions (4) were made. (For some examples, see fig. 3.)

For the three processes considered in (1), the counter-terms' contributions as a function of the δ's are:

$$\mathcal{A}_a^{counter} = \frac{g^2}{32\pi^4 i M^2 m^2}(\delta_M - \delta_m + \delta_g + 2\delta_c)(t^2 + s^2) \qquad (6)$$

$$\mathcal{A}_b^{counter} = \frac{g^2}{64\pi^4 i M^2 m^2}(\delta_M - \delta_m + \delta_g + \delta_c + \delta_0 - \delta_{c_\theta})s^2 \qquad (7)$$

$$\mathcal{A}_c^{counter} = \frac{g^2}{32\pi^4 i M^2 m^2}(\delta_M - \delta_m + \delta_g + 2\delta_0 - 2\delta_{c_\theta})(s^2 + t^2 + u^2) \qquad (8)$$

We notice that only the m^2 terms (if any) contained in the δ's affect the one-loop order four-W amplitude. These terms were previously calculated by van der Bij and Veltman [3] and were shown to be unobservable at the one-loop level for processes not involving external Higgs lines. Thus, in the limit $m^2 \gg s,t,u \gg M^2$, the one-loop order amplitude does not depend at all on the physical processes chosen to fix the free parameters g,M, etc. The quantities δ_g, δ_c, δ_0, δ_{c_θ} are free from m^2 dependence. The only relevant quantity is:

$$\delta_M - \delta_m = -2\pi^2 i \frac{m^2}{M^2}\left[\frac{3}{4}\frac{1}{n-4} + \frac{3}{8}\log(m^2) - \frac{25}{32} + \frac{9\pi}{32\sqrt{3}}\right] \qquad (9)$$

The heart of the calculations is, of course, the evaluation of approximately 1000 loop diagrams using the algebraic manipulation program Schoonship [6]. We have kept the leading terms which are quadratic in s, t, u with or without logarithms such as $\log(s/m^2)$, etc..

These diagrams can essentially be grouped into two sets. The first set consists of diagrams with no Higgs propagators inside the loop. We have found that the contributions from these diagrams essentially reduce to two-point functions and completely cancel in all the three processes at which we have looked. The second set consists of diagrams with Higgs propagators occurring in the loop. After appropriately expanding all the integrands, again, only two-point functions are left. (For calculational details, see (refs. [4].)

5 Results

The radiative corrections for the processes considered are:

$$\mathcal{A}_a^1 = -\frac{g^4}{16\pi^2 M^4}\left\{\frac{1}{18}ts + \left(\frac{5}{9} - \frac{9\pi}{32\sqrt{3}}\right)(t^2 + s^2)\right.$$
$$+\left(\frac{1}{96}ts + \frac{5}{96}s^2\right)\log(s/m^2)$$
$$+\left(\frac{1}{96}ts + \frac{5}{96}t^2\right)\log(t/m^2)$$
$$\left. + \frac{1}{32}u^2 \log(u/m^2)\right\}$$

$$\mathcal{A}_b^1 = -\frac{g^4}{16\pi^2 M^4}\left\{\left(\frac{37}{72} - \frac{9\pi}{32\sqrt{3}}\right)s^2 + \frac{1}{72}t^2 + \frac{1}{72}u^2\right.$$
$$+\frac{1}{32}s^2 \log(s/m^2)$$
$$+\left(\frac{1}{64}t^2 + \frac{1}{192}u^2 - \frac{1}{192}s^2\right)\log(t/m^2)$$
$$\left.+\left(\frac{1}{64}u^2 + \frac{1}{192}t^2 - \frac{1}{192}s^2\right)\log(u/m^2)\right\}$$

$$\mathcal{A}_c^1 = -\frac{g^4}{16\pi^2 M^4}\left\{\left(\frac{13}{24} - \frac{9\pi}{32\sqrt{3}}\right)(s^2 + t^2 + u^2)\right.$$
$$\left.+\frac{1}{16}s^2 \log(s/m^2) + \frac{1}{16}t^2 \log(t/m^2) + \frac{1}{16}u^2 \log(u/m^2)\right\},$$

The corresponding J=0 partial-wave amplitudes are:

$$a_a = \frac{g^2 s}{128\pi^2 M^2}\left[1 + \frac{g^2 s}{64\pi^2 M^2}\left(-\frac{20}{9}\log(s/m^2) - \frac{2441}{108} + \frac{12\pi}{\sqrt{3}}\right)\right]$$
$$a_b = \frac{g^2 s}{64\pi^2 M^2}\left[1 + \frac{g^2 s}{64\pi^2 M^2}\left(-\frac{5}{9}\log(s/m^2) - \frac{905}{108} + \frac{9\pi}{2\sqrt{3}}\right)\right]$$
$$a_c = \frac{g^2 s}{128\pi^2 M^2}\left[0. + \frac{g^2 s}{64\pi^2 M^2}\left(-\frac{10}{3}\log(s/m^2) - \frac{256}{9} + \frac{15\pi}{\sqrt{3}}\right)\right] \quad (10)$$

These results are plotted in fig. 3. In fig. 4, the critical energy $\sqrt{s^*}$ obtained by coupling the channels involving $W_L^+ W_L^-$ and $(\frac{1}{\sqrt{2}})W_L^0 W_L^0$, with amplitudes given by (10), is plotted as a function of the Higgs mass (fig. 5). These results agree with [5].

6 Conclusion

The one-loop corrections are positive throughout the energy range of interest. Numerically speaking, for a Higgs mass of 1 TeV, the radiative corrections to the tree amplitude ratio remain within 5% for energies below $500 GeV$. For a heavier Higgs mass, the results we have presented should be taken only as an indication because perturbation theory is no longer valid. Nevertheless, for a very heavy Higgs (5-10 TeV), the critical energy at which the J=0 partial-wave reaches unity is reduced by (20-30%) when including the one-loop corrections.

References

[1] B. Lee, C. Quigg, and H. Thacker, Phys. Rev. D16, 1519 (1977)

[2] M. Chanowitz and M. Gaillard, Nucl. Phys. B261, 379 (1985)

[3] J. Van der Bij and M. Veltman, Nucl. Phys. B231, 205 (1984)

[4] M. Veltman and F.J. Yndurain, *preprint* UM-TH-89-04
R. Bouamrane, PhD thesis (1989), *unpublished*

[5] S. Dawson and S. Willenbrock, Phys. Rev. Let. 62, 1232 (1989)

[6] SCHOONSHIP, 68000 version Jan 1 1989.

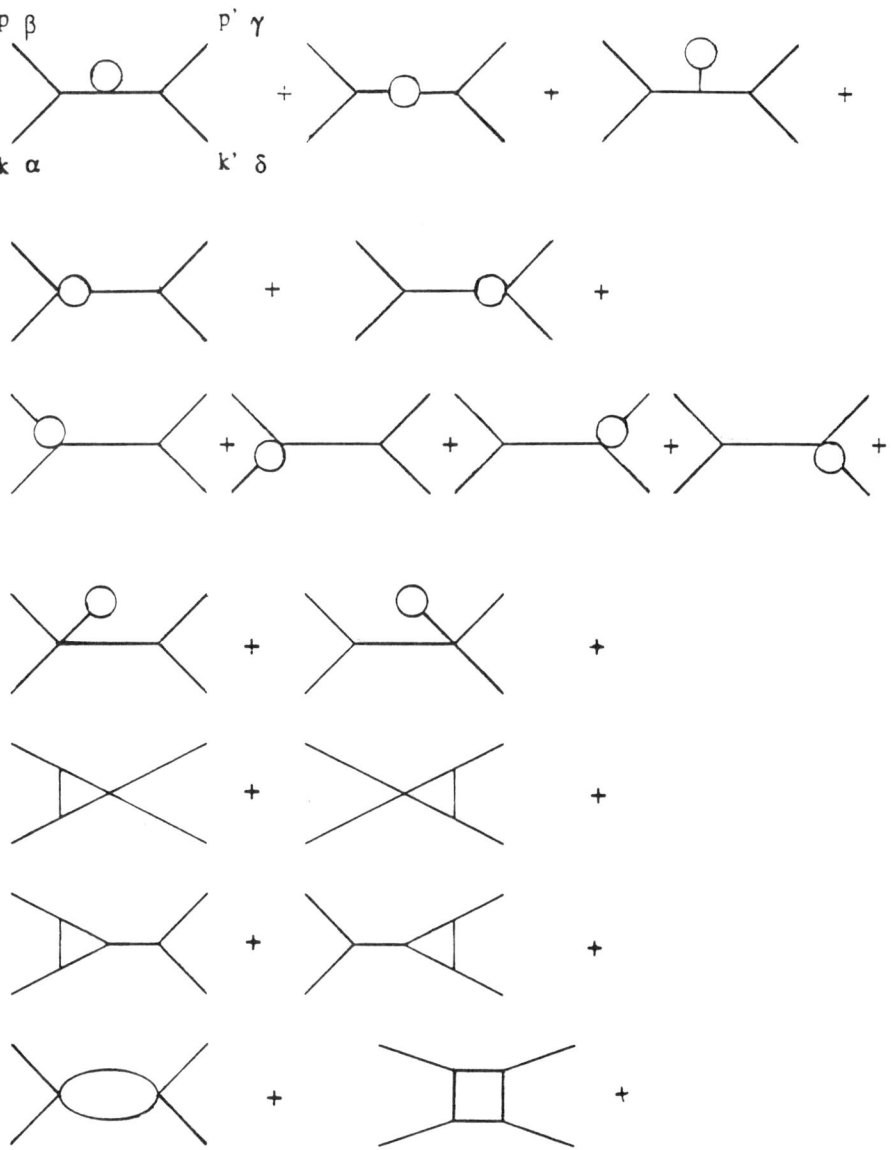

Figure 1: Topologies contributing to the one–loop corrections for WW scattering

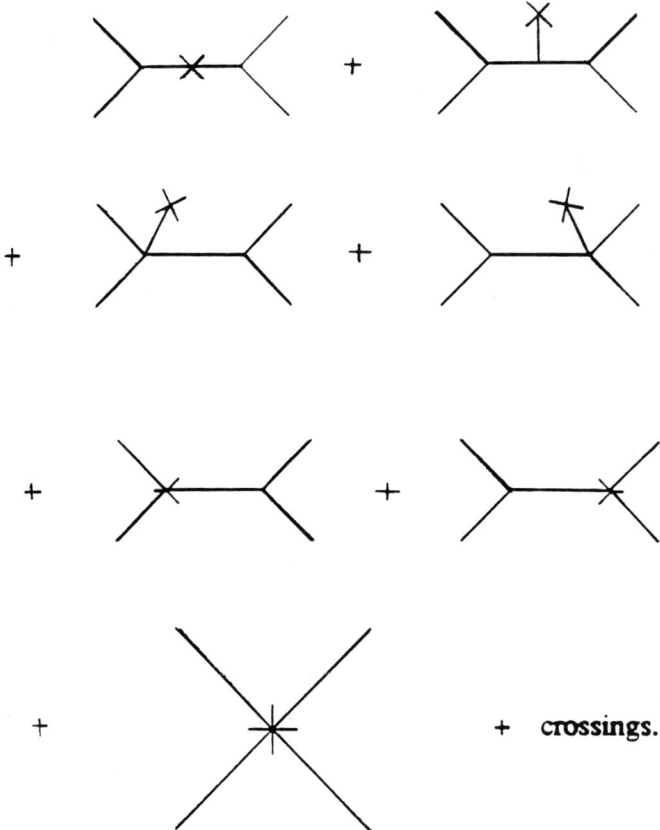

Figure 2: Counter-terms contribution to WW scattering

$H \;\text{-------}\!\!\times\qquad -m^2 M \delta_t$

$W_\mu^+ \;\text{---}\!\!\times\!\!\text{---}\; W_\nu^- \qquad -M^2(2\delta_c + 2\delta_M + \delta_t) + 2\delta_c\,(p.q\,\delta_{\mu\nu} - p_\mu q_\nu)$

$$W^0 \to (W^+, W^-) \;=\; \left[\delta_g + 2\delta_c + \tfrac{s_\theta}{c_\theta}\delta_{A0} + \delta_{c_\theta}\right] \times W^0 \to (W^-, W^-)$$

$$(W^0, W^0, W^+, W^-) \;=\; \left[2\delta_g + 2\delta_c + 2\delta_0 + 2\tfrac{s_\theta}{c_\theta}\delta_{A0} + 2\delta_{c_\theta}\right] \times (W^0, W^0, W^+, W^-)$$

Figure 3: Examples of Feynman rules

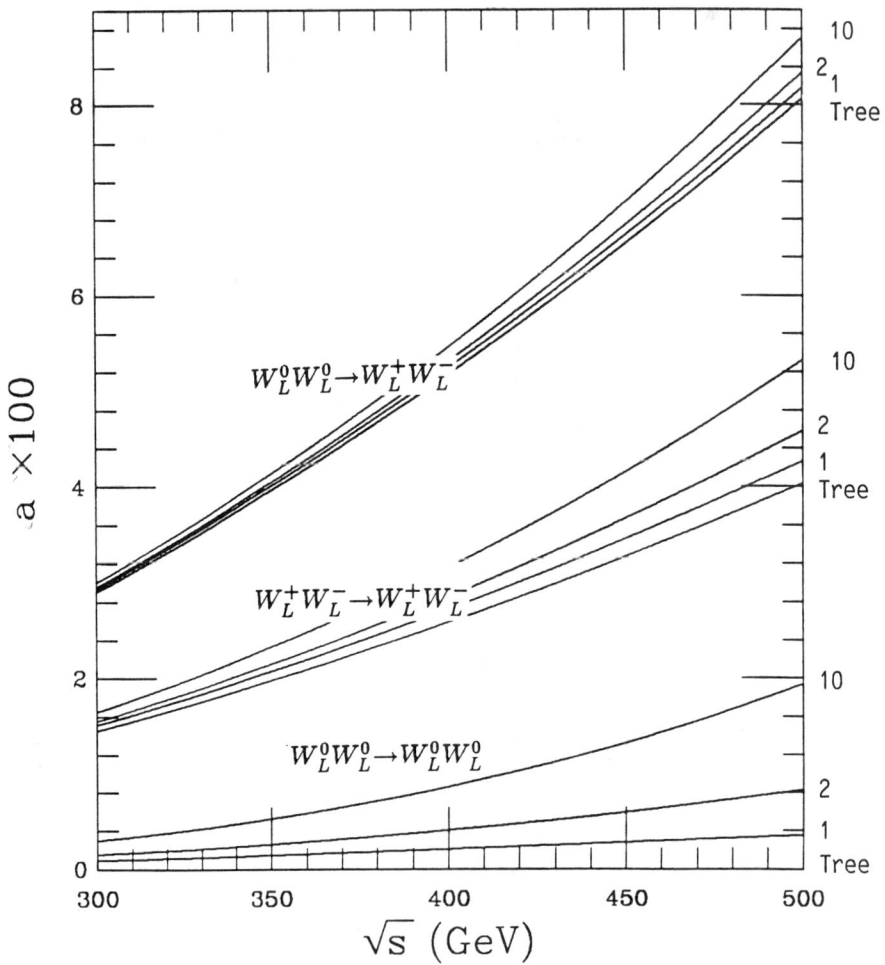

Figure 4: J=0 partial-wave amplitude in the energy window 300 to 500 GeV.

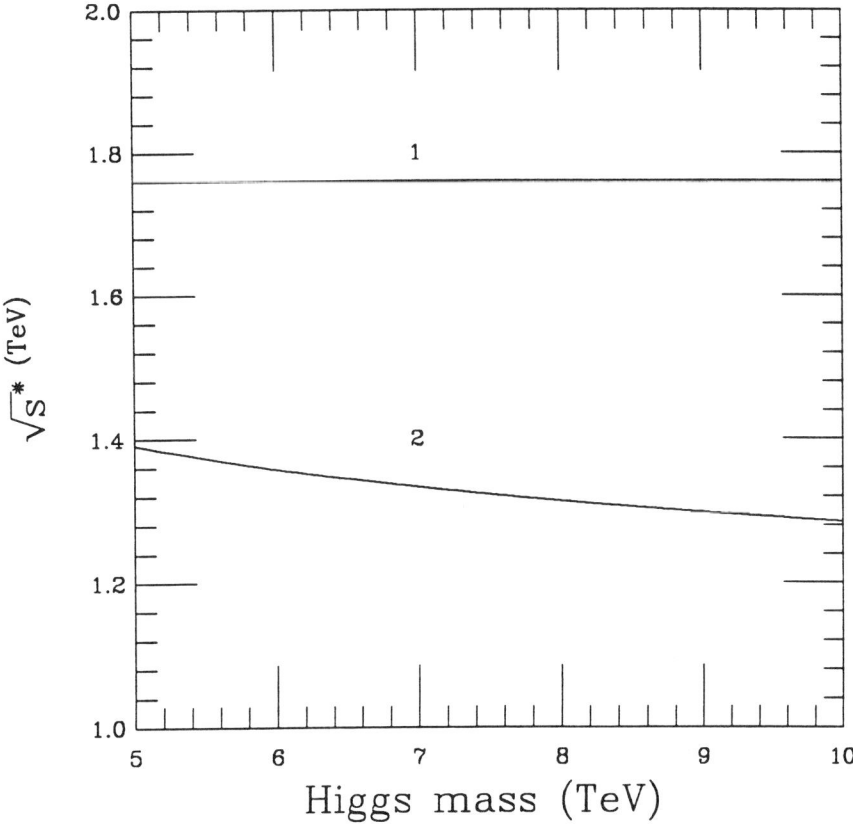

Figure 5: Critical energy as a function of the Higgs mass. Tree level (1). Tree level + one–loop corrections (2).

Algorithmic Solutions in QED and QCD

ALGORITHMIC SOLUTIONS FOR QED AND QCD

Michelangelo MANGANO[1]

To Giorgio Gamberini, who was happily among us when I started writing this, and now is no more.

This report is divided into two parts. In the first part I will discuss some problems arising when trying to describe multi-jet production with QCD Monte-Carlo's, while in the second part I will shortly summarize the results presented during this Workshop related to Algorithmic Solutions for QED and QCD. The plan of the first half is as follows: after a short Introduction, I will discuss *i)* Drell-Yan production accompanied by large amounts of transverse energy and *ii)* some problems related to the introduction of higher order corrections into hadronic QCD Monte-Carlo's. I refer to Kato's talks during this Workshop for an extensive introduction to QCD shower Monte-Carlo's [1,2].

1 Introduction

Since the main underlying issue of this Workshop is high precision tests and measurements, a word of warning is probably in order: contrarily to the standards of accuracy which are achievable in QED, both experimentally and theoretically, measurements and calculations in QCD provide a much lower precision. Higher order corrections in QCD are large, and calculations are complicated by the non-abelian nature of the theory, thus making precise predictions very difficult. In addition to this, the final states which are calculable in perturbative QCD are not exactly the final states which are observable experimentally, because of confinement, and a comparison of theory and experiment in QCD has very often to be mediated by a more or less phenomenological description of the hadronization process [2].

QCD Monte-Carlo's are always multi-particle generators, and, because of confinement and of the mass gap ($m_\pi > 0$), the multiplicity is a priori a well defined quantity. This contrasts with QED Monte-Carlo's, where the multiplicity is only defined thanks to a cut-off and is otherwise, strictly speaking, infinite. However, while QED Structure Function Monte-Carlo's have the advantage of being testable against the predictions of esclusive exact calculations, as we have

---

[1]Istituto Nazionale di Fisica Nucleare, Scuola Normale Superiore and Dipartimento di Fisica, Pisa, ITALY

seen in various occasions during this Workshop, there are no fully exclusive exact calculations to perform the same tests on QCD shower Monte-Carlo's. The reason for this is again the high complexity and large multiplicity of a hadronic final state, which prevents exact analytic calculations to be made. The study of the leading logarithmic singularities of the theory provides the analytic framework to carry out, in a given approximation, the Monte-Carlo simulation of these processes.

Experimentally, hadronic final states in, say, $p\bar{p}$ collisions, are often analized in an inclusive way, *e.g.* performing calorimetric measurements of jet energy distributions. Since energy measurements are not very sensitive to the infrared behaviour of the theory, one hopes that exact calculations done at the parton level (*i.e.* omitting the development of the shower and the hadronization of quarks and gluons) will be appropriate to describe these infrared-safe variables, and can therefore be used to test – even though only partially – the shower Monte-Carlo's.

In studying the performances of a hadronic Monte-Carlo, two issues are of particular importance: the first one, obviously, is the reliability of the approximations involved, how well these approximations work, and where do they break down. The second issue is the efficiency of the generation itself: to perform statistically significant studies, the generation of a large number of events is necessary, and it is important to be able to bias the generation towards the class of events that we are interested in in an efficient way, to avoid high rejection rates and waste of CPU/real time. In the following I will address these two issues by discussing few examples, namely large p_t Drell-Yan production and higher orders in QCD parton-parton scattering.

2 Large p_t Drell-Yan

As was explained during his introductory lecture by Kato, the generation of hadronic collisions starts from the hard scattering process, and evolves *outward* with the emission of gluon radiation: in other words, the initial-state hard partons are evolved *backward* and the final-state partons are evolved *forward* towards smaller and smaller values of Q^2. The probabilities for the emission of radiation are determined by the Altarelli-Parisi splitting functions and the proper Sudakov form-factors. These probabilities are furthermore corrected by imposing phase-space constraints on the possible momenta of the emitted gluons, constraints necessary to enforce within the deterministic development of the shower the effects of quantum interference due to the emission of large-angle gluons from different branches of the shower. For time-like evolution, for example, these interference effects amount to constraining the emission of a gluon from a given parton i with an angle θ_{ig} between gluon and parton which must be smaller than the angle θ_{ij} between parton i and the parton j which is *color connected* to it. The radiation process then takes place within cones of decreasing radius, whose boundary is given by connected color lines.

For the process $p\bar{p} \to W + X$, with the W eventually decaying into leptons, the hard scattering process is given by the simple $q\bar{q} \to W \to e\nu$ matrix elements. The only QCD radiation comes from the initial-state evolution, which generates a transverse momentum for the W. If the p_t of the W is not too large, say $p_t < M_W$, the leading-log approximation (LLA) which is used in the Monte-Carlo is appropriate, and the Monte-Carlo [3] correctly reproduces the result of an exact analytic calculation [4] for the W p_t distribution at the order α_s in which the leading soft singularities have been resummed (see Fig.1). If p_t is larger, however, the leading-log approximation is not good anymore, since terms of the order (p_t^2/M_W^2), which in the LLA are neglected, become important.

This is illustrated in Fig.2, where I show the inclusive W p_t distribution in $p\bar{p}$ collisions at $\sqrt{s}$=1.8 TeV obtained from the LLA Monte-Carlo and the one obtained by an exact order α_s parton-level calculation. As we see, the agreement between the two predictions is good only up to transverse momenta of the order of $M_W/2$. In the same Figure I plot the transverse energy distribution of the leading jet possibly present in each event, defining *jet* a cluster of transverse energy contained within a cone of radius 0.7 in $\eta-\phi$ space (η and ϕ being respectively pseudo-rapidity and azimuth). I require the axis of the jet to be confined to a central rapidity region with $|\eta| \leq 1$. In the case of the exact parton-level calculation the jet is simply given by the parton recoiling against the W, with the same central rapidity constraint. A comparison between the W and the jet p_t distributions for the two calculations, suggests that the transverse energy which compensates for the W transverse momentum is rather broadly spread, and possible jets recoiling against the W usually have a smaller transverse momentum than the W itself.

The proper simulation of W bosons accompanied by large amounts of transverse hadronic energy is extremely important for the study of backgrounds to interesting signals of exotic physics. This is the case of heavy top pair production, for example, with the W from one of the top quarks decaying leptonically and the other decaying hadronically, thus simulating events with a W and two or more jets. The same would apply to any other exotic source of W pairs, like a heavy Higgs or some technicolor excited state.

We thus need an efficient and reliable way of generating events with large-p_t W's. The Monte-Carlo scheme fails this goal in two ways: on one side, as we saw from the plots in Fig.2, the LLA is not sufficiently good at large p_t. On the other side, even if the approximation were good, since the p_t spectrum is rapidly falling, and since in the *backward* evolution scheme there is no way to force by hand the generation of events with large p_t, one would have to generate many millions of events before having few hundreds of interesting events in the tail of the p_t distribution.

The simplest way to get around these problems is to generate directly events with a large-p_t W, by using as hard process not the usual Drell-Yan, but the matrix elements for the $q\bar{q} \to Wg$ and $qg \to Wq$ processes. In this way the large p_t of the W comes directly from the emission of a single hard parton. This two-to-two hard scattering is then fed into the usual branching evolution routines for the development of the full initial state radiation (and now final as well, from the large-p_t parton recoiling against the W).

This procedure allows to efficiently generate events directly in the large-p_t tail of the distribution, by forcing the p_t to be always larger than a given threshold. Furthermore, the inclusive p_t distribution of the W will now agree 'by construction' with the result of the order α_s calculation. Few comments are however in order. First of all, the matrix elements for the production of a large-p_t W are not sufficient to give a full description of the Drell-Yan process at order α_s. This is because at this order we also have virtual corrections to the simple $q\bar{q} \to W$ diagrams. These corrections regularize the infrared and collinear singularities arising when the p_t of the W goes to 0. To completely implement these corrections in the Monte-Carlo is a hard task, and will require more studies [2]. Until then, one cannot lower the p_t threshold to values too small, where one can use, nevertheless, the standard Drell-Yan Monte-Carlo.

Secondly, in order to mantain the capability of the Monte-Carlo to describe correctly the quantum interference effects, one has to calculate the 'W + jet' matrix elements in such a way as to keep track of the color flows in the diagrams. In other words, one has to break up the calculation into the sum of the contributions from the various color flow patterns that contribute to a given hard process and generate event by event different color configurations with the proper weights. In this way one knows which are the pairs of color connected partons that will give rise to the cones within which the emission of radiation will take place. This problem can be easily solved for this process, in analogy with what is contained in Herwig for the treatment of hard photon processes, and I will indicate how to solve it in general when discussing QCD multi-jets in the next Section.

Finally I want to mention that even though using this recipe we generate the right inclusive W p_t distribution, we are still not guaranteed that at a more exclusive level we are doing the right thing. To show an example of how differences can possibly arise, in Fig.3 I plot the jet inclusive E_t distribution obtained by generating events using the order α_s matrix elements. One should notice that

---

[2] More on this problem in the next Section

now the distribution is much closer to the parton-level result compared to the result displayed in Fig.2, even in the p_t region (below 40 GeV) where the inclusive W p_t spectrum was properly described by the ordinary Drell-Yan Monte-Carlo. This suggests that jets generated by the LLA initial-state evolution are slightly fatter than those initiated by the order α_s calculation. A comparison with experimental data would be very important in this case, and remains the ultimate real check of the Monte-Carlo itself.

As was mentioned above, configurations where the W is accompanied by two or more jets are important backgrounds to the detection of top pair production. It is therefore important to check whether the Monte-Carlo – with the possible introduction of the order α_s matrix elements – properly simulates this background. In the order α_s plus LLA evolution described above, the second jet arises either from initial-state radiation or from the splitting of the final-state hard parton. In this latter case the two jets will tend to be close to one another, due to the collinear enhancement, while in the former case the two jets will tend to be back-to-back. In Fig.4 I show the distribution of the difference of azimuthal angle between the two jets in $W + 2$ jet events generated using the order α_s plus LLA Monte-Carlo and using a parton level calculation at the order α_s^2 [5]. Also shown are preliminary data from the CDF experiment at Fermilab [6]. It appears that while the data and the parton level calculation agree quite well, the Monte-Carlo tends to produce slightly fewer events than expected in the back-to-back configuration.

Again one could cure this problem by adding to the Monte-Carlo the order α_s^2 parton-level matrix elements. In this case, however, one would have to face matching problems arising when two hard partons become collinear or one becomes soft. These configurations are well described by the order α_s Monte-Carlo, which contains the proper Sudakov form factors, and one should make sure that the results of the two approaches match.

3 QCD Multi-Jets

A complete calculation of QCD matrix elements for parton-parton scattering to the order α_s^3 was recently completed by K.Ellis and J.Sexton [7]. It is being used by two groups [8] for the calculation of inclusive jet cross sections at the next-to-leading-log order (NLO). The phenomenological importance of this calculation is that the full knowledge of the order α_s^3 radiative corrections allows a more precise determination of the cross sections, reducing the systematic theoretical uncertainties to a level of possibly 20%. If this precision could me met by the experiments, it would provide a stringent test of QCD and it could allow to discriminate between various sets of structure functions. At this level of precision, however, it is important to make sure that higher order effects like the fragmentation process do not affect the predictions [9]. It then becomes important to be able to merge these higher order calculations into a shower Monte-Carlo. In this section I will shortly discuss some of the problems that arise when trying to carry out this program.

First of all NLO calculations are usually too inclusive to be used in a shower Monte-Carlo. This is because the parton level calculation, in order to be gauge invariant, requires the use of massless gluons (and often massless quarks for 'simplicity'), and the KLN theorem forces specific sums over final and initial states in order to cancel IR and mass singularities. These sums would have to be 'deconvoluted' by the Monte-Carlo in a way analogous to the way the Monte-Carlo performs the backward evolution of the initial state, generating event by event the various configurations of initial state radiation which at the inclusive level will reconstruct the hadron structure functions. I believe this will be the hardest problem to solve.

Initial-final state interferences at the NLO (*e.g.* box diagrams) are hard to implement into a shower Monte-Carlo even in QED. Some suggestions on how to deal with this problem in QED were given by O. Nicrosini in his talk, but their extension to non-abelian theories is not straightforward.

To be consistent with NLO calculations one should be able to describe the branching evolution at the same order. A first step in this direction was described by Kato in his talk. Kato described a Monte-Carlo with NLO branching evolution that can be used for e^+e^- QCD final states and for Deep Inelastic Scattering. Here the problems mentioned above are less severe, since a proper gauge choice forces the emission to take place from one leg of the final state only, and there is no initial state radiation. More work will be necessary before this scheme can be applied to purely hadronic reactions.

Another important problem is related to the Monte-Carlo treatment of the quantum interference effects in the shower. As was mentioned above, to properly implement these effects it is necessary to keep track of the pattern of color flow in the process under consideration. Right now, these NLO calculations are performed by summing over the various color configurations, and this important information is lost. The same problem would arise when trying to implement into the shower Monte-Carlo higher order tree level hard sub-processes needed for a correct generation of multi-jet events. To maintain the color structure of the process one should try to use the techniques described by S. Parke in his talk, namely the *dual expansion* of the amplitude [10]. In this technique amplitudes are calculated by decomposing them into *sub-amplitudes* determined by the inequivalent patterns of color flows in the diagrams describing the given process. These sub-amplitudes enjoy the following properties:

- They are gauge-invariant;
- They sum up incoherently to the leading order in $1/N$ when summing over colors;
- They contain collinear singularities only for collinear pairs which are color connected; near these poles, the required factorization properties (the Altarelli-Parisi equations) are satisfied.

This last property (collinear singularities for color connected pairs), is closely related to the idea of preconfinement, which is almost automatically achieved in shower Monte-Carlo's with color-coherence implemented in the described way.

Notice that this last comment about large-angle gluon coherence should also apply to the NLO Monte-Carlo described by Kato, since there the order α_s^2 splitting functions are calculated by summing over colors, thus loosing the possibility of implementing some of the interference effects.

4 "Algorithmic Solutions for QED and QCD": Workshop Summary

This was just meant to be a short review of the talks that were presented during this Workshop and were related to *algorithmic* issues, by which I mean Monte-Carlo implementations, new/old tricks, approximations and bookkeeping devices to get *numbers*. The more formal presentations were reviewed by G. Sterman in his Summary Talk. For the details of the various talks and the relevant bibliography I refer the reader to the papers contained in these Proceedings.

4.1 QED Monte-Carlos

I will start from the QED Monte-Carlo's. We had presentations by G. Bonvicini, B. Ward, A. Courau and R. Miquel. The first two described shower-like Monte-Carlo's, while the second two described matrix element Monte-Carlo's.

Bonvicini presented MOE, the QED structure function Monte-Carlo developed in collaboration with L. Trentadue [11] and based on the theoretical developments described in Ref.[12]. MOE describes the *photon* shower through the evolution of both the longitudinal momentum fraction z (the Altarelli-Parisi equations) and the evolution of the transverse momentum of the emitted photons with respect to the emitting charged particle (the Bassetto-Ciafaloni-Marchesini equations [13]). In addition, the order p_t^2/s corrections to the usual Altarelli-Parisi splitting functions are included in the calculation of the branching probabilities. These two features, BCM kernels and exact splitting functions guarantee the equivalence of this Monte-Carlo with an exact order α calculation. The multiplicities are chosen event by event according to the usual Poisson distribution, and the order α^2 deviations from Poissonian behaviour are partly accounted for by a redefinition of the allowed phase space (*e.g.* infrared cut-off) for the emission of the photons following the first. In this way MOE describes exactly not only all of the leading log's at any order in α, but also resums some subleading ones (see [11] and Bonvicini's talk for a detailed *accounting* of resummed *log*'s).

Similar results are achieved by YFS, the Monte-Carlo by Jadach and Ward [14] based on the seminal work by Yennie Frautschi and Suura [15]. The authors implement the YFS techniques for the handling of soft divergencies with renormalization group evolution (a' la Weinberg and t'Hooft) to describe the emission of an arbitrary number of photons in Bhabha and in $e^+e^- \to \mu^+\mu^-$ scattering. This Monte-Carlo accounts for corrections induced by W exchange, and resums log's of the following form: $(\alpha/\pi L^2)^n$, $(\alpha/\pi Ll)^{n \leq 2}$ and $(\alpha/\pi L)^{n \leq 2}$. As usual,

$$L = \log(s/m^2) - 1 \qquad l = \log(E/\epsilon), \qquad (4.1)$$

ϵ being the infrared cut-off.

A. Courau described his *library* of matrix-element Monte-Carlo's, built out of the Equivalent Photon Approximation, a.k.a. the Weiszäker-Williams method. One looks at kinematical configurations dominated at the dynamical level by poles in some of the propagators, and in these regions one can well describe the exact matrix elements with relatively simple analytic expressions, which can be integrated to give rise to cross-sections. Coureau showed how this can be done even in presence of constraints on the phase space, such as those dictated by the acceptance of an experimental apparatus. As examples, the following processes were described in detail: $ee \to eel^+l^-$ and $ee \to ee\gamma$. Whenever available, the agreement with experimental data is always good.

R. Miquel, finally, presented full exact calculations for the following processes: $e^+e^- \to \gamma\nu\nu$ (to the order α^4), and $ee \to ee\gamma$ with massive electrons. All of the electro-weak effects were included. The calculation employed the helicity amplitude methods (described during the Workshop by R. Gastmans), and was realized through the use of a Fortran code carrying out numerically the spinor algebra.

Whenever comparisons are possible, these Monte-Carlo's all agree very well among themselves and with others not presented directly here, but available in the literature and partly discussed by P. Rankin in her introductory talk (see also [16]). The level of agreement which is required by today's experimental capabilities is below 1%, and it is at this level that the Monte-Carlo's should be probed. It is important to realize, however, that 1% agreement for some inclusive variable (σ^{tot}, σ^{peak}, Z shape,...) does not guarantee a 1% agreement at the level of some more exclusive quantity, or in rare processes which do not count at the level of total cross-sections, but which are important for determining the possible presence of new physics. This could be the case, for example, of the so-called neutrino counting experiments, described during a parallel session of this workshop. Here one looks for events of the kind $e^+e^- \to \gamma + nothing$, 'nothing' representing presumably the decay of the Z into undetected neutrinos. The number of events of this kind is clearly proportional to the number of neutrino species. A small contribution to these final states could come, however, from some exotic processes, such as, for example, photino pair pruduction $e^+e^- \to \gamma + \tilde{\gamma}\tilde{\gamma}$ via exchange in the t-channel of a heavy scalar electron, supersymmetric partner of the electron. The spectrum and angular distribution of these photons is similar to those due to the W-exchange corrections to the standard Z annihilation process and scales with the scalar electron mass like $(M_W/m_{\tilde{e}})^4$; to properly estimate the standard model background to this supersymmetric process, and to possibly detect a weak signal, one needs a simulation which should be much more reliable than 1% at the inclusive level, since these processes would be much rarer than 1% of the total cross-section. Therefore it is very important to also test the various Monte-Carlo's (both of the structure-function type and exact matrix element calculations) on these more exclusive variables, like large angle distributions of hard photons. Needless to say, at the exclusive level and for these 'exotic' channels the precision required is not at the 1% level, but probably around the 10%. It would then be advisable to have Monte-Carlo's working in two modes, one for high precision (semi-)inclusive measurements, and the other for efficient and reliable (say, 10%) generation of configurations which have a small cross-section but which are important backgrounds to exotic phenomena.

4.2 QCD Monte Carlos

An important effort to incorporate QCD next-to-leading order effects into a shower Monte-Carlo was described by Kato. He presented the results of his collaboration with T. Munehisa [17], implemented into two Monte-Carlo's, one for $e^+e^- \to$ jets and the other for Deep Inelastic Scattering. For the sake of definiteness, let us concentrate on the e^+e^- case. $q\bar{q}$ final states are generated using parton level matrix elements, and the evolution is carried out using the order α_s^2 splitting functions describing 1-to-2 and 1-to-3 decays. If after the first branching the reconstructed virtual mass of the quark coming from the color singlet vertex is larger than a given cut-off, the weight of the event is recalculated using the order α_s matrix elements for the $e^+e^- \to q\bar{q}g$. The use of NLO vertices and branchings should make it possible, by comparison with data, the independent determination of α_s and Λ_{QCD}. Pairs of intersecting rungs of a ladder are reabsorbed into the 1-to-3 branchings, thus incorporating in the ladder evolution some of the quantum coherence effects. However, since the Monte-Carlo does not keep track of the color flow, preconfinement is not built into the evolution, and some independent fragmentation model has to be put in at the end to simulate the hadronization.

As a final comment, it would be interesting to exploit the relation between this approach and the approach employed by Bonvicini and Trentadue in their QED Monte-Carlo , where the exact Altarelli-Parisi 1-to-2 splitting functions ($P(z,p_t^2)$) are used together with the BCM evolution kernels.

4.3 Progresses in Perturbative Techniques

The need to carry out more and more precise theoretical predictions leads to the development of new calculational techniques. During the Workshop we had talks by Gastmans and S. Parke presenting the Helicity Amplitude technique and the Dual Amplitude technique.

The HA technique described by Gastmans was initially developed to allow analytical calculations of QED processes with many particles in the final state. Instead of calculating Feynman diagrams for a given process in a generic helicity configuration, performing the sum over polarizations only after having squared up the amplitude, it turns out to be useful to calculate directly at the matrix element level amplitudes for fixed helicity configurations. Gauge invariance allows then the choice of specific polarization vectors to describe the given helicities , and these can be chosen in such a way as to make the expression for the amplitude as simple as possible. The final result is just obtained by summing the square of the various HA's, which obviously do not interfere with each other. Gastmans described in detail how this works, and gave a few examples of the effectiveness of this method.

When calculating many-parton processes in a non-abelian theory, it is fundamental to organize the color algebra in an effective way. The Dual Amplitude technique [10], described by S. Parke, provides the solution to this problem. Parke proved that the most general tree-level amplitude for a Yang-Mills $SU(N)$ theory can be decomposed in the following way:

$$A^{n-gluons} = \sum_{perm} \text{tr}(\lambda^{a_1}\lambda^{a_2}...\lambda^{a_n})\mathcal{A}(1,2,...n) \qquad (4.2)$$

where the sum is over the non-cyclic permutations of (1,2,..,n) and the λ's are the matrices of $SU(N)$ in the fundamental representation. The sub-amplitudes (*Dual Amplitudes*) $\mathcal{A}(1,2,..,n)$ enjoy the remarkable properties mentioned above in Section 2. Using this representation of the amplitude important properties of the perturbative expansion emerge, and wih the help of the HA method very complex calculations become possible. The DA technique can be easily generalized to include processes with fermions and different gauge groups as well. The main theoretical challenge in this field remains though the extension of the HA technique to loop amplitudes.

New interesting calculations were presented by H. Veltman and R. Bouamrane. Veltman described the full calculation of radiative corrections to Polarized Compton Scattering, with massive fermions. This calculation is important, *e.g.* , for luminosity monitoring in polarized e^+e^- scattering experiments, such as those that will be performed in the next future at SLC.

Boumrane discussed the EWK corrections to WW scattering, in the attempt to understand the structure of perturbative unitarity of the $SU(2) \times U(1)$ theory in presence of a very massive Higgs.

Finally, J. Im introduced us to the *Noodle Method*, which is an integration technique designed to provide efficient and fast integration of many particle cross-sections over complicated phase-space volumes.

Acknowledgements: I want to thank Giovanni Bonvicini for the wonderful job he did organizing this successful Workshop, and for the insistence with which he encouraged me to attend it. I want to thank the participants for their active presence, and Laura Phillips for her competent collaboration with us.

References

[1] COJETS: R. Odorico, Computer Phys. Comm. 32 (1984) 139;
BIGWIG: G. Marchesini and B. Webber, Nucl. Phys. B 238 (1984) 1; B. Webber, Nucl. Phys. B238 (1984) 492;
PYTHIA: H.-U. Bengtsson and G. Ingelman, Computer Phys. Comm. 34 (1985) 251;
FIELDAJET: R.D. Field, Nucl. Phys. B264 (1986) 687;
ISAJET: F. Paige and S.D. Protopopescu, Brookhaven report BNL-38034 (1986);
JETSET: T. Sjöstrand and M. Bengtsson, Computer Phys. Comm. 43 (1987) 367;
EUROJET: A. Ali, B. van Eijk and E. Pietarinen, see B. van Eijk thesis, Amsterdam (1987).

[2] See also: B.R. Webber, Ann. Rev. of Nucl. Sci. 36 (1986) 253.

[3] All the simulations discussed in this work were carried out using the HERWIG Monte-Carlo : G. Marchesini and B.R. Webber, Nucl. Phys. B310 (1988) 461.

[4] G. Altarelli, R.K. Ellis, M. Greco and G. Martinelli, Nucl. Phys. B246 (1984) 12;
C.T.H. Davies, B.R. Webber and W.J. Stirling, Nucl. Phys. B256 (1985) 413.

[5] M. Mangano and S.J. Parke, FERMILAB-Pub-89/106-T (1989).

[6] T. Kamon, presented at the 8th Topical Workshop on Proton-Antiproton Collider Physics, Castiglione della Pescaia (ITALY), September 1-6, 1989. To appear in the Proceedings.

[7] R.K. Ellis and J.C. Sexton, Nucl.Phys. B269 (1986) 445.

[8] M. Aversa, P. Chiappetta, M. Greco and J.-M. Guillet, Phys. Lett. 210B (1988) 225; *ibid.* 211B (1988) 465;
S. Ellis, Z. Kunszt and D. Soper, Phys. Rev. Lett. 62 (1989) 726.

[9] M. Mangano, presented at the 8th Topical Workshop on Proton-Antiproton Collider Physics, Castiglione della Pescaia (ITALY), September 1-6, 1989. To appear in the Proceedings.

[10] M. Mangano, S. Parke and Z. Xu, *in* Proc. of "Les Rencontres de Physique de la Vallee d'Aoste", La Thuile, Italy, (1987), ed. M. Greco, Editions Frontières, p.513;
F. A. Berends and W. Giele, Nucl. Phys. B294 (1987) 700;
M. Mangano, S. Parke and Z. Xu, Nucl. Phys. B298 (1988) 653.

[11] G. Bonvicini and L. Trentadue, Nucl. Phys. B323 (1989) 253.

[12] E.A. Kuraev and V.S. Fadin, Sov. Journ. Nucl. Phys. 41 (1985) 466;
G. Altarelli and G. Martinelli, in the CERN Yellow Report "Physics at LEP", CERN 86-02, p.47;
O. Nicrosini and L. Trentadue, Phys. Lett. 196B (1987) 551.

[13] A. Bassetto, M. Ciafaloni and G. Marchesini, Nucl. Phys. B 163 (1980) 477.

[14] S. Jadach and B.F.L. Ward, Phys. Rev. D38 (1988) 2897.

[15] D.R. Yennie, S.C. Frautschi and H. Suura, Ann. of Phys. 13 (1961) 379.

[16] R. Kleiss, CERN-TH 5439/89.

[17] K. Kato and T. Munehisa, Phys. Rev. D36 (1987) 61.

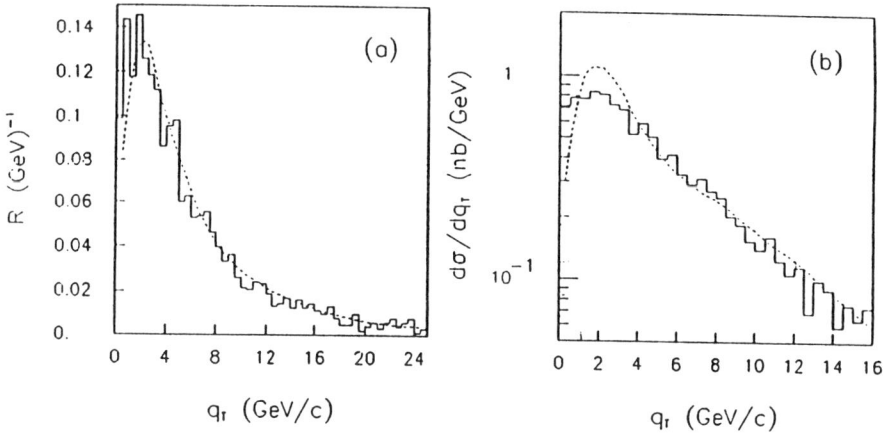

Figure 1: Monte-Carlo transverse momentum distribution of bosons as obtained in [3]: (a) at zero rapidity, compared with Altarelli et al. [4] (dashed); (b) at all rapidities, compared with Davies et al. [4] (dashed)

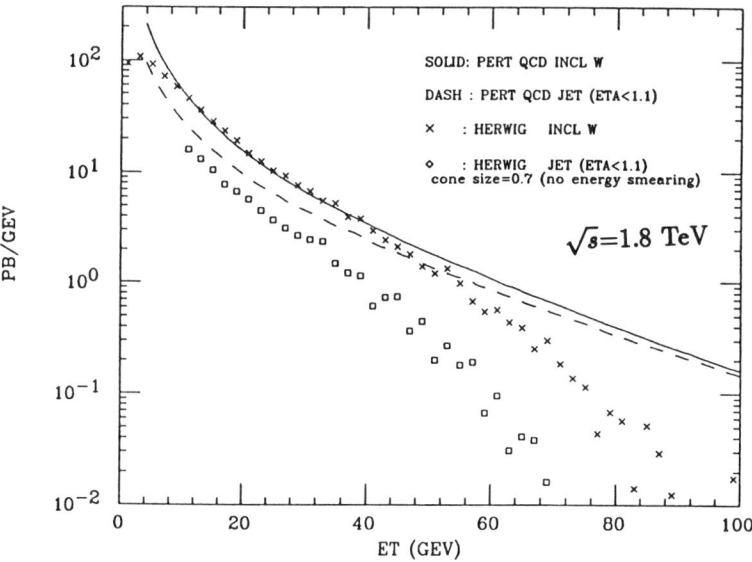

Figure 2: Monte-Carlo inclusive transverse momentum distribution of W bosons (solid) and inclusive E_t distribution of central jets (dashed), compared with the results of an analytic tree level order α_s calculation (crosses and squares, resp.).

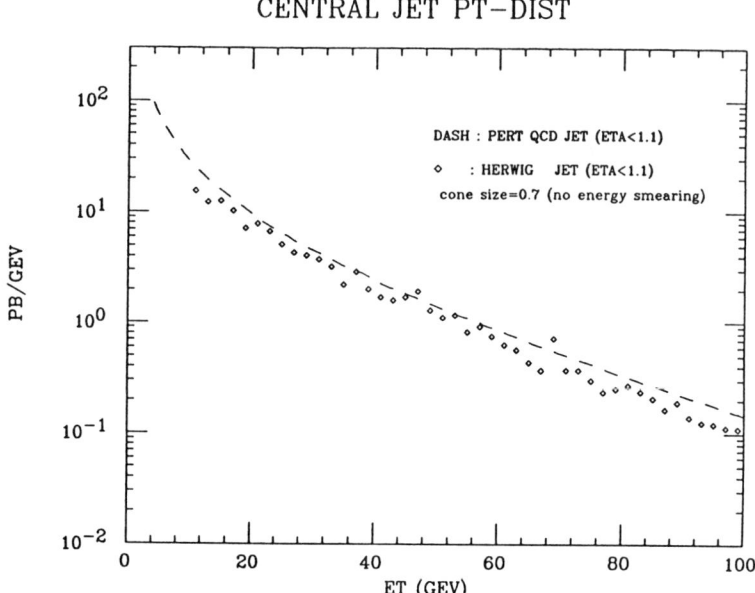

Figure 3: Monte-Carlo inclusive central jet E_t distribution, with order α_s hard sub-process matrix elements (diamonds), compared to the result of the analytic parton-level calculation (dashes).

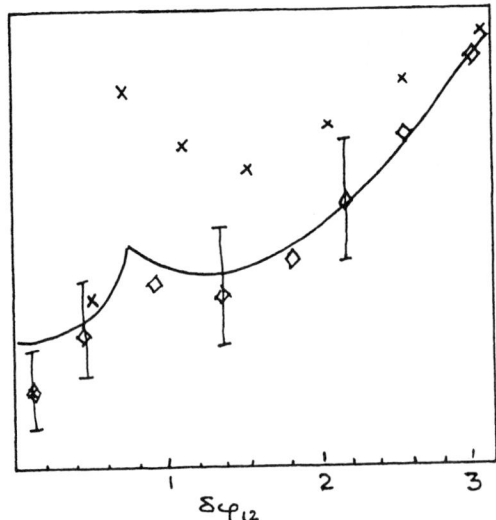

Figure 4: Jet-jet azimuthal difference, in W +2 jet events. Monte-Carlo with order α_s hard matrix elements (crosses), parton level order α_s^2 (solid) and CDF preliminary data (diamonds).

Review of QCD generators

Kiyoshi KATO

Department of Physics, Kogakuin University

Shinjuku, Tokyo 160, Japan

Talk at the Workshop on QED Structure Function, May 22-25, 1989, Ann Arbor, Michigan

Contents

1. Introduction
    1.1 Introduction to QCD jets
    1.2 Short history of QCD cascade
    1.3 Plan of this paper
2. QCD Cascade in LL order
    2.1 Basic idea
    2.2 Time-like cascade in e^+e^- annihilation
    2.3 Angular ordering
    2.4 Double cascade scheme
    2.5 Space-like cascade
3. QCD cascade in NLL order
4. Comments on radiative correction for QCD generator
5. Summary

Abstract

Monte Carlo simulation of QCD cascade is reviewed including the recent development to incorporate next-to-leading order corrections.

1. Introduction

1.1 Introduction to QCD jets

The discovery of jets in e^+e^- annihilation [1] showed us a dynamical activity of quarks. By use of perturbative QCD jet phenomena are studied extensively. Since the observed structure of jets is complex, Monte Carlo simulation based on QCD is an essential tool for the study of jets[2]. Since QCD is an asymptotic-free theory, the effective coupling constant, $\alpha_s(Q^2)$ is small in large Q^2 region where Q stands for a momentum scale of the process in question. We apply perturbative method in large Q^2 region while we cannot solve QCD in small Q^2 region. Due to this situation, the standard strategy to generate jets is as follows. First, we generate partons based on the perturbative QCD. Next, we use a model of hadronization to convert partons into hadrons which are observed in detectors. In this talk, since this workshop is for the study on electrons and photons which do not hadronize, we focus our attention on the first part and the discussion on models of hadronization [3] is skipped. Though the jet is an universal phenomenon in high energy hadronic interaction, we concentrate on e^+e^- annihilation to make concrete discussion throughout this paper except for section 2.5.

We use some approximation to calculate QCD perturbative series since it is not yet possible to calculate all order diagrams. There are two methods to do this. One method is the matrix element (MA) method in which we calculate diagrams exactly up to a fixed order of matrix elements. Another one is based on the renormalization group method and we sum up only leading terms of all order diagrams.

In MA method, we first integrate matrix elements and obtain the portion of n-jet production. According to the fraction we select the multiplicity of jets and generate 4-momenta of partons by the matrix element. In the second method, the jet is formulated as a stochastic branching process of quarks and gluons. A parton produced first can have a virtual mass and can branch into child-partons. The probability distribution of the virtual mass and that of energy distribution for child-partons are given by perturbative QCD. Iteration of this 'decay' processes produces a number of partons and we can employ Monte Carlo method to generate partons which is called as QCD cascade.

The MA method has some shortcomings. Due to technical difficulty we only calculate up to four-jet distribution (and higher order tree cross sections) while through the analysis of experimental data by a cluster method it is found that the method is insufficient to describe multi-jet distribution [4]. Further, MA method unavoidablly gives a discontinuous distribution to a class of physical quantities. To avoid mass singularity, we must introduce a dimensionless cutoff to define jets, e.g., ϵ and δ introduced by Sterman and Weinberg [5] or y_{min}. As an example, we consider the cross section of 2 jets and that of 3 jets. At $W = 100$GeV the cross section is saturated by 3 jets when we take the minimum mass of jets to be 7GeV which is resolvable by a practical detector. This breaks the perturbation. We can compare the defects of MA method with conventional 1-loop calculation in QED. In QCD experimentalists have observed multi-jets while in QED events with multi-photons, e.g., $e^+e^- \to \ell\bar{\ell}\gamma\gamma$ have been observed. In QCD we get wrong prediction for small jet-mass due to mass singularity while in QED non-radiative cross section becomes negative for a small photon energy cutoff. These facts lead us to the cascade method in QCD and to the introduction of exponentiation in QED.

1.2 Short history of QCD cascade

In the early age of perturbative QCD, its main interest was on inclusive quantities. The first important step to handle jets as a system of QCD quanta was the jet calculus by Konishi, Ukawa and Veneziano [6]. Following their work QCD cascade in LL was developed by Odorico and others [7]. Later, through the detailed study of soft gluons [8] the effect of coherence or so-called angular ordering was included in QCD cascade by Marchesini and Webber [9]. Based on these ideas the generator of LUND [10] and that of Webber [11] were constructed.

While these works are in LL order, the formulation of QCD cascade beyond LL order was developed by Kalinowski, Konishi, Scharbach and Taylor[12]. They also calculated quantities in quark sector. The calculation of functions of NLL order in gluon sector was done by Gunion, Kalinowski and Szymanowski [13]. The construction of Monte Carlo model in NLL order was still a non-trivial task after these works and it was done by Munehisa and the author as the development of a new QCD generator named NLLjet [14,15,16].

1.3 Plan of this paper

In the next section, the QCD cascade in the leading-logarithmic (LL) order is discussed. First, we present basic tools of perturbative QCD. Next, time-like cascade is formulated for e^+e^- process. The effect of coherence is also discussed. The double cascade scheme is studied in the next subsection to handle overlap of phase space. LL cascade for space-like evolution is discussed in the following subsection. In section 3, the formulation of cascade including the next-to-leading logarithmic (NLL) correction is presented with comparison of its results with experimental data. After section 4 in which the role of QED correction for QCD generators is discussed, we present summary of QCD cascade in section 5.

2. QCD cascade in LL order

2.1 Basic idea

In LL order we sum all terms of $\Sigma \alpha_s(Q^2)^n (\log Q^2)^n$ because $\alpha_s(Q^2) \log Q^2$ becomes large even if $\alpha_s(Q^2)$ is small. The large logarithm originates from the collinear singularity: When a parton branches into two partons it gives a term as $\sim \int dp_T^2/p_T^2$ where p_T is the transverse momentum between two partons. An important observation for the formulation of QCD cascade is that when we work in physical gauge, large logarithms appear only in ladder diagrams. This allows us to treat jets as a stochastic branching process since we can neglect interference terms. It is also noted that there is no ghost in physical gauge. A convenient

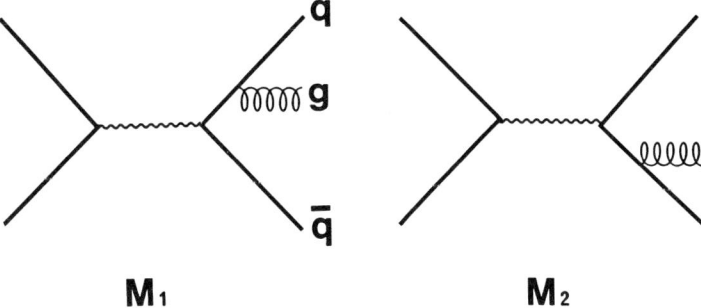

Fig.1 Feynman diagrams for $e^+e^- \to q\bar{q}g$.

choice of physical gauge is the light cone gauge. The polarization tensor of a gluon is

$$d^{\mu\nu} = -g^{\mu\nu} + \frac{k^\mu n^\nu + n^\mu k^\nu}{k \cdot n} \tag{1}$$

where n^μ is a light-like($n^2 = 0$) vector to specify gauge and k^μ is the momentum of the gluon.

As an example we consider the process $e^+e^- \to q\bar{q}g$. As is shown in Fig.1, this process is expressed by two Feynman diagrams in the lowest order. The cross section is

$$\frac{d\sigma}{dz_1 dz_2} = \sigma_0 \frac{\alpha_s}{2\pi} C_F \frac{z_1^2 + z_2^2}{(1-z_1)(1-z_2)} \tag{2}$$

where z_1 and z_2 are energy fractions of quark and antiquark, σ_0 is the lowest order cross section for $e^+e^- \to q\bar{q}$ and $C_F = 4/3$.

To show the origin of collinear singularity, we take the direction of gauge vector to that of anti-quark and boost the system opposite to the direction of the gauge vector. In this infinite momentum frame, we use variables $x = (p_q n)/(qn)$ and p_T, the light-cone momentum fraction of quark and the transverse momentum of quark with respect to the direction of boost. (Fig.2) In this frame the Eq.(2) becomes

$$\frac{d\sigma}{dx\,dt} = \sigma_0 \frac{\alpha_s}{2\pi} C_F \left(\frac{1}{t} \frac{1+x^2}{1-x} + 2(x - \frac{1}{1-x}) + t(1 - x + \frac{1}{1-x}) \right) \tag{3}$$

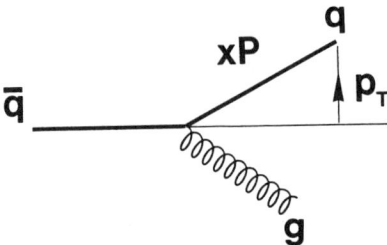

Fig.2 $q\bar{q}g$ system in the infinite momentum frame.

where $t = p_T^2/(x(1-x)s)$ and $s = q^2 =$ (the squared center-of-mass energy). The collinear singularity is expressed by $1/t$. Using the notation of Fig.1, the contribution of each term is as follows:

	Feynman gauge	Light − cone gauge
$M_1 M_1^*$	$\frac{(1-t)^2}{t}(1-x)$	$\frac{(1-t)^2}{t}\frac{1+x^2}{1-x}$
$2M_1 M_2^*$	$\frac{1-t}{t}\frac{2x}{1-x}$	$(1-t)\frac{2x}{1-x}$
$M_2 M_2^*$	$t\frac{1}{1-x}$	$t\frac{1}{1-x}$

Above calculation shows that in the physical gauge we can neglect the interference term if we sum up only leading terms. It should be noted that an antiquark does not evolve for the present choice of gauge. Asymmetry between quark and antiquark will be discussed later.

The above result is extended to general case. We define inclusive distribution of quark and gluon, $D_q(x, Q^2)$ and $D_g(x, Q^2)$, for $e^+e^- \to q+$ anything and $g+$ anything. We write down the general equation for ladder diagrams (Fig.3) and we get integral-differential equation, Altarelli-Parisi equation [17], which is QCD version of Lipatov equation in QED[18].

$$\frac{d}{dt}D_a(x, Q^2) = \frac{\alpha_s(Q^2)}{2\pi}\sum_b P_{ab} \otimes D_b(x, Q^2), \quad a, b = q, g. \quad (4)$$

Altarelli-Parisi function P_{ab} represents a branch producing a parton a from a

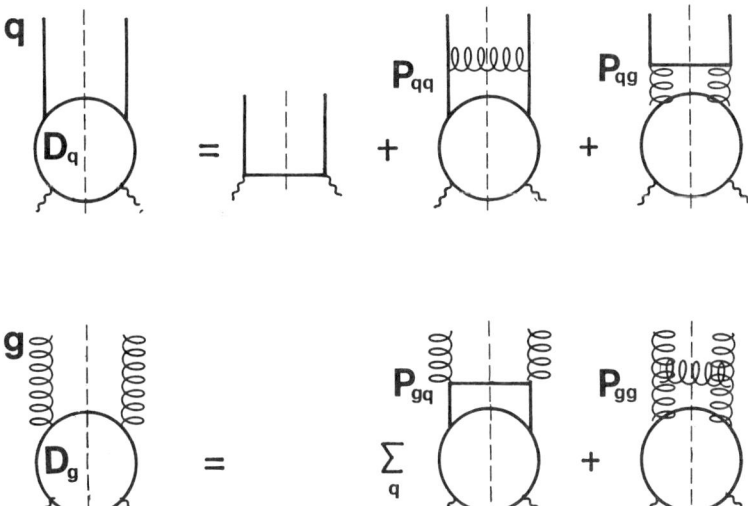

Fig.3 Diagramatical representation of Altarelli-Parisi equation, Eq.(4), and P-functions, Eq.(5). A solid (curly) line line stands for a quark (gluon).

parton b as in Fig.3.

$$\begin{aligned}
P_{qq}(x) &= C_F \frac{1+x^2}{1-x}, \\
P_{gq}(x) &= P_{qq}(1-x), \\
P_{qg}(x) &= T_R(x^2 + (1-x)^2), \\
P_{gg}(x) &= C_A(\frac{1-x}{x} + \frac{x}{1-x} + x(1-x))
\end{aligned} \qquad (5)$$

where $t = \log(Q^2/\Lambda^2)$, $C_A = 3$, $T_R = 1/2$ and $\otimes$ stands for the convolution integral as

$$f \otimes g(z) = \int_0^1 dx dy \delta(z - xy) f(x) g(y) . \qquad (6)$$

To define P-functions, regularization at $x = 0, 1$ is not explicitly shown in Eq.(5).

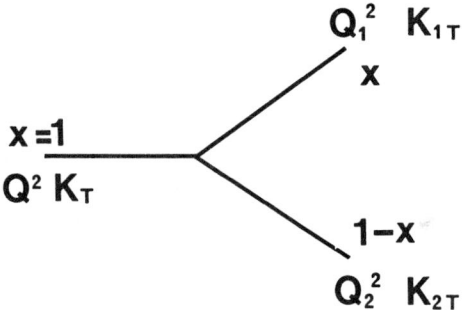

Fig.4 Kinematics of a two-body branching is described by momentum fraction, virtuality and transverse momentum.

2.2 Time-like cascade in e^+e^- annihilation

The evolution QCD cascade is controlled by equations in the last subsection. Kinematical consideration is important here. For the two-body branching shown in Fig.4 we get an equation as the conservation of $-$ component of momentum(p^-),

$$Q^2 + \mathbf{K}_T^2 = \frac{Q_1^2 + \mathbf{K}_{1T}^2}{x} + \frac{Q_2^2 + \mathbf{K}_{2T}^2}{1-x} \qquad (7)$$

where x is the fraction of p^+ and $\mathbf{K}$ is the transverse momentum which are decomposed as $\mathbf{K}_{1T} = x\mathbf{K}_T + \mathbf{k}$ and $\mathbf{K}_{2T} = (1-x)\mathbf{K}_T - \mathbf{k}$. From this equation we get a condition,

$$\mathbf{k}^2 = x(1-x)\left(Q^2 - \frac{Q_1^2}{x} - \frac{Q_2^2}{1-x}\right) \geq 0 \qquad (8)$$

which constrains the range of x as $x_- < x < x_+$ where $x_\pm$ are roots of Eq.(8). We introduce a cutoff Q_0^2 as

$$Q^2 \geq Q_0^2. \qquad (9)$$

This cutoff also regularizes infrared singularity. It is important to note that masses of produced partons, Q_1^2 and Q_2^2, are dependent on each other through Eq.(8). Integrating Eq.(4) we get the Sudakov factor [19] or the non-emission probability, $\Pi_{NE}^a(Q^2, Q_1^2)$ following the paper of Odorico[7]. It is the probability that a parton a does not emit a parton in the phase volume between Q^2 and Q_1^2 and it is given by

$$\Pi_{NE}^q(Q^2, Q_1^2) = \exp\left[-\int_{Q_1^2}^{Q^2} \frac{dK^2}{K^2} \int_\epsilon^{1-\epsilon} dx \frac{\alpha_s(K^2)}{2\pi} P_{qq}(x)\right],$$

$$\Pi_{NE}^g(Q^2, Q_1^2) = \exp\left[-\int_{Q_1^2}^{Q^2} \frac{dK^2}{K^2} \int_\epsilon^{1-\epsilon} dx \frac{\alpha_s(K^2)}{2\pi}(P_{gg}(x) + \sum_q P_{qg}(x))\right]. \quad (10)$$

where $\epsilon = Q_0^2/K^2$ is the kinematical boundary defined by Eq.(9).

Using these quantities, the Monte Carlo simulation of jets goes as follows:

(1) There is a parton a which will evolve. It has a maximum virtuality Q^2 which is to be determined by kinematics.

(2) We determine whether the parton a branches or does not branch. If it branches, we determine the mass of parton a and if it does not branch, the parton a has a mass of Q_0 and it will not branch any more. To do this step, we generate a uniform random number r between 0 and 1. If $r < \Pi_{NE}^a(Q^2, Q_0^2)$, it does not branch. In the reversed case the parton a branches and its squared mass K^2 is determined by $r = \Pi_{NE}^a(Q^2, K^2)$.

(3) When a parton a branches into partons b and c, we determine their momentum fraction and their species(q or g). For the quark branching, momentum fraction is determined by $P_{qq}(x)$. For the gluon branching, we first decide the decay mode, $g \to gg$ or $g \to q\bar{q}$, by weights $\int P_{gg}(x)dx$ and $\int P_{qg}(x)dx$ and determine the momentum fraction by $P_{gg}(x)$ or by $P_{qg}(x)$.

(4) Above procedures are repeated until all partons have the mass of Q_0 and are inactive. Finally, we calculate 4-momenta of partons from their momentum fraction and virtuality. The azimuthal rotation angle at each two-body branching is chosen at random.

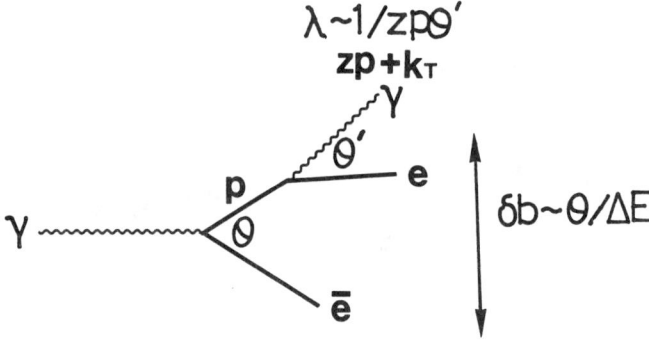

Fig.5 Angular ordering in QED for $\gamma \to e^+e^-\gamma$.

2.3 Angular ordering

As was noted before, maxima of available virtuality of two partons interfere through Eq.(8) when they have a common mother-parton. One of the merit of angular ordering is that we can escape from this complication of kinematics. The angular ordering is known in QED many years ago as Chudakov effect [20]. As is shown in Fig.5, we consider a process $\gamma \to e^+e^-\gamma$. When the photon transverse wavelength λ is greater than the transverse emission distance δb of e^+e^- pair, the process is suppressed since the radiated photon 'sees' a neutral system. The condition $\lambda < \delta b$ implies $\theta > \theta'$, i.e., branching angles are ordered.

In the study of infrared structure of QCD [8], it is found that interference diagrams give large logarithms as much as ladder diagrams. This fact threatens the formulation of QCD cascade. However, the large logarithms from interference terms cancel non-ordered configuration in ladder terms and the remaining contribution comes from only the ladder diagrams with the restriction to phase space.(Fig.6)

The angular ordering was first implemented by Marchesini and Webber into a Monte Carlo program[9,11]. They use an angular variable, $\xi = (p_1 p_2)/E_1 E_2$, as an evolution variable and require the angular ordering at every branching. Due to this the overlap of phase space is absent. Since their angular variable is dependent on a frame, the center-of-mass energy of the whole system is not

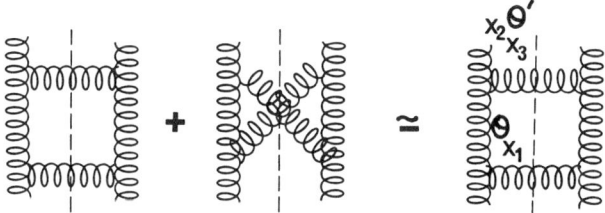

Fig.6 Angular ordering in QCD, $\theta > \theta'$ for $g \to ggg$ in case of $x_1 \gg x_2 \gg x_3$.

determined until the evolution process finishes.

2.4 Double cascade scheme

As is discussed in subsection 2.1, in the light-cone gauge the antiquark does not radiate gluons when n is parallel to $p_{\bar{q}}$. In this sense jets in this gauge have the structure of single cascade. In the center-of-mass frame, it results in small asymmetry between q and $\bar{q}$. In Webber's model, the interference of phase spaces is suppressed and the relation,

$$(\Pi_{NE}^{angular\ ordering})^2 \simeq \Pi_{NE}, \qquad (11)$$

allows to let both q and $\bar{q}$ evolve.

To treat quark and antiquark jets symmetrically, we have introduced 'double cascade scheme'[15]. The singular part of Eq.(3) is symmetric for the interchange of $t \leftrightarrow z$ where $z = 1 - x$. In the center-of-mass system

$$\begin{aligned} z_1 &= 1 - z + tz \sim 1 - z \quad \text{(for } t, z \ll 1\text{)} \\ z_2 &= 1 - t. \end{aligned} \qquad (12)$$

Using an identity $1 = \theta(z-t) + \theta(t-z)$, we write the inside of bracket of Eq.(3)

as

$$(\ ) \text{ of Eq.(3)} \simeq (\theta(z-t) + \theta(t-z))(\frac{2}{tz} - \frac{2-z}{t} - \frac{2-t}{z})$$
$$\simeq \theta(z-t)(\frac{2}{tz} - \frac{2-z}{t}) \quad (q-\text{side cascade}) \quad (13)$$
$$+ \theta(t-z)(\frac{2}{tz} - \frac{2-t}{z}) \quad (\bar{q}-\text{side cascade})$$

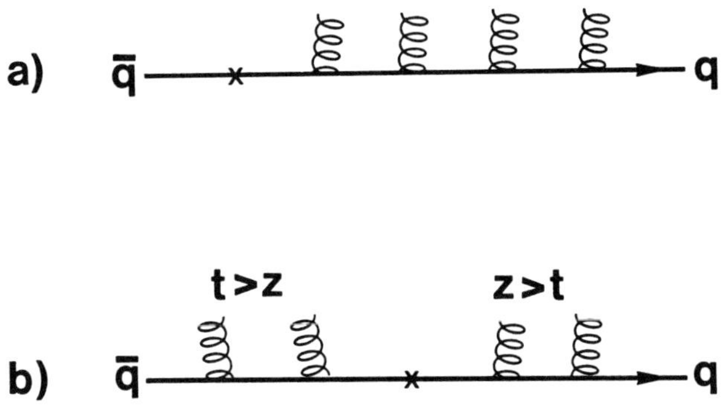

Fig.7 (a) Single cascade and (b) double cascade in light-cone gauge.

We obtain the relation

$$(\Pi_{NE}^{double})^2 \simeq \Pi_{NE} \quad (14)$$

for

$$\Pi_{NE}^{double}(Q^2, Q_1^2) = \exp\left[-\int_{Q_1^2}^{Q^2} \frac{dK^2}{K^2} \int_{\epsilon}^{1-\epsilon} dx \frac{\alpha_s(K^2)}{2\pi} P(x)\theta(z-t)\right]. \quad (15)$$

Thus we can let a quark and an antiquark evolve symmetrically restricting their phase space to $z > t$ ($t > z$) for q ($\bar{q}$). (Fig.7) And the consistency of this treatment is assured by the Eq.(14).

As is discussed above, the choice of gauge is important to the formulation of QCD cascade since the kinematical consideration is non-trivial for a Monte Carlo model. The reason why we use light-cone gauge is that we can use covariant variables, x and Q^2, for evolution and that quantities in NLL order, which are presented in the next section, are calculated only in this gauge.

2.5 Space-like cascade

In this subsection, we consider the cascade by partons in the initial state. The kinematical condition is different since virtuality of parton is space-like. Another point is that there exists no parton beam, i.e., partons come from a hadron. Partons in a hadron are described by distribution functions which represent the structure of hadron. The prediction of cascade must be consistent with the Q^2-dependence of distribution functions predicted by conventional QCD. The formulation of QCD cascade for space-like case is done by many authors [21,22]. Here we present the formulation of Munehisa and Tanaka [22]. We use z instead of x in this subsection. As is shown in Fig.8, the kinematical constraint is

$$\mathbf{k}^2 = z(1-z)\left(-|Q^2| + \frac{|Q_1^2|}{z} - \frac{|Q_2^2|}{1-z}\right) \geq 0. \tag{16}$$

In contrast to the case of Eq.(8), the above equation gives only a condition of $z < z_-$.

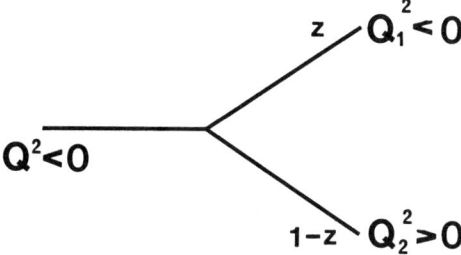

Fig.8 Kinematics of a two-body branching in space-like cascade.

To make concrete discussion, we consider the deep-inelastic lepton-hadron scattering. The cascade is formulated either as forward evolution or as backward evolution. (Fig.9)

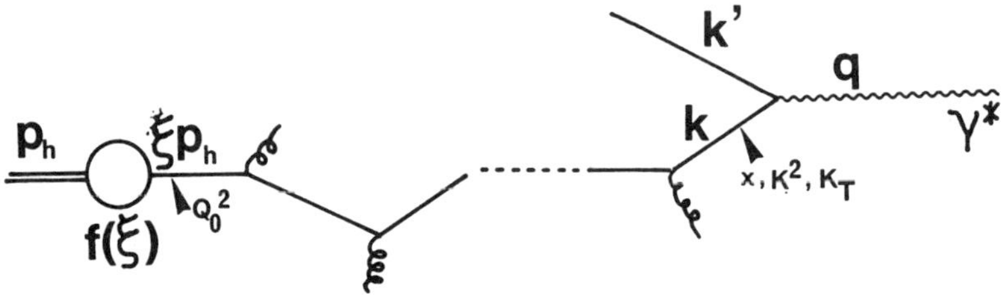

Fig.9 Schematic diagram of space-like cascade.

In the former, a parton of momentum fraction ξ is chosen by $f(\xi)$, the distribution in a hadron at $Q^2 = Q_0^2$. It evolves forward as the time-like cascade using Π_{NE} and $P(z)$. When a selected virtuality exceeds $-q^2$, the evolution stops. Then the Bjorken variable of this event, x_{BJ} is calculated solving the kinematical condition at the right-most vertex in Fig.9, i.e., $(-|k^2|/x - q^2/x_{BJ})(x - x_{BJ}) = (x_{BJ}/x)\mathbf{K}_T^2$. In the backward evolution, the parton that collides the current is determined first by the condition that its maximum allowed virtuality to be $-q^2$. It evolves backward toward the hadron and it stops when the determined virtuality is Q_0^2. The Bjorken variable is determined by $x_{BJ}^{-1} = (1/\xi \Pi z_i) - (\mathbf{P}_{hT}^2/-q^2)\xi \Pi z_i$ where $\mathbf{P}_{hT}$ is the transverse component of P_h. Here we define the transverse momentum perpendicular to the direction of current (q) and we use the frame in which the transverse momentum of colliding parton, i.e., the parton of momentum k in Fig.9, is zero. In both cases, $-q^2$ of the scattering is given as input and the Bjorken variable $x_{BJ} = -q^2/2P_h q$ is determined after the evolution. Radiated partons also develop jets as time-like cascade. By both methods we have calculated the x_{BJ} distribution of non-singlet quark. Here the d_V distribution of set-1 by Owens and Reya [23] at $Q_0^2 = 4 \text{GeV}^2$ is used as $f(\xi)$. The result for

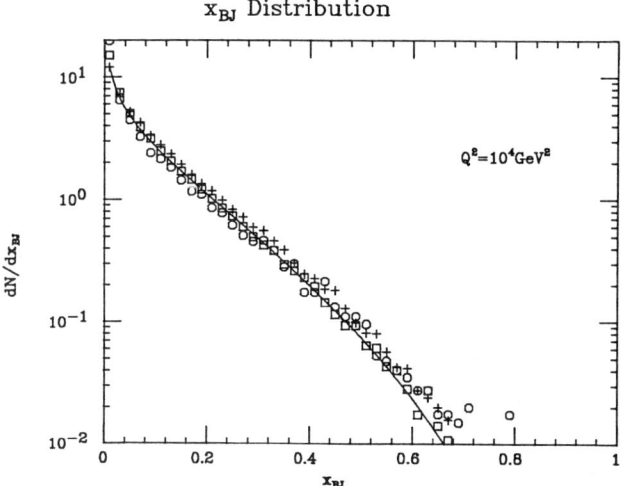

Fig.10 Comparison of x_{BJ} distribution of non-singlet quark by forward evolution(cross) and that by backward evolution(circle). The solid line is the fit by conventional Q^2-evolution of structure function[23]. The square points are the result of forward evolution when all radiated gluons are kept on-mass-shell.

$Q^2 = 10^4 \text{GeV}^2$ is shown in Fig.10 and two methods seem to agree well. In the Monte Carlo simulation of parton shower, it is not possible to avoid 'formulation dependence' and the slight difference in these figures suggests us the magnitude of this ambiguity.

3. QCD cascade in NLL order

The QCD cascade in LL order is discussed in the last section. Though it is useful, there are reasons why we must go beyond the leading order. The most important motivation is that Λ_{QCD}, which is the fundamental parameter of QCD, is determined only in NLL order. Another point is that LL cascade is not good for large p_T region while in NLL order it is possible to include 3 jets contribution in a consistent manner. Also study in NLL gives a way to resolve ambiguous points in LL.

We have constructed a QCD cascade model named **NLLjet** with following guiding principles.

(1) The higher order contribution is smaller than that in the lower order so as to keep the perturbation valid.

(2) Distribution of generated partons is consistent with the prediction by conventional QCD for inclusive quantities.

(3) Functions those represent branching probability are to be positive.

These conditions are not trivial and we need detailed study of kinematics and careful assembly of components in NLL order as is shown in the following. First, we review shortly the formulation of QCD cascade beyond the LL order[12]. In this section we do not discriminate between quark and gluon in its diagramatical display.

step $-$ 1 We write the off-shell unitarity equation of Polyakov in axial gauge for $e^+e^- \to anything$. In the diagram below a solid line stands for a full propagator, a vertex is one-particle irreducible and a vertical dashed line represents the cut to take imaginary part of full propagators.

step $-$ 2 We divide phase spaces so that virtuality of a particle on the cut is less than a cutoff, Q_0^2. A cut propagator is replaced in a following manner:

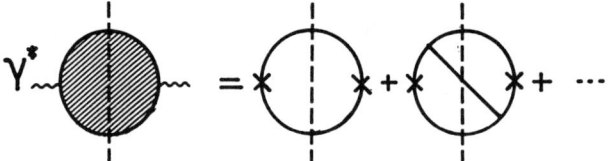

step $-$ 3 Next we contract lines which are making a diagram as a non-diagonal one (interference terms) and create effective vertices. An example of this procedure is shown below. After this step, the diagram is ladder-like.

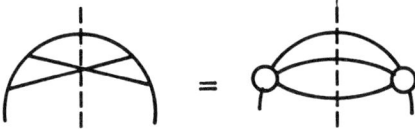

step − 4 We introduce a projection operator **P** to separate an uncut propagator into a peace with mass singularity and that without singularity. Again we define a new effective vertex contracting non-singular part. An example is below.

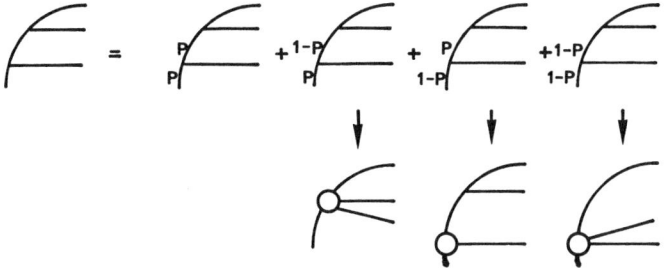

After these operations, the diagram has the following properties.

(1) It is diagonal and we can make stochastic treatment.

(2) All the particles have virtuality Q_0^2 on the cut. This works as a cutoff for jet evolution.

(3) Vertices are finite and a large logarithm is assigned to a conjugated pair of propagators.

In NLL order we sum all terms of

$$\sum a_n \alpha_s^n (\log Q^2)^n + \sum a'_n \alpha_s^n (\log Q^2)^{(n-1)} . \tag{17}$$

We need the following components:

· Primary vertex.

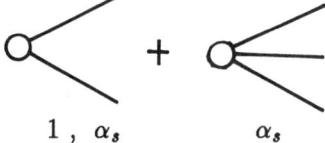

$1, \alpha_s \qquad \alpha_s$

· Two-body branch.

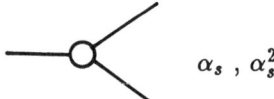

α_s, α_s^2

· Three-body branch.

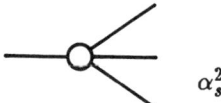

α_s^2

· Off-shell propagator.

    Π_{NE} by $P = P^{(0)} + \alpha_s/2\pi P^{(1)}$

These factors are dependent on others and we need detailed study of kinematics. On the other hand we can use the freedom to define these functions suitable for Monte Carlo simulation. Only the sum of *all* terms of NLL order has physical meaning. In order to produce large p_T jets one can introduce a primary three-jet cross section into LL cascade model. However, such a prescription has little sense in view of QCD perturbation.

Primary three-jet vertex is represented by Fig.11. For this we use on-shell $q\bar{q}g$ cross section $\sim (z_1^2 + z_2^2)/(1-z_1)(1-z_2)$ with a cutoff δ. In Fig.12, non-hatched region of the triangle is the mass singular region and the contribution from this region is included into $P_{qq}^{(1)}(z)$ where we denote the Altarelli-Parisi function in NLL order as $P = P^{(0)} + \alpha_s/2\pi P^{(1)}$. Three-jets are produced in the hatched

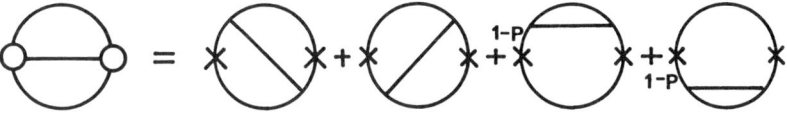

Fig.11 Primary three-jet vertex.

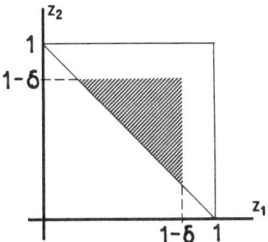

Fig.12 Phase space of primary three jets. z_1 and z_2 is the energy fraction of a quark and an antiquark.

phase space. This term corresponds to the coefficient function C_n in terms of conventional QCD and $P-$ function corresponds to anomalous dimension γ_n. As is well known, only the combination of $C_n - 2\gamma_n^{(1)}/\beta_0$ is independent of scheme. In this sense δ is a technical parameter and physical result is almost independent of δ. Our standard value is $\delta = 0.1$ and we have checked that the change of δ around this value little affects distribution of partons.

The structure of primary vertex in NLLjet is shown below where a box represents a cascade.

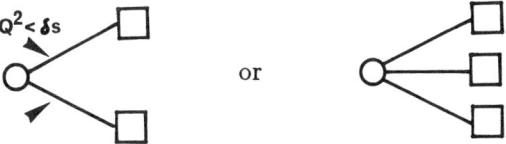

The two channels are selected at random by a ratio $\sigma_2(\delta) : \sigma_3(\delta)$. δ−dependent σ_3 is obtained by integrating the hatched region of Fig.12. σ_2 is calculated by subtracting σ_3 from the total cross section. According to the guiding principle (1), the value of δ is chosen so as that $\sigma_2 > \sigma_3$. It should be noted that actual 'three jets' are produced from both channels. Double counting of phase space is avoided naturally by imposing that available maximum virtuality of primary two-jet channel is less than δs since squared mass of quark or antiquark and gluon in the hatched area of Fig.12 is greater than δs.

There still remains an ambiguity. After the evolution has finished we calculate 4-vectors of partons. Since we use on-shell cross section, only two degrees of freedom exist whereas the three partons at primary three-jet generally acquire masses by evolution. Resultingly it is possible to assign only either of energy or angle of these massive jets in a consistent way with the variables at the primary vertex from where the cascade started. The resolution is beyond the NLL order. We have introduced the order α_s^2 effective three-jet cross section of Gottschalk and Shatz[24] for the primary three-jet generation to keep up the off-mass-shell nature.

We take $p_T^2 = z(1-z)Q^2$ for the argument of α_s which is already suggested through the study of infrared structure[8]. In NLL order, we can explicitly show the reason of this choice. Since we redefine $P_{qq}^{(1)}(z)$ as above, it has a term that behaves $\sim -\log(1-z)/(1-z)$ for $z \to 1$. This breaks our principle (1). When we choose p_T^2 for the argument of α_s, above unwanted term is canceled.

For the two-body and three-body branches detailed discussion and definition are found in ref.[14]. Here we briefly explain our prescription through a few examples. Some of three-body branching functions are negative in their original form[12,13]. However, they turns to be positive to meet our principle (3) after we impose momentum conservation and also improve definition of angular ordering.

The meaning of 'momentum conservation' is as follows. For instance, in Fig.4 the upper limit of Q_1^2 is about zQ^2 from Eq.(8) neglecting Q_2^2. Integration by transverse momentum gives a large logarithm $\log zQ^2$. For a moment we break momentum conservation: If we take the upper limit of the integral to be Q^2, we get $\log Q^2$. In LL order both have no difference since $\log zQ^2 - \log Q^2$ contributes only to NLL order. Of course in a practical Monte Carlo generator momentum should be conserved. However, there is no difference in formulation of LL cascade whether we keep momentum conservation or not. Conversely, we can control quantities of NLL order by imposing momentum conservation. By this way, we can define probability functions for three-body branch to be positive.

The angular ordering also works on quantities in NLL order in a similar manner since it restricts phase space. Gunion and Kalinowski studied the function for $g \to ggg$ and they made it positive by angular ordering[25]. However, their function was too large to break our principle (1). At this point one should note that the condition of angular ordering holds in a very limiting case. If we denote the momentum fraction of three gluons as x_1, x_2, x_3, it only holds for $x_1 \gg x_2 \gg x_3$ or alike. This means that any condition is valid if it is equivalent to the condition of angular ordering at the limiting region[14].

Above consideration is implemented into NLLjet. First, we have checked that it is self-consistent, i.e., our principle (2). We generate events by NLLjet and compute moments of inclusive distribution at various energies. Q^2- dependence of moments is predicted by conventional perturbative QCD. We have shown that input value of Λ to NLLjet is recovered by analysis of moments within a good accuracy as is shown in ref.[14]. Thus it will be possible to determine Λ_{QCD} in a definite scheme by NLLjet.

In order to study experimental data and to determine Λ_{QCD} a hadronization model must be linked with NLLjet. Analysis of MARKII data at $W = 29$GeV and other set of data is done by Kamae and Shirahashi using LUND string fragmentation[26]. They get $\Lambda_{\overline{MS}} = 235 \pm 52$MeV. Analysis of VENUS data at TRISTAN is also done using EPOCS hadronization[27] and a preliminary value of $\Lambda_{\overline{MS}} = 208^{+80}_{-62}$MeV is obtained[28]. In Figs.13-14, we show a few figures obtained by their analyses to demonstrate how NLLjet fits data and how it determines Λ.

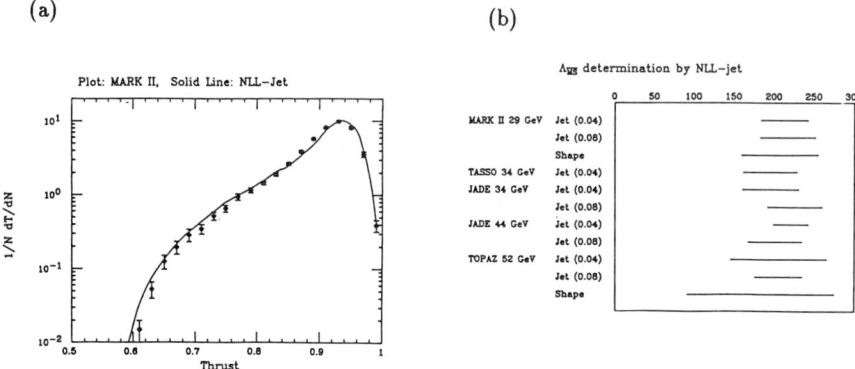

Fig.13 Analysis and determination of Λ by MARKII data[26]. (a) Fit for the thrust distribution of MARKII data and (b) determined $\Lambda_{\overline{MS}}$ for various sets of data.

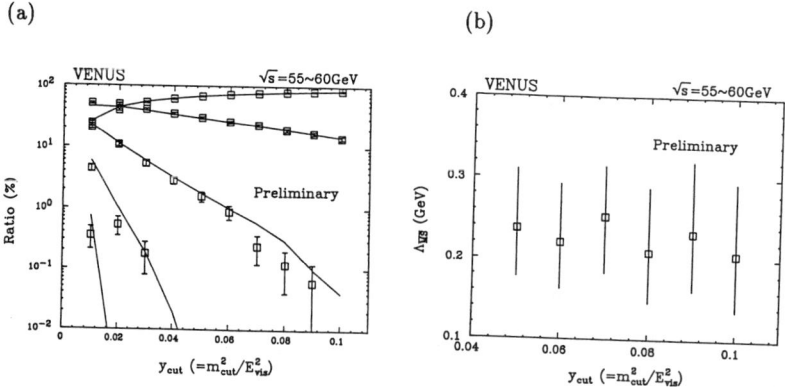

Fig.14 Analysis and determination of Λ by VENUS preliminary data[28]. (a) Fit for the jet-multiplicity by a cluster method and (b) $\Lambda_{\overline{MS}}$ for various y_{cut}.

4. Comments on radiative correction for QCD generator

This section is not on the discussion on QCD cascade itself and it is to draw attention of QED experts on a delicate point in QCD generators. At present, most of QCD generators for e^+e^- annihilation includes a package of the radiative correction as a part of system. Mostly, it is the 'initial state radiation'(ISR) based on the paper by Berends, Kleiss and Jadach[29]. In ISR we consider only the photon radiation from e^+e^-, the correction for electron-current vertex and the vacuum polarization. The radiation from $q\bar{q}$ and the vertex correction for quark is neglected here.

A question is whether the prescription of ISR is sufficient or not. For QED correction, ISR works. However, for electro-weak theory correction only to initial state has no sense since it depends on renormalization scheme[30]. From a practical point of view, the factorized treatment of ISR is convenient for QCD generators. Given the beam energy, the package of ISR generates a momentum of photon and it calculates modified center-of-mass energy W' and beam direction $\hat{n}'$. The QCD part receives only W' and $\hat{n}'$ and generates QCD jets. If we try to include final state radiation, hard work will be required: In MA method, matrix elements for $e^+e^- \to q\bar{q}\gamma$, $q\bar{q}g\gamma$, $q\bar{q}gg\gamma$, etc. must be calculated. In parton shower method, we must formulate a cascade for q, $\bar{q}$, g and γ. A plausible approximation is as follows: First, we generate $e^+e^- \to q\bar{q}$ or $q\bar{q}\gamma$ according to the electro-weak calculation including both initial and final radiation[31]. Around Z^0 region, we need improvement including exponentiation. Second, we start QCD cascade from the generated $q\bar{q}$ system.

For other related issues, $e^+e^- \to$ *hadrons* in the region of resonances (Z^0, toponia), and the correction for R−ratio, I skip discussion since they are covered by other talks in the workshop.

5. Summary

QCD cascade is formulated through the study of mass singularity in perturbative QCD. The choice of gauge and the kinematical consideration are important, which is little stressed in the conventional perturbative QCD. Monte Carlo event generators based on QCD cascade are developed and they are widely used for the study of jets in high energy reactions.

QCD cascade beyond the LL order is formulated and a Monte Carlo generator NLLjet is constructed on this formulation. NLLjet will work as a new tool for the analysis of hadronic events. It is essential to go beyond the LL order for the determination of Λ_{QCD}. Using the analysis of moments of inclusive distribution, we have checked that the distribution of partons in NLLjet gives correct Λ. Including hadronization, we get $\Lambda_{\overline{MS}} \simeq 200\text{MeV}$ analysing MARKII data and preliminary VENUS data.

The author acknowledges Y.Shimizu, T.Munehisa, H.Tanaka, T.Kamae, A.Shirahashi, J.Fujimoto and T.Watanabe for useful discussion. He also thanks to G.Bonvicini and Michigan University for warm hospitality at the workshop. He acknowledges H.Sugawara, K.Takahashi and M.Iwata for encouragement and the National Laboratory for High Energy Physics(KEK) for financial support. This work is also supported financially by the Special Research Fund of Kogakuin University.

References

[1] G.Hanson et al., Phys. Rev. Lett. **35**, 1609 (1975).

[2] B.Bambah et al., CERN-TH 5466/89, to appear in *the Proc. of 1989 LEP Physics Workshop*, (1989).

[3] T.Sjöstrand, Int. Jour. of Mod. Phys. **A3**, 751 (1988); H.Yamamoto, in *Proc. of 1985 Int. Symp. on Lepton and Photon Int. at High Energy,*

edited by M.Konuma and K.Takahashi, Kyoto (1986).

[4] W.Bartel *et al.*,(JADE Collaboration), Z.Physik. **C33**, 23 (1986).

[5] G.Sterman and S.Weinberg, Phys. Rev. Lett. **39**, 1436 (1977).

[6] K.Konishi, A.Ukawa, and G.Veneziano, Nucl. Phys. **B181**, 221 (1981).

[7] R.Odorico, Nucl. Phys. **B172**, 157 (1980); G.C.Fox and S.Wolfram, Nucl. Phys. **B168**, 285 (1980); T.D.Gottschalk, Nucl. Phys. **B214**, 201 (1983), *ibid.* **B227**, 413 (1983).

[8] A.Bassetto, M.Ciafaloni, and G.Marchesini, Phys. Rep. **100**, 201 (1983) *and references therein*.

[9] G.Marchesini and B.R.Webber, Nucl. Phys. **B238**, 1 (1984); B.R.Webber, Nucl. Phys. **B238**, 492 (1984).

[10] T.Sjöstrand, Comput. Phys. Commun. **39**, 347 (1986); T.Sjöstrand and M.Bengtsson, Comput. Phys. Commun. **43**, 492 (1987).

[11] G.Marchesini and B.R.Webber, Cavendish-HEP-87/9, (1987).

[12] J.Kalinowski, K.Konishi, and T.R.Taylor, Nucl. Phys. **B181**, 221 (1981); J.Kalinowski, K.Konishi, P.N.Scharbach, and T.R.Taylor, Nucl. Phys. **B181**, 253 (1981).

[13] J.F.Gunion and J.Kalinowski, Phys. Rev. **D29**, 1545 (1984); J.F.Gunion, J.Kalinowski and L.Szymanowski, Phys. Rev. **D32**, 2303 (1985).

[14] K.Kato and T.Munehisa, Mod. Phys. Lett. **A1**, 345 (1986); Phys. Rev. **D36**, 61 (1987).

[15] K.Kato and T.Munehisa, Phys. Rev. **D39**, 156 (1989).

[16] K.Kato, in *Perspectives on Particle Physics*, edited by S.Matsuda, T.Muta and R.Sasaki, World Scientific (1989); T.Munehisa, Talk at the Topical conf. on e^+e^- Collision Physics, Tsukuba (1989).

[17] G.Altarelli and G.Parisi, Nucl. Phys. **B126**, 298 (1977); J.F.Owens, Phys. Lett. **76B**, 85 (1978); T.Uematsu, Phys. Lett. **79B**, 97 (1978).

[18] L.N.Lipatov, Sov. J. Nucl. Phys. **20**, 94 (1975).

[19] V.Sudakov, Sov. Phys. JETP **3**, 65 (1956).

[20] A.E.Chudakov, Izvestia AN USSR(Fizika) **19**, 650 (1955).

[21] T.Sjöstrand, Phys. Lett. **157B**, 321 (1985); M.Bengtsson and T.Sjöstrand, Z.Physik. **C37**, 465 (1988); G.Marchesini and B.R.Webber, Nucl. Phys. **B310**, 561 (1988); H.R.Wilson, Phys. Lett. **201B**, 361 (1988); K.Kato and T.Munehisa, Z.Physik. **C40**, 85 (1988).

[22] T.Munehisa and H.Tanaka, in preparation.

[23] D.W.Duke and J.F.Owens, Phys. Rev. **D30**, 49 (1984).

[24] T.D.Gottschalk and M.P.Shatz, Caltech preprint, CALT-68-1172; Phys. Lett. **150B**, 451 (1985)

[25] J.F.Gunion and J.Kalinowski, Phys. Lett. **163B**, 379 (1985).

[26] T.Kamae, in *Proc. of 24th Int. Conf. on High Energy Physics*, Munchen (1989); K.Kato, T.Kamae, T.Munehisa, and A.Shirahashi, in preparation.

[27] K.Kato and T.Munehisa, KEK Report 84-18(1984), 87-5(1987).

[28] VENUS Collaboration, talk at the annual meeting of Japanese Physical Society, Tokai University, (1989); T.Watanabe, private communication.

[29] F.A.Berends, R.Kleiss, and S.Jadach, Nucl. Phys. **B202**, 62 (1982).

[30] J.Fujimoto, talk at this conference.

[31] J.Fujimoto and Y.Shimizu, Mod.Phys.Lett. **3A**, 581 (1988).

THE RENORMALIZATION GROUP IMPROVED YFS METHOD IN QED*,**

S. Jadach
CERN, Geneva 23, Switzerland
and
Institute of Physics, Jagellonian University,
Krakow, Poland

and

B. F. L. Ward
CERN, Geneva 23, Switzerland
SLAC, Stanford University, Stanford, CA 94309, USA
and
Department of Physics and Astronomy
University of Tennessee, Knoxville, TN 37996-1200, USA

ABSTRACT

We review the recently introduced Monte Carlo renormalization group improved Yennie-Frautschi-Suura method as it applies to higher order QED radiative corrections. Explicit Monte Carlo realizations are presented. Sample Monte Carlo data are illustrated which are of immediate interest to SLC and LEP scenarios. In this way the effects of multiple photon final states on high precision $Z°$ physics are assessed.

*Invited talk presented at the Ann Arbor Workshop on QED Structure Functions, May, 1989.

**Work supported by the DOE, contracts DE-AC03-76SF00515 and DE-AS05-76ER03956

Introduction

It is now true that $Z°$ physics has in fact started at the SLC and LEP[1]. Thus, the primary objective of these two e^+e^- colliders, that of testing the standard $SU_{2L} \times U_1$ at or below the level of 1%, becomes closer and closer to reality as each day passes. It is clear that the higher order radiative effects in such SLC/LEP environments must be controlled at or below the level of .3% if the true physics potential of these devices is to be realized in near term impending observations. Here, we wish to discuss one approach[2] which we have developed for exercising such a control over the radiative corrections even on an event-by-event basis. This approach we have called the renormalization group improved Yennie-Frautschi-Suura method, since it is based on a union of the renormalization group equation[3] of 't Hooft and Weinberg with the infrared summation methods of Yennie, Frautschi and Suura (YFS).[4]

For definiteness, we will illustrate the method using the QED part of the standard $SU_{2L} \times U_1$ theory. It is well-known[5] that such QED corrections are in fact the dominant part of the radiative corrections to the $Z°$ physics of interest to us here. It will be evident that our method extends naturally to the entire $SU_{2L} \times U_1$ theory. And, in our examples, we show this in a more direct way.

Indeed, the separation of the so-called pure Q.E.D. part of the $SU_{2L} \times U_1$ in $e^+e^- \to X$ is strictly valid only for certain processes and, for these processes, only for a fixed loop order.[6] For example, the process $e^+e^- \to \nu\bar{\nu} + \gamma$ has, already at the lowest observable order, the graphs in Fig. 1, where we see in Fig. 1 that

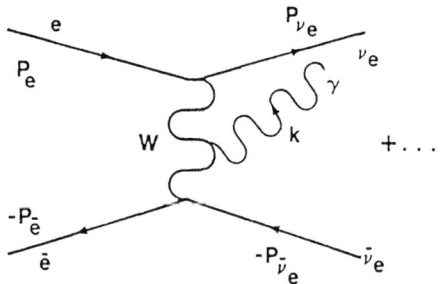

Fig. 1. Radiative $\nu\bar{\nu}$ production.

the $WW\gamma$ vertex, which mixes the QED effects with the SU_{2L} part of the theory, is already present at this lowest observable order. Thus, for this process, the separation of its pure QED part is not strictly possible.

Similarly, if we restrict ourselves to one-loop -$O(\alpha)$ corrections to $e^+e^- \to f\bar{f}$, we may separate out the QED part of the corrections as the diagrams in Fig. 2; the remaining corrections are shown in Fig. 3. This is the classic[7] isolation of the $O(\alpha)$ QED corrections to this process. We emphasize, however, that, in the higher orders, graphs such as those in Fig. 4 make such a separation again invalid in the strict sense. Thus, one must exercise caution in using the phrase "QED corrections."

Quite independent of the strictness of the concept of isolating the Q.E.D. corrections, what one clearly wants to do is to implement these Q.E.D. effects in the radiative correction calculations in such a way that the data may be corrected for them; this would then make the identification of possible new physics beyond the $SU_{2L} \times U_1$ theory more readily accessible. Thus, our strategy has been to develop a methodology which implements all large higher order Q.E.D. related effects at the level of the events themselves, i.e., at the level of an event generator. We present and illustrate the basis of our methods in what follows.

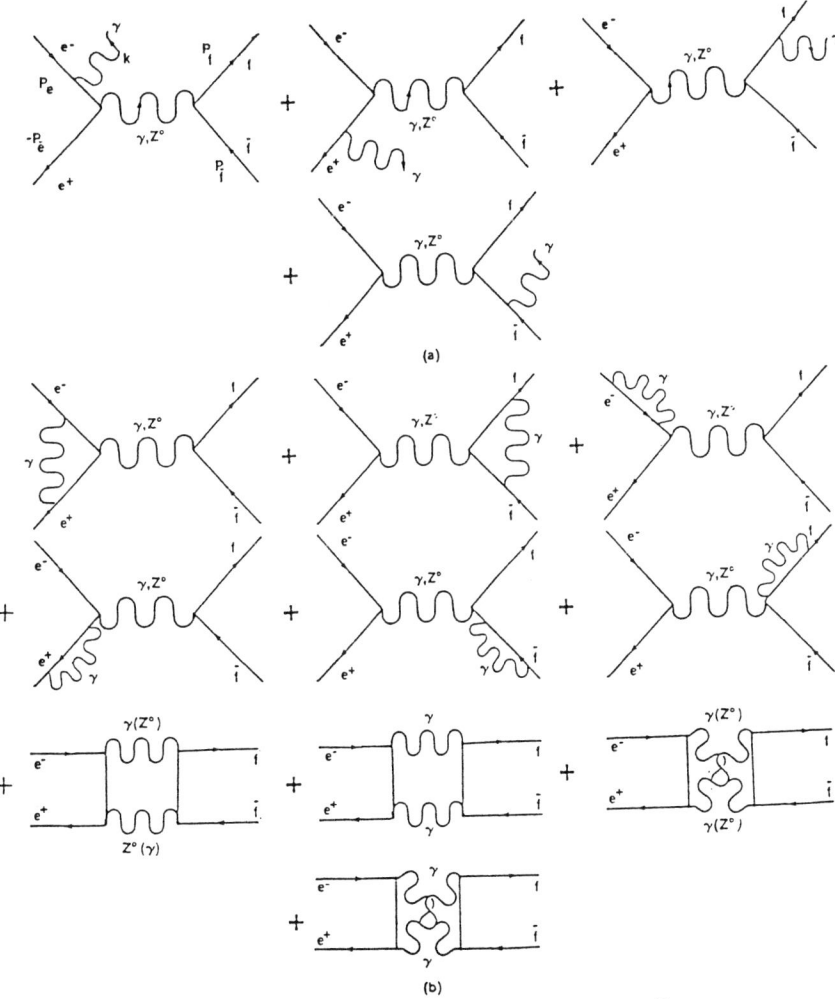

Fig. 2. "Pure Q.E.D." corrections to $e^+e^- \to f\bar{f}$ to $O(\alpha)$; (a), real bremsstrahlung; (b), virtual effects.

Our discussion proceeds as follows. In the next section, we review the methods of Yennie, Frautschi and Suura. In Section 3, we discuss the renormalization group improvement of the YFS theory. In Section 4, we discuss our Monte realizations of the resulting renormalization group improved YFS theory. In the ensuing section, we illustrate our methods with some sample Monte Carlo data. The final section, Section 6, contains our summary remarks.

Review of YFS Theory

In this section, we review the theory of Yennie, Frautschi and Suura with an eye toward the ultimate objective of realizing the theory via Monte Carlo methods. We start with the basic defining theorems in the theory.

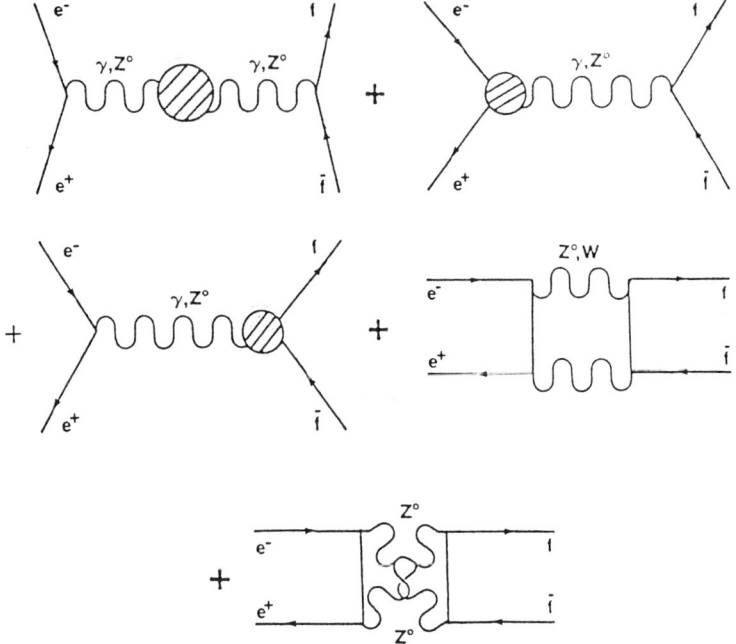

Fig. 3. Remaining electroweak corrections to $e^+e^- \to f\bar{f}$ at $O(\alpha)$.

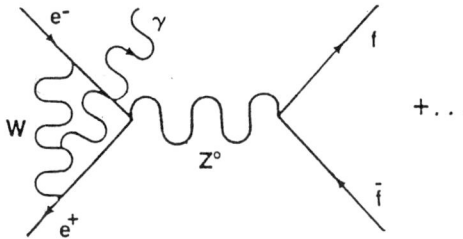

Fig. 4. Mixed Q.E.D. electroweak correction to $e^+e^- \to f\bar{f}$ at $O(\alpha^2)$.

Specifically, our objective is to sum up the large infrared effects to all orders in perturbation theory in the Q.E.D. fine structure constant α. Hence, in the $SU_{2L} \times U_1$ theory in the context of $Z°$ physics, a typical scenario is that illustrated in Fig. 5. The attendant infrared effects naturally arrange themselves according to their origin: they may be due to real photon emission or to virtual photon exchange.

Focussing first on virtual photon exchange, we recall that, by the first theorem of YFS, the amplitude $M^{(m)}$ in Fig. 5 may be represented as

$$M^{(m)} = \exp(\alpha B) \sum_{n=0}^{\infty} m_n^{(m)} \qquad (1)$$

where the residual amplitudes $m_n^{(m)}$ have no virtual infrared effects (n represents the number of γ loops in the respective contribution to $M^{(m)}$ as it is illustrated in Fig. 5) and B is the

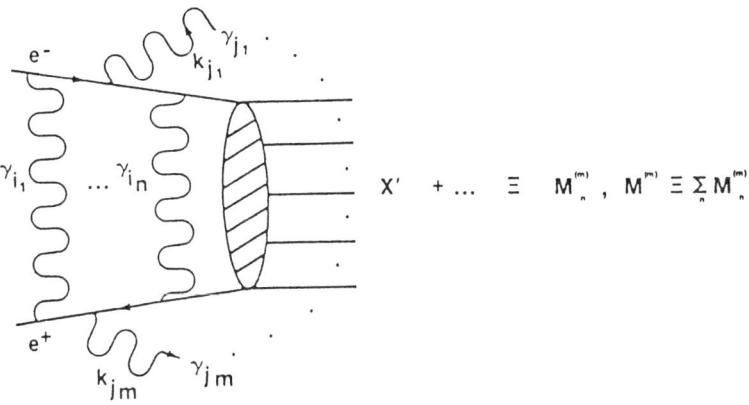

Fig. 5. n photon loop, m real γ emission amplitude for $e^+e^- \to f\bar{f}$.

famous virtual infrared function[2,4] (m_γ is our photon mass infrared regulator)

$$B = \frac{i}{8\pi^2} \int \frac{d^4k}{[k^2 - m_\gamma^2 + i\epsilon]} \left[\left\{ \frac{2P_e + k}{k^2 + 2kP_e + i\epsilon} - \frac{2P_{\bar{e}} - k}{k^2 - 2kP_{\bar{e}} + i\epsilon} \right\}^2 + \ldots \right] \qquad (2)$$

so that B is known.

Similarly, the second theorem of YFS allows us to write

$$\left| \sum_{n=0}^{\infty} m_n^{(m)} \right|^2 = \tilde{S}(k_1) \ldots \tilde{S}(k_m) \beta_0 + \sum_{i=1}^{m} \tilde{S}(k_1) \ldots \tilde{S}(k_{i-1}) \tilde{S}(k_{i+1}) \ldots \tilde{S}(k_m) \beta_1(k_i) + \ldots + \beta_m(k_1, \ldots, k_m) \qquad (3)$$

where $\tilde{S}(k)$ is the famous well-known real emission infrared singular factor,

$$\tilde{S}(k) = \frac{-\alpha}{4\pi^2} \left(\frac{P_{\bar{e}}}{P_{\bar{e}} k} - \frac{P_e}{P_e k} \right)^2 + \ldots \qquad (4)$$

and, hence, the residuals β_n are free of all virtual and real infrared singularities to all orders in α.

These two YFS theorems then allow us to write the classic YFS result for $e^+e^- \to X' + n(\gamma)$

$$d\sigma = \exp(2\alpha\,(\text{ReB} + \tilde{B})) \int \frac{d^4y}{(2\pi)^4} \exp[i(P_e + P_{\bar{e}} - P_{X'})y + D]$$

$$\{\beta_0 + \sum_{n=1}^{\infty} \int \prod_{j=1}^{n} \frac{d^3k_j}{k_j} \exp(-iyk_j)\,\beta_n\}\,d^3P_{X'}\cdot dE_{X'}\,, \tag{5}$$

where we have introduced the functions

$$2\alpha\,\tilde{B} = \int^{K_{max}} \frac{d^3k\,\tilde{S}(k)}{k^0}\,, \quad D = \int \frac{d^3k}{k}\,(e^{-iky} - \theta(K_{max}-k))\,\tilde{S}(k) \tag{6}$$

in such a way that (5) does not depend on K_{max}. In (5), the infrared singularities are cancelled to all orders in α. Using the source theory methods of Schwinger, Mahanthappa[8] has obtained a similar result to (5) for a generic cross section in Q.E.D.

A few comments about (5) are in order from the standpoint of its use in radiative correction simulations via Monte Carlo methods. First, we note that the 4-momentum is conserved rigorously by the formula. Secondly, the differential cross-sections with multiphoton final states have the four-vectors for these photons explicitly available. Finally, we note that the large UV (ultraviolet) effects are not in general addressed explicitly in (5).

As consequence of the first two observations, we may observe that (5) lends itself to a new type of radiative correction simulation via Monte Carlo methods with the rigorous event-by-event realization of the all orders (in α) infrared singularity cancellation via the exponential factor $\exp(2\alpha(\text{ReB} + \tilde{B}))$. To achieve the kind of accuracy desired at SLC and LEP ultimately, one should, as the third comment indicates, improve (5) via renormalization group[3] arguments to account for the respective large UV effects. As we shall show, such improvement does not affect our ability to realize (5) via Monte Carlo methods. Hence, we turn now to this renormalization group improvement of (5) in our next section.

Renormalization Group Improved YFS Theory

In this section, we present the elements of the renormalization group improvement of the YFS theory as it relates to (5). We begin with a review of the renormalization group itself.

Specifically, we wish to employ the methods of Weinberg and 't Hooft in realizing the renormalization group. Thus, if we let {Γ} denote the 1PI proper vertices of a renormalizable theory, then 't Hooft and Weinberg have shown that the independence of the unrenormalized theory of the normalization point $-\mu^2$ requires

$$(\mu\frac{\partial}{\partial\mu} + \beta(g_R)\frac{\partial}{\partial g_R} - \gamma(g_R)\,m_R\frac{\partial}{\partial m_R} - \gamma_\Gamma(g_R))\Gamma = 0 \tag{7}$$

if we image for simplicity that the theory involves one renormalized coupling g_R and one renormalized mass m_R. Here, the coefficient functions β, γ_θ and γ_Γ are all computable order by order in renormalized perturbation theory. The generalization to a theory with more than one coupling and more than one mass is then the immediate one: one includes the analog of $\beta \partial/\partial g_R$ ($-\gamma_\theta m_R \partial/\partial m_R$) for each coupling (mass), etc., on the lefthand side of (7). The solution of (7) may be represented as

$$\Gamma(\lambda\{P_j\}; g_R, m_R, \mu) = \lambda^{D_\Gamma} \Gamma(\{P_j\}; g_R(\lambda), m_R(\lambda), \mu)$$
$$\exp\left[-\int_1^\lambda \frac{d\lambda'}{\lambda'} \gamma_\Gamma(g_R(\lambda'))\right] \quad (8)$$

where $\{\lambda P_j\}$ are the external 4-momenta of Γ, D_Γ is the engineering dimension of Γ and $g_R(\lambda)$ and $m_R(\lambda)$ are the respective running coupling and mass of the theory. From (8), we see that the solution of (7) may be viewed as giving the large λ behavior of $\Gamma(\lambda\{P_j\})$ in terms of the value of Γ at $\lambda = 1$. This view is the view we will take then, since the respective λ behavior is the desired UV behavior of interest to us here. We should note that the idea to use the renormalization group in this connection of higher order radiative corrections to the $SU_{2L} \times U_1$ theory is a generalization to $SU_{2L} \times U_1$ of an idea for such an use by Tsai[9] for Q.E.D. radiative corrections; in that Q.E.D. application, Tsai used the Gell-Mann-Low formulation of the renormalization group. Here, we avoid the need to assume the Gell-Mann-Low limit by using the new improved renormalization group of Weinberg and 't Hooft.

A natural question[10] which arises is that the $\{\Gamma\}$ which one usually considers, the proper 1PI vertices, are amplitudes whereas we are interested in (5), which is a cross section. We note that what we do is to apply (7) to the 1PI vertices which comprise the amplitudes whose squares are proportional to the respective differential cross sections in (5). In this way, we realize the respective renormalization group improvement of (5). We have found[2] that, under this improvement, (5) retains its algebraic structure but that β_m, D, B and $\tilde{B}$ all are replaced by their renormalization group improved forms so that, for example, for β_n we have the replacement, in the Q.E.D. subgroup of the $SU_{2L} \times U_1$ theory,

$$\beta_n \to \beta_n' = \lambda^{2D_{M(n)}} \beta_n(\{P_i\}; m_R(\lambda), \alpha(\lambda), \mu) (e_R(\lambda)/e_R(1))^{-2n} \quad (9)$$

where $D_{M(n)}$ is the respective engineering dimension of the amplitude $M^{(n)}$ in Fig. 5 and $e_R(\lambda)$ is the running electric charge with $\alpha(\lambda) = e_R^2(\lambda)/(4\pi)$. There are analogous formulas for the renormalization group improved forms of D, B and $\tilde{B}$ (See Ref. 2.). In this way, we arrive at the renormalization group improved (RGI) YFS theory.

We have applied this theory at the level of the Monte Carlo event generator for $e^+e^- \to f \bar{f} + n(\gamma)$, $f \neq e$, and for $e^+e^- \to e^+e^- +$

n(γ) in the fortran programs[2] YFS(2) FORTRAN and BHLUMI FORTRAN respectively. Let us now illustrate, in the next section, how this is done.

MONTE CARLO REALIZATIONS

We have recently succeeded[2] in applying the basic renormalization group improved YFS theory to several processes in the SLC/LEP type scenario. Here, we illustrate how one proceeds to do this in practice with the once popular MMG1 (Mustral) type Monte Carlo event generator of Berends, Kleiss and Jadach[11] for the process $e^+e^- \to \mu^+\mu^- + (\gamma)$.

Specifically, given the formulae of an $O(\alpha)$ Monte Carlo like MMG1, one follows the recipe which we have given in Refs. 2 for constructing the respective $\beta_{0,1}$ for use in (5). For example, the respective β_0 is

$$\beta_0 = \frac{d\sigma}{d\Omega_\mu}(1\ loop) - 2\ Re\ (\alpha(1)\ B)\frac{d\sigma_0}{d\Omega_\mu}, \qquad (10)$$

where $d\sigma(1\ loop)/d\Omega_\mu$ is the $O(\alpha)$ corrected differential cross section for $e^+e^- \to \mu^+\mu^-$, $d\sigma_0/d\Omega_\mu$ is the corresponding Born differential cross section, and $d\Omega\mu$ is the respective μ^- differential solid angle relative to the incoming e^- direction in the e^+e^- c.m.s. system, for example. There is an analogous formula for β_1 (See Ref. 2.).

Having constructed $\beta_{0,1}$, one may now realize the formula (5) via Monte Carlo methods. To this end, one needs an efficient method of realizing the e^D factor in (5). This realization involves the event-by-event generation of multiple photon states so that one has in this way a new approach to the simulation of $SU_{2L} \times U_1$ higher order corrections in which the actual 4-momenta of the multiple photons in the respective events are also generated in the simulation. The key observation is the following. Upon integrating over the photon angular variables in (5) we find that their energies and their number n realize a restricted Poisson distribution

$$P_{n-1} = e^{-\bar{n}}\ \bar{n}^{(n-1)}/(n-1)!\ \text{where}\ \bar{n} = \beta\log(\epsilon/K_{max}),$$
$$\beta = (2\alpha/\pi)(\log(s/m_e^2) - 1),\ \text{for}\ s = (P_e + P_{\bar{e}})^2. \qquad (11)$$

Here, ϵ is the total energy in photons in the respective event so that it is distributed according to the differential distribution

$$d\sigma = \exp\{2\alpha(1)[ReB + \tilde{B}(P_i(1), m_{iR}(\lambda), E_{\gamma,max}/\lambda)]\}(\delta(\sqrt{s} - E_{\chi'})$$
$$\tilde{\beta}_0(\sqrt{s})\int_0^{K_{max}}\rho(\epsilon')\ d\epsilon' + \theta(\epsilon - K_{max})\ \tilde{\beta}_0(\sqrt{s})\rho(\epsilon))dE_{\chi'} + \exp\{2\alpha(1) \qquad (12)$$

$$[ReB + \tilde{B}(P_i(1), m_{iR}(\lambda), E'_{\gamma,max}/\lambda)]\}\frac{\tilde{\beta}_1(k')d^3k'}{k'}[\delta(\epsilon - k')$$

$$\int_0^{K_{max}} \rho'(\epsilon'-k')d(\epsilon'-k') + \theta(\epsilon-k'-K_{max})\,\rho'(\epsilon-k')]dE_X.$$

where we define $\rho^{(\cdot)}(\epsilon) = (\alpha(1)A/\epsilon)\,(\epsilon/E_{\gamma,max})^{\alpha(1)A}$ for $\epsilon = \sqrt{s} - E_{X'}$, $\alpha(1)A = \tilde{B}(K_{max}/\lambda)/\ln(2K_{max}/m_\gamma)$ with $E_{\gamma,max} = \sqrt{s}/2 - 2m_f^2/\sqrt{s}$ and $E'_{\gamma,max} = (s-2k'\sqrt{s} - 4m_f^2)/(2(\sqrt{s}-2k'))$; we effect this distribution using the usual uniform pseudo-random number techniques. Hence, the numerically delicate issue is the realization of the restricted Poisson distribution. This may be handled as we now indicate using a method first proposed by one of us (S.J.) in Refs. 2.

More precisely, we have used the mathematical result that, if $\{r_j\}$, $r_i \in (0,1)$ are uniformly distributed and we form the sequence $\{R_N\}$ as

$$R_{N+1} = -\sum_{i=1}^{N} \log r_i \quad , \quad 1 \le N \le n-1 \qquad (13)$$

where we take n-1 to be the value of N for which R_{N+2} first exceeds $\bar{n}$, then the n photon energies are identified as

$$k_i = \epsilon\, e^{z_i}/\sum_j e^{z_j} \qquad (14)$$

with $z_i = \log\epsilon + (\log K_{max} - \log\epsilon)\,R_i/\bar{n}$, $\qquad(15)$

where we agree to reject the entire event if we violate $k_n > K_{max}$. It may be verified that this indeed realizes our restricted Poisson distribution for $\epsilon > K_{max}$. For $\epsilon \le K_{max}$, we take n=0. Since (5) does not depend on K_{max}, we arrive in this way at a rigorous realization of all mutliple photon final states in a given experimental scenario - for we may set K_{max} arbitrarily low; it is simply a reference point for defining the numerical work in the Monte Carlo. No predictions depend on it. In this way, we realize the effect of e^D in (5), and, hence (5) itself by restoring the angular variables of the photons according to the distributions in (5) in the standard manner.

The resulting programs are YFS(2) FORTRAN and BHLUMI FORTRAN and are described in more detail in Refs. 2. Here, we will now illustrate them with some typical results in the next section.

RESULTS

In this section, we wish to illustrate our MC realizations of (5) via YFS2 and BHLUMI with some typical results. For definiteness, we will always use the MkII detector scenario as our prototypical detector case.

Considering first the total cross section for $e^+e^- \to \mu^+\mu^- + n(\gamma)$, we recall that in Refs. 2 we have shown that YFS2 agrees with the analytic result of Berends et al.[5] at the level of .1%. Thus, we will now focus on the situation regarding multiple photon effects in the angular character of the events in a process of the type $e^+e^- \to f\bar{f} + n(\gamma)$, $f \neq e$. In this connection, we consider the standard asymmetries

$$A_{FB} = (\sigma_F - \sigma_B)/(\sigma_F + \sigma_B), \quad A_{LR} = (\sigma_L - \sigma_R)/(\sigma_L + \sigma_R) \text{ and } A_{FB,pol.}$$

$$= (\sigma_{LF} - \sigma_{LB} - (\sigma_{RF} - \sigma_{RB}))/(\sigma_{LF} + \sigma_{LB} + \sigma_{RF} + \sigma_{RB}), \quad (16)$$

where $\sigma_{F(B)}$ is the cross section for $e^+e^- \to f\bar{f} + n(\gamma)$ with $\cos\theta_f > (<)0$ and $\sigma_{L(R)}$ is the cross section for this process when the electron is left (right) handed. $\theta_f \equiv \theta_{cm}(f)$ is the C.M.S. production angle of f relative to e. We take the cases $f = \mu$ and $f = b$ and, in both cases, we impose MkII-like μ-cuts to facilitate the comparison of the various effects:

$$E_\mu > 2\text{GeV}, \quad E_\gamma > .2\text{GeV}, \quad E_{VIS} > .1\sqrt{s}, \quad (17)$$

$$|\cos\theta_\mu| < .8 \quad , \quad |\cos\theta_\gamma| < .95$$

Further, we always evaluate the respective asymmetry at $\sqrt{s} = M_{Z^*}$. Our interest is in the percentage change in the respective asymmetry under the effect of multiple photon radiation.

Considering first the process $e^+e^- \to \mu^+\mu^- + n(\gamma)$ we obtain the results in Table I for the model $M_{Z^*} = 92\text{GeV}$, and $\sin^2\theta_W = .23$.

TABLE I: $\mu\bar{\mu}$ Asymmetries

- NO RADIATIVE CORRECTIONS

    $A_{LR} = .159$, $A_{FB,pol.} = .1023$, $A_{FB} = .0167$

- RADIATIVE CORRECTIONS

$1\gamma \equiv$ (B-K-J, $k_o = 3 \times 10^{-3}$)		$A_{LR} = .155 \pm .001$
YFS		$A_{LR} = .155 \pm .0009$
$1\gamma \equiv$ (B-K-J, $k_o = 3 \times 10^{-3}$)		$A_{FB,pol.} = .1009 \pm .005$
YFS		$A_{FB,pol.} = .1016 \pm .005$
1γ	(NO ACOL. CUT)	$A_{FB} = -.003 \pm .002$
YFS	(NO ACOL. CUT)	$A_{FB} = .0021 \pm .002$
1γ	10° ACOL. CUT	$A_{FB} = -.00065 \pm .0037$
YFS	10° ACOL. CUT	$A_{FB} = .0024 \pm .0037$
1γ	3° ACOL. CUT	$A_{FB} = .0023 \pm .0037$
YFS	3° ACOL. CUT	$A_{FB} = .0048 \pm .0037$

(This value of M_Z. has recently been improved[1] to a value close to 91 GeV; hence, Table I, for $\sqrt{s} = M_{Z'}$, should not be too far off in relating to us the expected percentage change in our asymmetries due to multiple photon radiation effects.) With this apology, we note that, as is well known, the asymmetries A_{LR} and $A_{FB,pol.}$, for muons, are indeed not affected very much by the initial state radiation, independent of whether the radiation is computed using 1γ Berends-Kleiss-Jadach type event generator (with $k_o = 3 \times 10^{-3}$) or using our YFS2 multiple photon event generator. The asymmetry A_{FB}, on the other hand, is evidently affected strongly by the initial state radiation. Indeed, the no acollinearity cut results indicate a correction ~100% and all 3 acol. cut scenarios suggest a significant difference between the 1γ and YFS2 results, although the level of our statistics precludes any definite conclusion in this last regard. We emphasize that, while the asymmetries appear quite similar in the presence of 1γ and multiple γ radiation, the actual cross sections differ by ~ 3-4% so that for precision k_o-independent cross sections one should use the multiple photon event generator. As we expect, a tighter acollinearity cut reduces the effect of the initial state radiation. The disappointing aspect of A_{FB} is that, even after 10^5 $\mu\bar{\mu}$ pairs, the statistical errors are still comparable to the size of the signal: 10^5 $\mu\bar{\mu}$ pairs corresponds to 3×10^6 Z°'s.

Thus, we turn next to same asymmetries (16) for the $b\bar{b} + n(\gamma)$ final state in $e^+e^- \to X$ at $\sqrt{s} = M_{Z'} = 91$ GeV and $\sin^2\theta_W = .2354$, for example. The situation appears to be improved significantly in this case regarding A_{FB}, as we show in Table II. The results in Table II for A_{LR} and $A_{FB,pol.}$ are consistent with their

TABLE II: $b\bar{b}$ Asymmetries

- NO RADIATIVE CORRECTIONS

$A_{FB} = .0708$, $A_{LR} = .116$, $A_{FB,pol.} = .609$

- RADIATIVE CORRECTIONS

5 X 10^5 EVENTS (YFS2)

$A_{LR} = .1126 \pm .0012$

$A_{FB,pol.} = .6143 \pm .0013$

$A_{FB} = .0672 \pm .0012$ NO ACOL. CUT

$A_{FB} = .0689 \pm .0012$ 1° ACOL. CUT

analogues in Table I - the percentage change due to radiation is small. However, for A_{FB}, we see a dramatic change in the nature of the radiative effect. The percentage change due to the radiation is -5.1% for no acol. cut and is -2.75% for a 1° acol. cut. Thus, unlike the case of muons, here the radiation leaves a relatively large and measurable (assuming appropriate tagging methods can be implemented) asymmetry. Hence, for unpolarized beams, this asymmetry appears attractive for relatively high luminosity scenarios like the LEP scenario.[12]

Having illustrated our YFS2 results, we now turn to BHLUMI results. In Refs. 2, we have shown how, for $k_o = .01$, a single bremsstrahlung Monte Carlo of the Berends-Kleiss[13] type and the YFS BHLUMI Monte Carlo are quite close to one another near $\sqrt{s} = M_{Z^\circ}$ for the MkII MINISAM cross section, wherein the cuts $16.2 \text{ mrad} \leq \theta_{cm}(e) \leq 24.5 \text{ mrad}$, $15.2 \text{ mrad} \leq \theta_{cm}(\bar{e}) \leq 25 \text{ mrad}$, $E_f' > .5(\sqrt{s}/2)$, $E_e' + E_{\bar{e}}' > .6\sqrt{s}$, and the superfluous "cut" acol. $< 10°$ are all imposed (if you wish, you may think of the last cut as a coarse check on the work, since it should have no effect for the MINISAM). Hence, here we concentrate on the actual degree of agreement of the two Monte Carlos (one can acutally move to the 1-γ Monte Carlo from the YFS option in BHLUMI by a simple radiation switch, which one specifies as an input parameter). The issue here is that the YFS option in BHLUMI has no k_o parameter whereas the 1-γ Monte Carlo most certainly does. At $M_{Z^\circ} = 92$ GeV, for example, we find, after 6×10^5 events, that

$$\sigma_{\text{MINISAM}} (1\gamma) \big|_{k_o=.01} = 246.8 \pm .3 \text{nb} \qquad (18)$$

$$\sigma_{\text{MINISAM}} (\text{YFS}) = 246.43 \pm .69 \text{ nb}$$

These results are fortuitously closer than $1\sigma \equiv .28\%$. However, the value of k_o is not really known so that the 1-γ result has an uncertainty which, on general grounds and by explicit numerical exercises[14], as we illustrate in Fig. 6, we have found to be $\sim k_o = 1\%$. Evidently, for the highest precision work, the YFS BHLUMI type results are indeed necessary; for, to repeat, such results do not have a k_o parameter.

Hence, we feel that our YFS methods, as realized by YFS2 FORTRAN and BHLUMI FORTRAN, do indeed provide a complete set of multiple photon event generators for the SLC/LEP $Z°$ physics scenario. Further applications of these methods will appear elsewhere.

Conclusions

The YFS methods discussed in this paper represent a new approach to high precision radiative corrections in the $SU_{2L} \times U_1$ theory in which the large Q.E.D. corrections from the infrared regime are treated to all orders. Such all orders work is already imperative at the level of the total cross section in $e^+e^- \rightarrow \mu\bar{\mu} + n(\gamma)$ near $\sqrt{s} = M_{Z^\circ}$ for high precision work-the simplest final fermion pair scenario. Our YFS methods allow such work to be accomplished on an event-by-event basis via Monte Carlo methods.

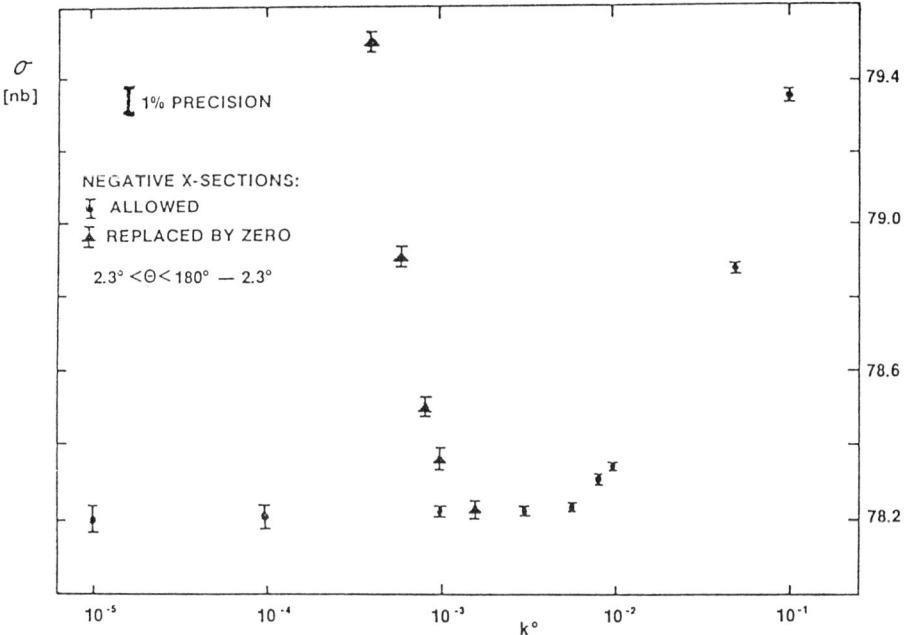

Fig. 6. k_o dependence of a 1γ Berends-Kleiss[13] type of Monte Carlo calculation for an SLC-LEP bhabha luminosity scenario.

In this way, arbitrary detector cuts are accommodated. To date, our detailed numerical results have been encouraging.

Specifically, we re-emphasize that, in Refs. 2, we have been able to show that, below the level of .1%, we reproduce the result of Berends et al.[5] for $e^+e^- \to \mu\bar{\mu} + (2\gamma)$. In addition, we have found that, while the multiphoton final states affect the values of σ_{HA}, A=F,B, H=L,R, by ~3-4% relative to a $k_o = 3 \times 10^{-3}$ 1γ Monte Carlo set of respective values, the respective asymmetries A_{LR} and $A_{FB,pol.}$ are essentially the same for the YFS and 1γ predictions. A_{FB} may be somewhat different for the YFS simulation when compared to the 1γ prediction; our statistics at this time preclude any statement. The analysis of a higher statistics Monte Carlo sample will appear elsewhere. We find that unlike $A_{FB}(\mu)$, the $A_{FB}(b)$ is still quite large after the multiple photon radiative effects are taken into account. This result is quite interesting from the standpoint of unpolarized beam e^+e^- collider physics at a relatively high luminosity.

Our methods have also been applied to the process $e^+e^- \to e^+e^- + n(\gamma)$. Here, building on Ref. 2, we emphasized that the resulting Monte Carlo event generator, BHLUMI FORTRAN, gives one an event-by-event realization of the multiple photon effects in this process in a manner which does not have any dependence on the famous $O(\alpha)$ k_o type parameter. The agreement between the YFS simulation and that of an $O(\alpha)$ type Monte Carlo with $k_o = .01$ is indeed quite good, however, for the MkII MINISAM cross section.

Hence, the multiphoton Monte Carlo assures us of the "reasonableness" of the specified value of k_0 for this <u>particular</u> observable.

In summary, the renormalization group improved YFS method is a theoretically rigorous method which has now been realized in Monte Carlo event generators which are available from the authors upon request. We thus have a practical way of handling the large IR and UV effects in the $SU_{2L} \times U_1$ theory in general and in its Q.E.D. subgroup theory in particular. And, indeed, such practical applications are now in progress at the SLC and LEP, both of which are now observing $Z°$'s.

Acknowledgements

The authors would like to thank the Organizers for giving them the opportunity to lecture on their methods to such a stimulating audience. The authors benefitted from discussions with the participants at the Workshop, especially those with Profs. V. Fadin, E. Kuraev, P. Rankin, L. Trentadue, G. Sterman, and G. Bonvicini. The authors thank Profs. J. Ellis, F. Dydak and G. Feldman for the support and kind hospitality of their Institutes while the research reported in this manuscript was completed.

References

1. See, for example, G. S. Abrams <u>et al</u>., Phys. Rev. Lett. <u>63</u>, 724 (1989); J. Steinberger, U. Amaldi, S. Ting and Michiletti, talks at CERN, 9/2/89; P. Piroue, in Proc. <u>ICTP Summer Workshop on Particle Physics and Cosmology, 1989</u>, to be published.

2. See for example, S. Jadach and B. F. L. Ward, Phys. Rev. D<u>38</u>, 2897 (1988); <u>ibid</u>. <u>39</u>, 1472 (1989); Phys. Lett B<u>220</u>, 611 (1989); UTHEP-88-1101, Phys. Rev. D., 1989, in press; SLAC-PUB-4834, Computer Phys. Comm., 1989, in press; S. Jadach, MPI-PAE/PTh 6/87; B. F. L. Ward, Phys. Rev. D<u>36</u>, 939 (1987); S. Jadach and B. F. L. Ward, CERN-TH. 5399/89; UTHEP-89-0703; and references therein.

3. S. Weinberg, Phys. Rev. D<u>8</u>, 3497 (1973); G. 't Hooft, Nucl. Phys. <u>B61</u>, 455 (1973).

4. D. R. Yennie, S. C. Frautschi and H. Suura, Ann. Phys. <u>13</u>, 379 (1961).

5. See, for example, G. Altarelli, in <u>Physics at LEP</u>, eds. J. Ellis and R. Peccei (CERN, Geneva, 1985); E. A. Kuraev and V. Fadin, Sov. J. Nucl. Phys. <u>41</u>, 466 (1985); F. Berends <u>et al</u>., Nucl. Phys. <u>B</u> <u>297</u>, 429 (1988), and references therein.

6. See, for example, S. Jadach and B. F. L. Ward, in "Radiative Corrections for Experimentalists," convenor R. Kleiss, CERN, 1989, and references therein.

7. See, for example, W. F. L. Hollik, DESY 88-188, 1988, and references therein.

8. K. T. Mahanthappa, Phys. Rev. <u>126</u>, 329 (1962).

9. Y. S. Tsai, SLAC-PUB-3129, 1983.

10. We thank Profs. V. Fadin and G. Sterman for discussions on this point.

11. F. A. Berends et al., Comput. Phys. Commun. <u>29</u>, 185 (1983).

12. See, for example, P. Zerwas and H. Kuhn, talk presented at the LEP100 Meeting, 9/5/89, CERN, Geneva, 1989; in <u>Proc. LEP100 Workshop</u>, ed. G. Altarelli (CERN, Geneva, 1989).

13. F. A. Berends and R. Kleiss, Nucl. Phys. <u>B228</u>, 537 (1983).

14. S. Jadach and B. F. L. Ward, to appear.

AN INCLUSIVE AND EXCLUSIVE ALGORITHM FOR QED EVOLUTION*

GIOVANNI BONVICINI

University of Michigan, Ann Arbor, MI 48109

and

LUCA TRENTADUE

Dipartimento di Fisica, Universita' di Parma, INFN,
Gruppo Collegato di parma, 43100 Parma, Italy

A novel algorithm to calculate radiative corrections to e^+e^- annihilation which is both inclusive and exclusive is proposed. The method uses the structure functions formalism and is based on the factorization of mass singularities in the QED perturbation expansion. The general approach is presented together with its ability to reproduce the infrared and collinear singularity structure of QED as well as the all-order multiple photon radiation. Some numerical results are given.

Contribution to the Ann Arbor Workshop on QED Structure Functions
May 22-25, 1989, University of Michigan, Ann Arbor

---

* Work supported by the Department of Energy, contract DE-AC02-84ER01112; and by the Istituto Nazionale di Fisica Nucleare.

© 1990 American Institute of Physics

1. Introduction

In the last years a powerful way to calculate QED radiative corrections (RC) in e^+e^- experiments [1] has been brought to the attention of the scientific community. Given the need of very high precision predictions at LEP/SLC, such an approach has already been used to calculate RC to inclusive quantities [2-5]. In this work we tackle the problem of a complete description of the evolution of a given QED process by using the structure function formalism. We build an algorithm which generates an arbitrary number of photons along with their individual 4-momenta, and which is suitable for a general application of RC to experimental data.

This algorithm is intended for both inclusive and exclusive processes. By inclusive, we mean those reactions where the final state is only partially reconstructed or computed; in an exclusive reaction the entire observable final state is reproduced.

Historically, inclusive calculations have always achieved high precision for a small subset of observables [6, 7]. At LEP/SLC, for example, analytic formulae describe precisely the inclusive shape of the Z^0. The entire class of weak corrections can be dealt with analytically too, the reason being that no extra weak bosons are radiated.

However, RC to data are in most cases an integral over a multidimensional space. There are three degrees of freedom for each observed particle. Extra dimensions and irregular integration boundaries are added by detector acceptances, resolutions, efficiencies and imposed cuts. As is well-known [8], a Monte Carlo procedure is the only way to perform such complex integrals.

Finite order perturbation methods, which are not precise in the most densely populated regions of the phase space, are strictly exclusive. This originates from allowing only a limited number of photons, which in turn forces the introduction of an unphysical cutoff. However, they were found suitable for Monte Carlo event generation [9] and have been far more extensively used in applications to experiments than inclusive methods.

Here we present a method that both has aspects of a correct infrared behaviour and which allows generation of exclusive final states. The former aspect, together with its main consequence of a correct inclusion of multiple all-order photon radiation, is now believed to be a natural ingredient of any radiative correction scheme. It has also the feature of keeping track explicitly of the transverse degrees of freedom. In this we differ from other QED structure function Monte Carlos, which have a fixed number of photons and only a part of the correct multiphoton 4–momentum [10–12].

The paper is organized as follows: In sec. 2, we discuss the Poissonian properties of evolution equations in longitudinal phase space. We limit ourselves to the evolution of the initial state in e^+e^- annihilation. In sec. 3, we implement the transverse degrees of freedom and add modifications as needed to push the precision of the algorithm beyond 1% . In sec. 4, we present some results and conclusions.

2. The Poissonian properties of evolution equations.

The analogy with QCD Drell–Yan type processes carried out in refs. [1–5] is based on a Lipatov–Altarelli–Parisi evolution equation for electron or positron states which in the nonsinglet channel, *i.e.*, for the evolution of a single valence electron or positron state without any sea contribution, is

$$D(x,s) = \delta(1-x) + \frac{\alpha}{2\pi} \int_{m^2}^{s} \frac{ds'}{s'} \int_{x}^{1} \frac{dz}{z} P(z) D(\frac{x}{z}, s') \quad . \tag{1}$$

$P(z)$ is the regularized $e \to e + \gamma$ vertex given by

$$P(z) = \frac{1+z^2}{1-z} - \delta(1-z) \int_0^1 dt \frac{1+t^2}{1-t} \quad . \tag{2}$$

The second (virtual) term is zero everywhere but at $z = 1$. The density $D(x,s)$ represents the probability of finding, inside a parent electron (positron) of mass m, at the scale s, an electron (positron) with a fraction x of a given kinematic variable. According to the majority of QCD–type applications, e^+ and e^- are evolved independently, and we define, for now,

$$x = \frac{p_L}{E} = 1 - k\cos\theta \quad . \tag{3}$$

Here k is the fractional energy of the emitted photon in units of the beam energy E, p_L is the longitudinal momentum and θ is the angle between the emitted photon and the radiating leg.

The cross section is derived from a convolution integral over the electron and positron legs

$$\sigma(s) = \int dx_1\, dx_2\, \sigma_0(s') D(x_1, s)\, D(x_2, s) \tag{4}$$

and $s' = x_1 x_2 s$. Iterated and truncated to first-order, eq. (1) is

$$D(x,s) = B(0)\, \delta(1-x) + B(1)\, P(x) \quad . \tag{5}$$

The $B(n)$ are the probability to radiate n photons above a *physical* cutoff (see sec. 4) during the evolution, and in the unitary approximation they play the role of branching ratios. In the following, we will implicitly assume the kernels multiplying the $B(n)$ to be unitary, so that all the information on the branching probability will be contained in the $B(n)$, which we will now determine. Iterated to third order, eq. (1) has the form:

$$D(x,s) = \delta(1-x)$$
$$+ \frac{\alpha}{2\pi}\left[\int \frac{ds'}{s'} P(x) \right.$$
$$+ \frac{\alpha}{2\pi}\left[\int \frac{ds''}{s''} \int \frac{dz}{z} P\left(\frac{x}{z}\right) P(z) \right.$$
$$\left.\left.+ \frac{\alpha}{2\pi}\left[\int \frac{ds'''}{s'''} \int \frac{dz}{z} \int \frac{dy}{y} P(\frac{x}{yz}) P(z) P(y) + \cdots \right]\right]\right] . \quad (6)$$

Upon integrating in x each term, above and below the dimensionless cutoff x_0, one extracts the coefficients $B(n)$. We drop nondominant terms which correspond to the sea electrons, and find

$$B(0) = 1 - \bar{n} + \frac{\bar{n}^2}{2} - \frac{\bar{n}^3}{6} + \cdots = e^{-\bar{n}} ,$$
$$B(1) = \bar{n} \, B(0) , \quad (7)$$
$$B(2) = \frac{\bar{n}^2}{2} B(0) , \quad etc.,$$

with

$$\bar{n} = \frac{\alpha}{\pi} L(-l - 3/4) , \quad L = \log(\frac{s}{m^2}) , \quad l = \log(x_0) . \quad (8)$$

The probability distribution is Poissonian [6, 13] with an average number of photons depending logarithmically on the cutoff. By forcing the emission probability to be binomial, zero or one emitted photon, and imposing the condition that both probabilities be positive, we find

$$B(0) = 1 - \bar{n} \; , \qquad B(1) = \bar{n} \; , \qquad 0 < \bar{n} < 1 \; . \tag{9}$$

This is the well-known limit on the cutoff that affects truncated expansion Monte Carlos [9], or the equivalent result given by eq. (5).

Finally, the substitution $L \to L-1$ in eq. (8) correctly normalizes the emission probability [14] and reproduces, by direct inspection of the matrix elements, all the dominant terms up to order α^2, and the l-dominant terms [2].

A calculus theorem can be used to transform the double emission term in eq. (6) (the proof for n photons is automatic) into an explicit probability density

$$\int_{x/1-x_0}^{1-x_0} \frac{dz}{z} P(z) \, P\left(\frac{x}{z}\right) = \frac{d}{dx} \int_{\Omega} dz' dz'' \, P(z') \, P(z'') \; , \tag{10}$$

where the integration area Ω is defined by:

$$z', z'' < 1 - x_0 \; , \qquad zz' > x \; .$$

From eqs. (7) and (10), it appears that the iterative solution of eq. (1) is an infinite sum of terms with Poissonian normalization, and each term is given by an iterated convolution of an elementary kernel. The details of the proof for eq. (10) can be found in the Appendix.

Having generated a Poissonian photon multiplicity, the algorithm proceeds to generate the longitudinal variable for each photon according to the kernel in eq. (2).

In the case of two photons, the steps are:

(1) Generate z_1, $P(z)$ as defined in eq. (2), and $0 < z < 1 - x_0$.

(2) Repeat the first step for z_2. (11)

(3) Let $x = z_1 z_2$, $k_1 = 1 - z_1$ and $k_2 = z_1(1 - z_2)$.

The second photon is allowed by the evolution to have an energy $z_1 x_0 < k_1 < z_1$, part of which is below the cutoff (fig. 1). As the phase space for the second photon is reduced, the cutoff is moved inward to maintain a Poissonian statistics. The probability of radiating a photon k_1 and a second photon k_2 below the cutoff is given by the integral of the $P(z)$ over the shaded area in fig. 1(b). The result is:

$$\left(\frac{\alpha L}{2\pi}\right)^2 \int_{1-x_0}^{1-z_1 x_0} \frac{2 dz_2}{1-z_2} \frac{1+z_1^2}{1-z_1} = \left(\frac{\alpha L}{\pi}\right)^2 \frac{1}{2} \log z_1 \frac{1+z_1^2}{1-z_1}, \quad (12)$$

which has to be compared with those non-Poissonian $\alpha^2 L^2$ terms [3] where 1 real photon plus 1 soft-virtual photon are radiated. The two expressions are identical and the virtual-virtual (0 real photons) and real-real (2 real photons) terms are easily proven to be factorizable [3] in the $\alpha^2 L^2$ terms. We conclude that our variable cutoff method achieves factorization of the L-dominant terms [2].

3. Inclusion of transverse degrees of freedom.

Equation (1) can be generalized [15] to also take into account transverse degrees of freedom. In QED, by keeping the masses and by using the correct kinematics, the evolution equation becomes:

$$\frac{\partial D(x, p_T, s')}{\partial s'} = \frac{\alpha}{2\pi} \frac{1}{s' - m^2} \int_x^1 \frac{dz}{z} P(z) \int \frac{d^2 q_T}{\pi} \delta_k^* D\left(\frac{x}{z}, p_T - \frac{x}{z} q_T, s'\right). \quad (13)$$

The $D(x, p_T, s)$ represents the probability to find inside the parent e^- (e^+) an e^- (e^+) with a fraction x of a longitudinal variable and a 2-vector transverse variable p_T with respect to the parent direction. We define, for now, the p_T^2 as:

$$p_T^2 = E^2 k^2 \sin^2\theta \quad , \tag{14}$$

so that the variables x and p_T are independent components of a 3-vector.

The δ_k^* function guarantees a correct kinematic splitting at the vertex. In the space-like (initial state) case [16], the argument of δ_k^* at the ith step of the evolution is:

$$(1-z) M_{i+1}^2 - z(1-z) M_i^2 - q_T^2 \quad , \tag{15}$$

where $M_{i+1}^2 < 0$ is the virtualness of the inner leg, $M_i^2 < 0$ is the virtualness of the outer leg, and $M_1^2 = -m^2$ (fig. 2).

We repeat the steps of the previous section. To $O(\alpha)$ we find [16]:

$$D(x, p_T, s) = B(0)\, \delta(1-x)\, \delta^2(p_T)$$
$$+ B(1)\, P(x)\, \frac{1}{p_T^2 - (1-x)^2 M_1^2} + O\left(\frac{p_T^2}{E_\gamma^2}\right) \quad , \tag{16}$$

and for the total cross section we have:

$$\sigma(s) = \int dx_1\, dx_2\, d^2 p_{T1}\, d^2 p_{T2}\, \sigma_0(s')\, D(x_1, p_{T1}, s)\, D(x_2, p_{T2}, s) \quad , \tag{17}$$

where

$$s' = 2\left(\sqrt{x_1^2 E^2 + p_{T1}^2}\sqrt{x_2^2 E^2 + p_{T2}^2} + x_1 x_2 E^2 - p_{T1} \cdot p_{T2}\right) \quad . \tag{18}$$

In the previous section, we discussed how to generate the photon multiplicity and longitudinal variables, and according to eqs. (15) and (16) the following complete set of formulae hold at the ith step of the iteration [see also Alg. (11)]:

$$x_i = \prod_{j=1}^{i} z_j \,, \qquad k_{Li} = \prod_{j=1}^{i-1} z_j(1-z_i) \,, \qquad (19)$$

$$p_{Ti} = q_{Ti} + x_i \, p_{Ti-1} \,, \qquad k_{Ti} = -q_{Ti} + (1-x_i) \, p_{Ti-1} \,.$$

The k_L, k_T are the x, p_T variables for the photon. Figure 3 shows that in eq. (19) the evolution variable q_T is the photon transverse momentum respect to the instantaneous direction of the leg and that eq. (19) simply conserves the 3-momentum at each step of the evolution. The evolution of the energy of the leg is trivially derived from the zero mass of the photon.

At the end of the evolution of the initial state, the electrons and positrons which are off mass shell are used to calculate the value of the annihilation cross section by evaluating the Born term at the reduced center-of-mass energy [eq. (18)].

This completes the description of the basic algorithm for the QED evolution.

The basic structure function approach was designed to carry up to leading logarithmic factorization of the evolution. Without losing any of the advantages of the main algorithm described above, we have introduced two modifications to achieve the second-order next-to-leading result and agreement with a first order calculation in the very hard region ($p_T^2 \approx s$), as needed for a powerful and high precision (less than 1%) prediction.

We compare with an exact first-order calculation [17] (which we have calculated using REDUCE [18]) and include further less dominant terms to achieve the $O(\alpha)$

matrix element up to $O(m^2/E^2)$. This procedure absorbs in the iteration many nondominant terms in the double emission exact matrix element (which we also calculated with REDUCE). The result is

$$P(z, q_T, s) = \frac{1}{1-z} \frac{1}{q_T^2} \left[1 + z^2 - 2 \frac{q_T^2}{s} \left(1 - z - \frac{q_T^2}{s} \right) - z(1-z) \frac{2M_i^2}{q_T^2} \right], \quad (20)$$

where new definitions are used (β is the velocity of the electron or positron)

$$z = 1 - k, \quad q_T^2 = |M_{i+1}^2 - M_i^2|, \quad |M_2^2 - M_1^2| = E^2 k(1 - \beta \cos\theta). \quad (21)$$

The effect of eq. (20) is to modify the probability density in phase space. However, the final algorithm retains the physical property of evolution of the mass, as described in eq. (15). Equations (18) and (19) describe 3–momentum conservation but now must be used with (p_L, p_T) as they can be extracted from (z, q_T) generated by eq. (20).

Finally, the first-order K–factor [1]:

$$K = 1 + \frac{\alpha}{\pi} \left(\frac{\pi^2}{3} - \frac{1}{2} \right) \quad (22)$$

is added as an overall normalization constant. This simple modification reproduces the exact next-to-leading result [1–3].

To summarize, this algorithm:

(1) uses special kernels, which are equivalent to the exact first order matrix element [eq. (20)];

(2) evolves independently the two legs and, on each leg, conserves 4-momentum [eq. (19)], evolves the mass [eq. (16)] and the cutoff [eq. (10)] in order to achieve factorization of the nonleading terms.

4. Results and conclusions.

The comparison between an analytic calculation for the Z^0 shape [19] and the Monte Carlo is shown in fig. 4, with the real-virtual pairs terms removed [10]. The agreement between the two predictions is completely consistent with the statistical error[*] of the Monte Carlo.

Figure 5 shows the radiative corrections to the process $e^+e^- \to \nu\nu\gamma$. The solid line is from an analytic calculation [16] and the data points are from our Monte Carlo. Events are counted if they have only one photon with energy $E_\gamma > 1$ GeV, zenith $\theta_\gamma > 20°$ and no other photons above a 1-GeV threshold and outside a $3°$ cone around the beam pipe. Two comments are in order.

First, the slight disagreement in the 1–4 GeV region ($\approx 5\%$) is due to the lack of logarithmic terms [see eq. (12)] in ref. [16], and to the independent emission modeling that does not reproduce the evolution of the mass and p_T [see eqs. (15) and (21)]. Both effects decrease the peak in fig. 5 less than 1%, but are more significant in the broad valley on the left. The first effect is estimated to be $\approx -2\%$ at 2 GeV. The second has been estimated by writing an independent emission (IE) Monte Carlo and by comparing it with the one described in the previous sections. The

---

[*] To our surprise, the disagreement between the fitting function and the Monte Carlo was nowhere worse than .3 σ, where $\sigma \approx .3\%$ near the peak and $\sigma \approx 1\%$ on the tail. The difference arises from using the Monte Carlo in a weighted mode for this particular set of runs. Weighted and unweighted mode have been checked to be equivalent.

effect is ≈ −2%. Our MC and the IE one differ significantly in the configuration described above only in what follows. Because our MC evolves p_T and mass, the unvetoed rate at large angle is 15% higher than in the IE case. However, the same effect increases the vetoing probability of about the same amount, leaving approximately the same total rate but different vetoing rates. On the peak no further hard photons are emitted and ref. [16] compares well with our algorithm. A comparison with $O(\alpha)$ calculations would not be significant because the peak/valley ratio, which dominates the corrections to the shape of the spectrum, has RC of order 50%. It is obvious that, near the Z peak, the neutrino counting experiment cannot be corrected properly with a first-order calculation or an independent emission model.

Second, in ref. [16] the $O(\alpha^2)$ calculation is a one-photon $[O(\alpha)]$ cross section, radiatively corrected. On the contrary, in our Monte Carlo the evolution starts from the pure annihilation Born term. The agreement (with the limitations discussed above) between the two predictions is a direct consequence of the capability of the algorithm to achieve the next-to-leading result, and its capability to predict RC both inclusively and exclusively.

In fig. 6, we show the detailed shape of the initial state radiator [16] in the large energy loss region as compared again with an analytic calculation.

In fig. 7, the photon multiplicity for the initial state radiator is finally shown. When the radiator is effectively damped by the resonance, both the hard part of the photon spectrum and the high multiplicity tail are suppressed.

We conclude by discussing the physical content of our method. The methods generally used to calculate RC to the data [9], which are truncated expansions, always put a lower limit on the cutoff [$\bar{n} < 1$, eq. (9)]. The optimum cutoff is then unrelated to the physical parameters of the experiment, and is neither the energy resolution, nor a bookkeeping parameter. Rather, the cutoff is an arbitrary external parameter which is tuned to optimize the approximation of the calculation and the results of the Monte Carlo are cutoff-dependent [9].

In the algorithm presented here, the cutoff can be set arbitrarily low [as introduced in eq. (8)]. The lower the cutoff, the more steps there will be in the evolution of the system. Results are independent of large variations of the cutoff (we have tried up to a minimum, computer-limited cutoff of 10^{-7}), as long as the cutoff is kept much less than all the experimental scales. Operatively this means that the cutoff, to be physical, should be always much less than the minimum of the beam energy spread, acollinearity and one photon resolution of the experiment.

In secs. 2 and 3, we have shown that the $\alpha^2 L^2$, $\alpha^2 Ll$ and $\alpha^2 l^2$ terms are correctly included in our recipe, as well as more dominant logarithms. Figure 4 in the soft region and fig. 6 in the hard region empirically support our statement.

Re-absorption of nonleading terms in the evolution of a few quantities (in our case the cutoff and the masses) has been long known to be a standard procedure in QCD [20]. In our case, this is particularly simple due to the much simpler branchings and quantum numbers in QED. Further reasons for the success of this approach are the following advantages over QCD:

(1) the initial state [the Dirac δ in eq. (1)] is known exactly;

(2) the evolution is chain-like as opposed to tree-like (that is, one or two of the three particles at the elementary vertex are on mass-shell). This also has the advantage of eliminating sea electrons from the picture. Four-fermions amplitudes are small ($\approx 1\%$) and can be dealt with in other ways.

(3) the integral on ds' [eq. (1)] is solvable exactly, with or without virtual pairs, therefore generating an exact statistics without loss of next-to-leading terms as in QCD.

As a last remark, chain-like evolution allows a predetermination of the number of steps of the evolution, and the evolution can start from the external relativistic invariant down to the inner line. Since it is the external relativistic invariant which enters in the exact matrix element [in the case of the initial state, see eqs. (20) and (8)], we have seen that one obtains a very natural inclusion of nonleading pieces in the evolution. Not surprisingly, this Monte Carlo evolves oppositely to well-known QCD examples [21] and, in the soft limit, explicitely calculates the Sudakov form factor [22].

We would like to thank M. Greco for constant encouragement and R. Frey for a careful reading of the manuscript. G. B. thanks S. J. Brodsky for many interesting discussions and L. T. thanks Stefano Catani and Oreste Nicrosini for many useful comments and discussions.

Appendix

We start from the following calculus theorem [23]:

$$\frac{d}{du} \int_{h(u)}^{g(u)} f(u,v)dv = \frac{dg}{du} f[u,g(u)] - \frac{dh}{du} f[u,h(u)] + \int_{h(u)}^{g(u)} \frac{\partial}{\partial u} f(u,v)dv \ .$$

Let us now consider the α^2 term of the expansion [eq. (6)], which describes the emission of two real photons

$$\int_{x/1-x_0}^{1-x_0} P(z) \, P\left(\frac{x}{z}\right) \, dz \text{ over} z \ .$$

The upper limit is the separation between the no-emission region and the real emission region [i.e., the quantity called $B(0)$ in the text is the sum of all pieces of the expansion integrated between $1 - x_0$ and 1]. The lower limit corresponds, in the Feynman diagram modeling, to the separation between two real photons and one real photon plus one virtual photon.

The integral function $Q(y)$ is defined as follows:

$$Q(y) = \int_y^{1-x_0} P(z')dz' \ , \qquad Q(1-x_0) = 0 \ .$$

We have

$$\int_{x/1-x_0}^{1-x_0} P(z) \, P\left(\frac{x}{z}\right) \frac{dz}{z} = \int_{x/1-x_0}^{1-x_0} P(z) \frac{\partial}{\partial x} Q\left(\frac{x}{z}\right) dz = \int_{x/1-x_0}^{1-x_0} \frac{\partial}{\partial x} \left[P(z) \, Q\left(\frac{x}{z}\right)\right] dz$$

and

$$\int_{x/1-x_0}^{1-x_0} \frac{\partial}{\partial x}\left[P(z)Q\left(\frac{x}{z}\right)\right] dz = \frac{d}{dx}\int_{x/1-x_0}^{1-x_0} P(z)\, dz \int_{x/z}^{1-x_0} P(z')\, dz'$$

$$- \frac{d(1-x_0)}{dx} P(z)\, Q\left(\frac{1-x_0}{z}\right) + \frac{d(x/1-x_0)}{dx} P(z)\, Q(1-x_0) \quad .$$

The second term in the right-hand side is zero because of the derivative of a constant, and the third term is zero because of $Q(1-x_0)$. The following identity holds:

$$\int_{x/1-x_0}^{1-x_0} P(z)\, P(\frac{x}{z})\, \frac{dz}{z} = \frac{d}{dx} \int_{x/1-x_0}^{1-x_0} \int_{x/z}^{1-x_0} P(z)\, P(z')\, dzdz' \quad ,$$

and the integration area in the (z, z') space can be rewritten as

$$z, z' < 1 - x_0 \quad , \quad zz' > x \quad .$$

References

[1] E. A. Kuraev and V. S. Fadin, SJNP 41(3) 466, 1985.

[2] O. Nicrosini and L. Trentadue, PL 196B: 551, 1987.

[3] F. A. Berends, G. J. H. Burgers and W. L. Van Neerven, NPB 297: 429, 1988.

[4] J. E. Campagne and R. Zitoun, LPNHEP-88-06.

[5] G. Altarelli and G. Martinelli, CERN/EP 86-02, p. 47.

[6] E. Etim, G. Pancheri and B. Touschek, Nuovo Cim. 51 (1967) 276.

[7] M. Greco *et al.*, PL 56B (1975) 367.

[8] F. James, Rept. Prog. Phys. 43: 1145, 1980.

[9] F. A. Berends and R. Kleiss, NPB 177: 237, 1981.

[10] J. Alexander *et al.*, PRD37: 56, 1988.

[11] D. Kennedy *et al.*, SLAC-PUB-4128.

[12] J. E. Campagne and R. Zitoun, private communication.

[13] F. Bloch and A. Nordsieck, PR 52 (1937) 54.

[14] G. Bonneau and F. Martin, NPB27: 381, 1971.

[15] A. Bassetto, M. Ciafaloni and G. Marchesini, NPB 163: 477, 1980.

[16] O. Nicrosini and L. Trentadue, to be published in Nucl. Phys. B.

[17] F. A. Berends and R. Kleiss, NPB 260: 32, 1985.

[18] A. C. Hearn, Rand Publication CP78.

[19] F. A. Berends, G. J. H. Burgers and W. L. Van Neerven, PL 185B: 395, 1987.

[20] G. Marchesini and B. R. Webber, NPB 238: 1, 1984.

[21] M. Bengtsson, T. Sjostrand and M. Van Zijl, ZPC 32: 67, 1986.

[22] V. V. Sudakov, ZHETF 30: 187, 1956.

[23] C. J. Everett and D. E. Cashwell, LA-9721-MS.

Figure Captions

Fig. 1. Evolution of the cutoff in the approximation that the electron and positron evolve independently: (a) phase space allowed to the first photon; (b) phase space allowed to the second photon.

Fig. 2. Four-momentum conservation at the elementary vertex. The quantities are defined in the text.

Fig. 3. Evolution of the leg. The quantities are defined in the text.

Fig. 4. Radiative corrections to the Z^0 line shape in the $\mu\mu$ channel as predicted by the Monte Carlo (data points), and by an analytic exact $O(\alpha^2)$ exponentiated calculation from ref. [19]. According to ref. [10], the predictions of refs. [1], [2] and [19] are indistinguishable. The Z^0 resonance parameters are: $M = 93$ GeV, $\Gamma(M) = 2.5$ GeV, and $\sin^2\theta_W = .223$. The Monte Carlo was run with 10^5 events per point, a cutoff x_0 of 10^{-5} and $s' > .1s$: (a) $87 \geq \sqrt{s} \leq 99$ GeV; (b) $92.3 \geq \sqrt{s} \leq 93.7$ GeV.

Fig. 5. Radiative corrections to $\gamma\nu\nu$ photon energy spectrum as predicted by the Monte Carlo (data points), and by an analytic exponentiated calculation, from ref. [16]. The relevant parameters are: $M = 93.2$ GeV, $\Gamma(M) = 2.6$ GeV, $\sin^2\theta_W = .223$, $\sqrt{s} = 98$ GeV, $x_0 = 10^{-5}$ and 3 species of neutrino. Photons are required to have an energy $E_\gamma > 1$ GeV, zenith $\theta_\gamma > 20°$ and are vetoed if another photon of energy greater than 1 GeV has $\theta > 3°$.

Fig. 6. The s' spectrum in the large energy loss region, as calculated by the Monte Carlo. The relevant parameters are the same of fig. 4 and the spectrum is not weighted for the Born term.

Fig. 7. Photon multiplicity as calculated by the Monte Carlo with the same parameters of fig. 4. Solid line: the multiplicity unweighted by the Born term. Dashed line: the multiplicity weighted by the Born term.

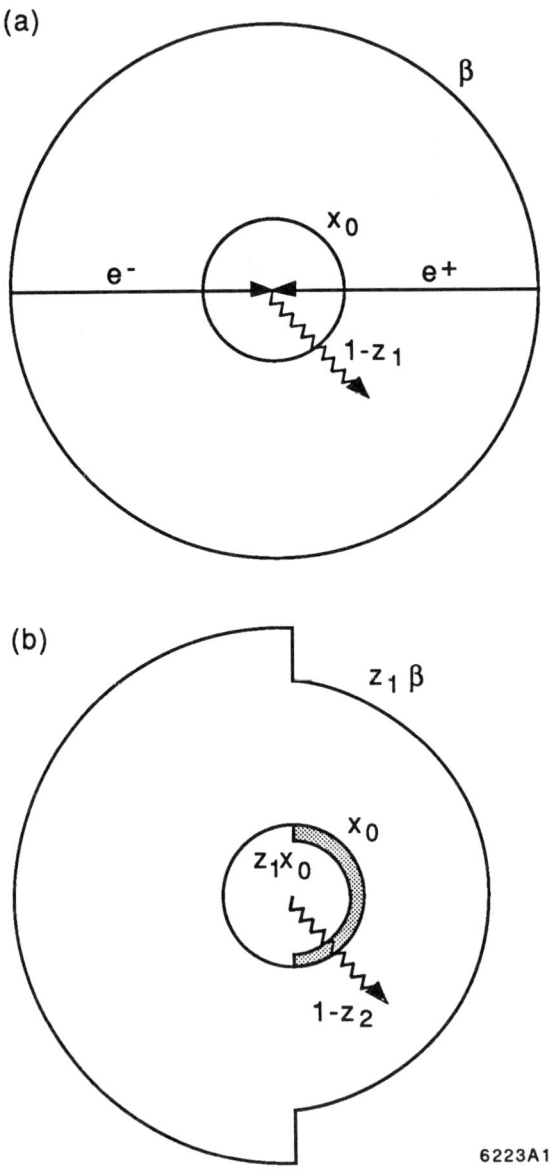

Fig. 1

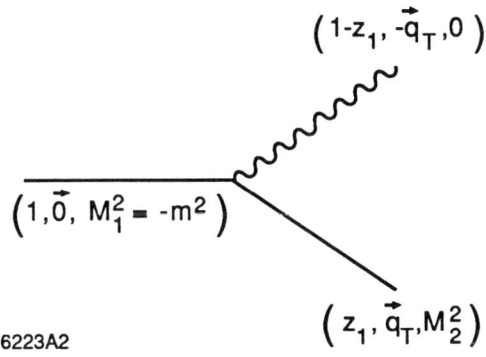

Fig. 2

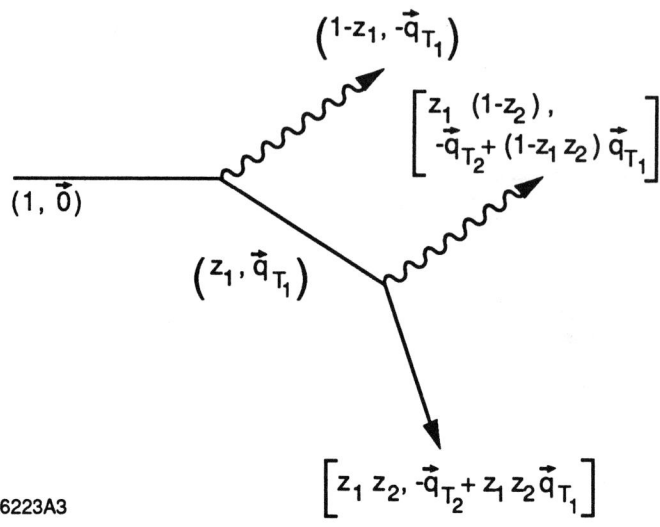

Fig. 3

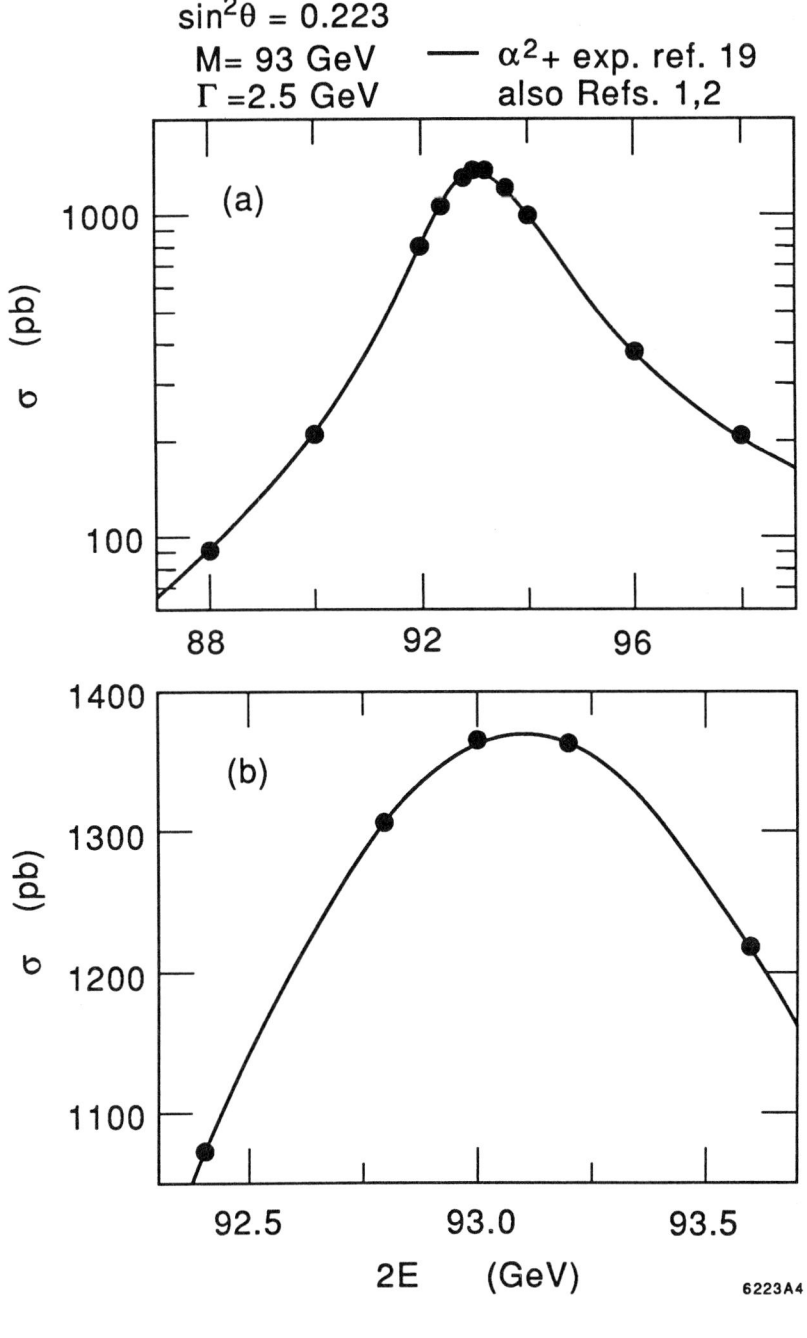

Fig. 4

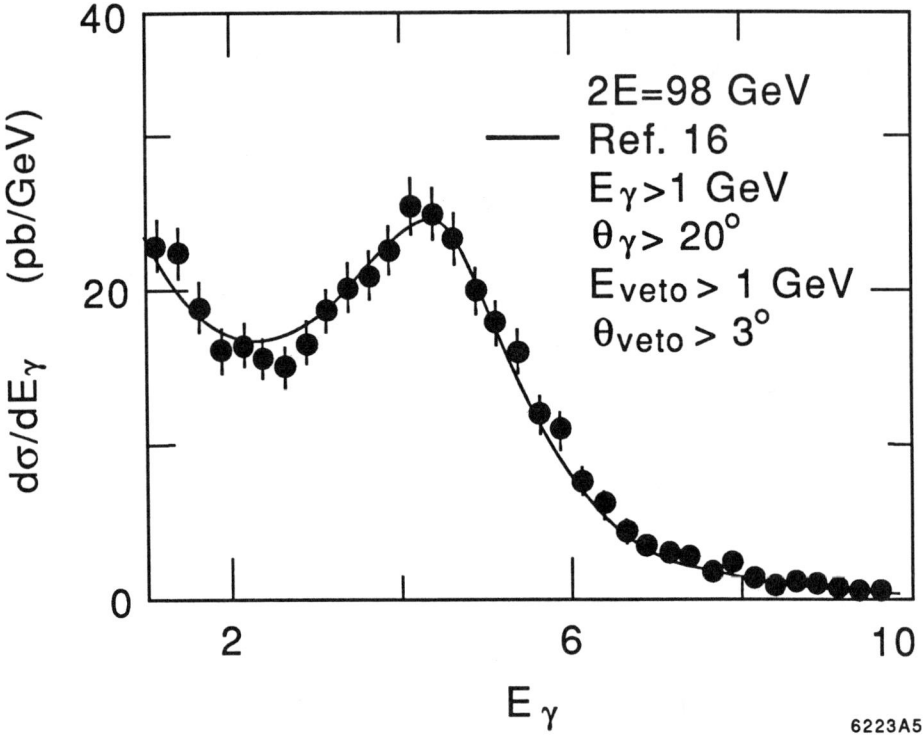

Fig. 5

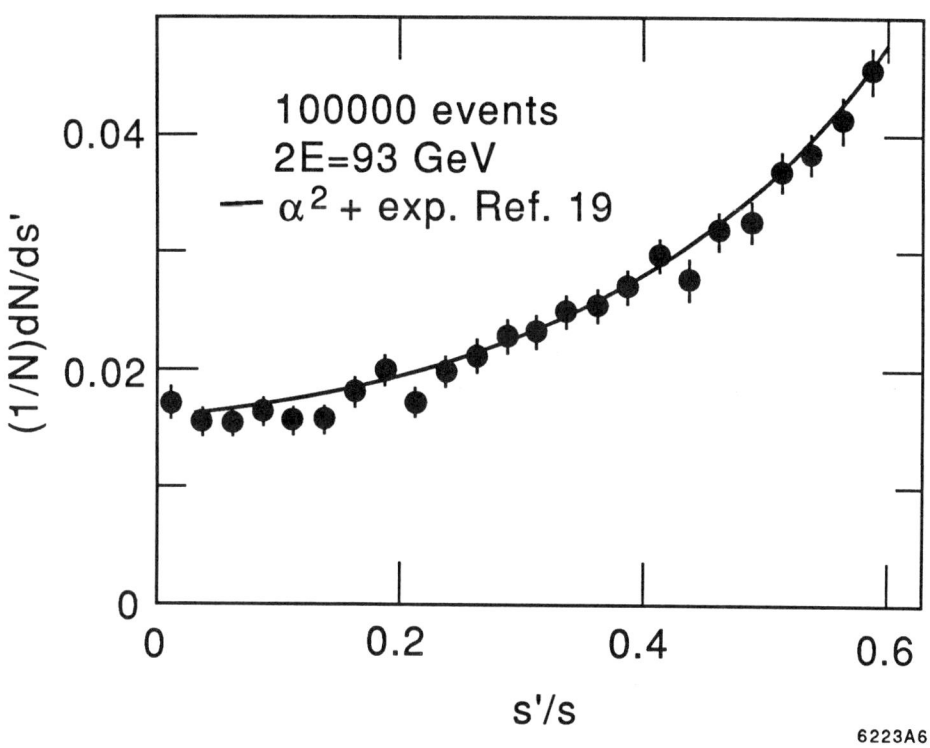

Fig. 6

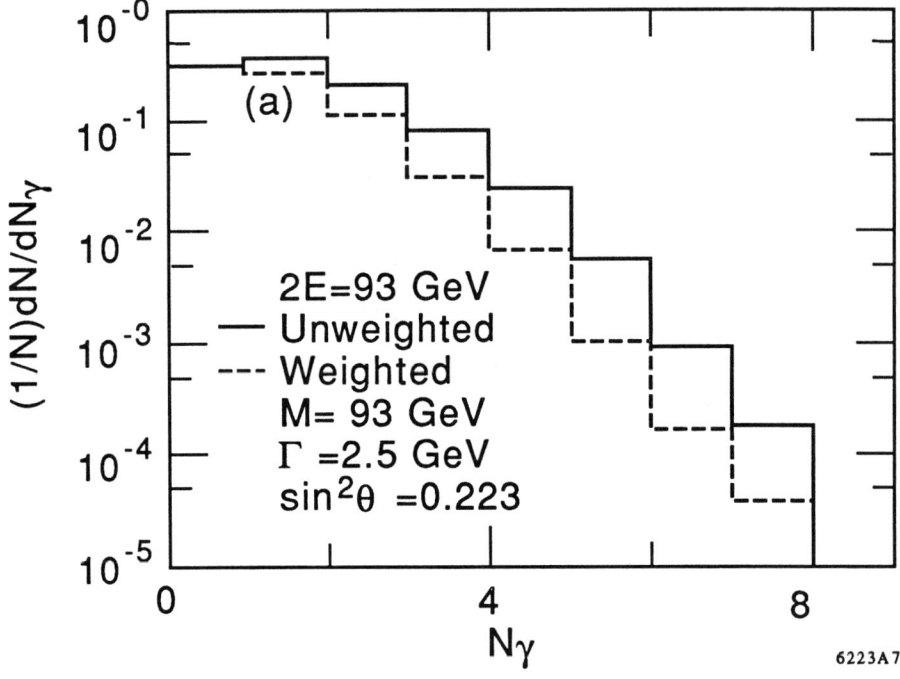

Fig. 7

Helicity Amplitudes Calculation *

M. Martinez, R. Miquel
Laboratori de Física d'Altes Energies
Universitat Autònoma de Barcelona, E-08193 Bellaterra (Barcelona) Spain

C. Mana
CERN, CH-1211 Geneva, Switzerland

August 18, 1989

Abstract

A method for computing Feynman diagrams at the tree level using an helicity amplitudes formalism is reviewed. This method allows the calculation of as many diagrams as needed and keeping all masses in a very simple and elegant way which fully exploits the symmetries of the problem. The method has been successfully applied in many processes and, in practice, the only actual limitation on the complexity of the problem treated is the actual computation power available for the user.

1 Introduction

The HELICITY AMPLITUDES method as we will describe it here, can be considered just as a practical way of computing Feynman diagrams at the amplitude level. This makes sense only if we specify the momenta and *spin state* of all the particles involved in the process since the amplitudes can be only added when the initial and final states are completely specified, that is, the matrix element squared is computed in the following way:

$$|iM|^2 = \sum_{\lambda} |\sum_j (iM_j)|^2 \qquad (1)$$

where

j runs over all possible diagrams
λ run over all the spin states into consideration

The particular choice we use to define the spin states is in the case of massless particles the helicity (equivalent in this case to the chirality) because of its easy representation and interesting properties.

The "standard method" used to compute Feynman diagrams is based on the spin sum properties:

$$\sum_{\lambda} \epsilon_\lambda^\mu \epsilon_\lambda^{*\nu} = -g^{\mu\nu} \qquad (2)$$

---

*Talk presented by M.Martinez

for massless spin 1 gauge bosons and

$$\sum_\lambda u_\lambda(p,m)\bar{u}_\lambda(p,m) = \not{p} + m \qquad (3)$$

for spin 1/2 particles, which are used when combining amplitudes to compute terms of the matrix element squared to close traces. This method in a natural way gets rid off the need of spin state information. The final result of the calculation usually can be expressed in terms of SCALAR products of the type

$$(p_i \cdot p_j) \qquad (4)$$

being p_i and p_j particle four-momenta.

In opposition, the basic idea of the Helicity Amplitudes method is that if chiral spinors are used then

$$u_\lambda(p)\bar{u}_\lambda(p) = \left(\frac{1+\lambda\gamma_5}{2}\right)\not{p} \qquad (5)$$

that is, no sum over λ has to be implemented and then each amplitude can be computed separately for a given helicity configuration. The final result can be expressed in terms of SPINOR products of the type

$$[\bar{u}_{\lambda_i}(p_i)u_{\lambda_j}(p_j)] \qquad (6)$$

which are complex functions of p_i, p_j, λ_i and λ_j.

At least a couple of reasons can be pointed out justifying the use of this second method, namely

i.- the fact that the amplitude of each diagram can be computed independently allows to fully exploit the symmetry of the diagrams. Usually all the Feynman diagrams contributing to a given process show enormous symmetries: several amplitudes can be obtained from another just by exchanging particle labels or "rotating" external legs. In practice, only few topologically different diagrams can be found for any process, and therefore is enough computing just the amplitude for the GENERIC DIAGRAMS. Doing so:

  a.- we are enabled to attack processes with a large number of diagrams (very unpractical with the standard trace-closing technique) being in the method we are going to explain here our computational power the only practical limitation.

  b.- it is trivial to obtain a simple and clear codification and also very compact expressions.

ii.- the fact that diagrams are added at the amplitude level provides an optimal control of the cancellations among them. In processes in which there are strong cancellations among the contributions coming from different diagrams this method allows the cancellation to occur directly between the relevant pieces instead of between terms of the matrix element squared (as is the case in the standard method), and therefore:

  a.- the needed accuracy is just half the number of digits required when working directly with matrix element squared.

  b.- there is no need for mixing different spin configurations (which actually correspond to different physical processes).

Additionally in our approach all particle masses are taken into account without any additional effort and if there are massless gauge bosons as external particles (which is most of the time the practical case) the check of the whole calculation is automatic.

Few different approaches have been suggested and successfully used in the last years [1-2]. We could classify them in two groups:

(a) the approaches foreseen to be used to produce analytical results (at the prize of neglecting particle masses) and

(b) the ones designed to be used in computer programs to lead to numerical results (keeping everything on).

As we are mainly interested in calculational techniques for Monte Carlo event generators, that is for numerical calculations, we are going to focus in the second approach which is, in fact, more general but we will show also how the first one can be easily deduced from it when neglecting the particle masses. We are going to follow the work of the CALKUL coll., with the important modifications introduced by Berends, Daverveldt and Kleiss [2] and further developments done by ourselves [3].

2 Feynman diagrams

The starting point for building helicity amplitudes is analyzing the structure of the amplitudes that we obtain when applying Feynman rules to write down the Feynman diagrams of a given process.

The external spin 1/2 particles are described by spinors $u(p,m)$ verifying

$$\begin{aligned} \slashed{p}\ u_\lambda(p,m) &= m\ u_\lambda(p,m) \\ \sum_\lambda u_\lambda(p,m)\bar{u}_\lambda(p,m) &= \slashed{p} + m \end{aligned} \quad (7)$$

and antiparticles show up as particles of negative mass and reversed "fermionic number" flow. These spinors show up paired delimiting the ends of fermionic lines with a well defined "fermionic number" flow direction. This actually represents bilinears of the form

$$[\bar{u}_{\lambda_i}(p_i,m_i)\ \Gamma\ u_{\lambda_j}(p_j,m_j)] \quad (8)$$

being Γ a set of γ matrices taking into account:

i.- the couplings with vector bosons

ii.- the propagator of virtual fermions

The external spin 1 massless gauge bosons [1] are described by polarization vectors $\epsilon_\lambda^\mu(k)$ verifying

$$\begin{aligned} \epsilon_\lambda \cdot k &= 0 & \epsilon_\lambda \cdot \epsilon_\lambda &= 0 \\ \epsilon_{-\lambda} &= \epsilon_\lambda^* & \epsilon_\lambda \cdot \epsilon_{-\lambda} &= 1 \end{aligned} \quad (9)$$

---
[1] Actually the only stable ones

in such a way that any objects verifying the above relations make an acceptable choice for $\epsilon_\lambda^\mu(k)$. For instance
$$\epsilon_\lambda^{\mu*} = N \ [\bar{u}_\lambda(k)\gamma^\mu u_\lambda(p)] \tag{10}$$
being

N a normalization factor (which we will compute later)

p^μ any 4-vector non-proportional to k^μ

Therefore an amplitude can be seen just as the product of a set of bilinears interconnected among them through Lorenz indices, namely
$$iM_i = \alpha \ [\bar{u}_1 \ \Gamma_A \ \gamma^\mu \ \Gamma_B \ u_2][\bar{u}_3 \ \Gamma_C \ \gamma_\mu \ \Gamma_D \ \gamma_\nu \ \Gamma_E \ u_4][\bar{u}_5 \ \Gamma_F \ \gamma^\nu \ \Gamma_G \ u_6] \tag{11}$$
where now the matrices Γ_i are basically products of

i.- chiral projectors $P_R = \frac{1+\gamma_5}{2}$ and $P_L = \frac{1-\gamma_5}{2}$

ii.- matrices $\slashed{p}$

Shifting the chiral projectors and substituting the matrices as
$$\slashed{p}_i = \left(\sum_\lambda [u_\lambda(p)\bar{u}_\lambda(p)]\right) - m_i \tag{12}$$
we end up essentially with bilinears of the following kind:
$$\begin{aligned}[\bar{u}_{\lambda_1}(p_1)(C_L P_L + C_R P_R)u_{\lambda_2}(p_2)] &\longrightarrow \text{Spinor products ("Y" functions)} \\ [\bar{u}_{\lambda_1}(p_1)\Gamma^\mu u_{\lambda_2}(p_2)][\bar{u}_{\lambda_3}(p_3)\Gamma'_\mu u_{\lambda_4}(p_4)] &\longrightarrow \text{"Z" functions}\end{aligned} \tag{13}$$
being
$$\Gamma^\mu \equiv \gamma^\mu(C_L P_L + C_R P_R) \tag{14}$$

3 Basic spinors

The standard way of building up spinors explicitly is solving Dirac's equation in the rest frame of the particle to obtain BASIC SPINORS.
$$\begin{aligned} m \ \gamma^0 \ u_i(0,m) &= m \ u_i(0,m) \\ m \ \gamma^0 \ v_i(0,m) &= -m \ v_i(0,m) \end{aligned} \tag{15}$$
taking an explicit representation of the γ matrices and afterwards boosting this solution to a frame with velocity $\vec{\beta} = \vec{p}/E$, namely
$$\begin{aligned} u_i(p,m) &= \frac{\slashed{p}+m}{\sqrt{2m(m+E)}} \ u_i(0,m) \\ v_i(p,m) &= \frac{-\slashed{p}+m}{\sqrt{2m(m+E)}} \ v_i(0,m) \end{aligned} \tag{16}$$
for $i = 1, 2$.

This spinors verify Dirac equation and obviously

$$\sum_i u_i(p)\bar{u}_i(p) = \slashed{p} + m$$
$$\sum_i v_i(p)\bar{v}_i(p) = \slashed{p} - m \qquad (17)$$

The problem is that this kind of procedure does not work for building spinors for massless particles. To do so we can instead start from the last equation and then for massless particles

$$\sum_i w_i(k_0)\bar{w}_i(k_0) = \slashed{k}_0 \qquad \text{being } k_0.k_0 = 0 \text{ (massless 4-momentum)} \qquad (18)$$

in such a way that defining the two spin states as the chiral ones (the most natural choice for massless particles) we can write

$$w_\lambda(k_0)\bar{w}_\lambda(k_0) = \frac{1+\lambda\gamma_5}{2}\slashed{k}_0 \qquad (19)$$

being λ the helicity $\lambda = \pm 1$. The main point is that such spinors can act as BASIC SPINORS too, that is, can be transformed to produce the spinor of a particle of momentum p in the following way:

$$u_\lambda(p,0) = \frac{\slashed{p}}{\eta}w_{-\lambda}(k_0) \qquad \text{being } \eta = \sqrt{2(p.k_0)} \qquad (20)$$

where u_λ has all the properties of a chiral spinor of helicity λ and the transformation applied can be regarded as a sort of boost. The result that it becomes singular for $p.k_0 = 0$ reflects the fact that no Lorenz transformation can "flip" the spin of a massless particle.

Moreover, we are enabled to produce also the spinor of a particle of momentum p and mass m in the following way:

$$u_\lambda(p,m) = \frac{\slashed{p}+m}{\eta}w_{-\lambda}(k_0) \qquad \text{being } \eta = \sqrt{2(p.k_0)} \text{ as before} \qquad (21)$$

where u_λ is a spinor for which λ denotes the two possible spin projections along

$$s^\mu = \frac{p^\mu}{m} - \frac{m}{(p.k_0)}k_0^\mu \qquad (22)$$

that is

$$u_\lambda(p,m)\bar{u}_\lambda(p,m) = \frac{1+\lambda\gamma_5\slashed{s}}{2}(\slashed{p}+m) \qquad (23)$$

Obviously now the transformation is not a boost but has a similar structure and produces the desired results. Thus it has to be considered as the natural extension of the massless boost to the case of massive particles.

Therefore $w_\lambda(k_0)$ can be regarded as the BASIC SPINORS we need to fix up the phase among all the spinors occurring in a process because we have a well defined method to build any spinor as a function of the same $w_\lambda(k_0)$. The defining properties of $w_\lambda(k_0)$ are

$$w_\lambda(k_0)\bar{w}_\lambda(k_0) = \frac{1+\lambda\gamma_5}{2}\slashed{k}_0 \qquad (24)$$

which defines λ as a chiral index and

$$w_\lambda(k_0) = \lambda \not{k}_1 w_{-\lambda}(k_0) \qquad (25)$$

which fixes up the phase between booth chiral states. k_0 and k_1 are any 4-vector satisfying:

$$\begin{aligned} k_0.k_0 &= 0 & &\text{massless vector} \\ k_1.k_1 &= -1 & &\text{space-like (rotation)} \\ k_0.k_1 &= 0 & &\text{orthogonality} \end{aligned} \qquad (26)$$

4 Spinor product

4.1 Massless case

For massless fermions with momenta p_1 and p_2 there are just two non-zero products

$$\begin{aligned} {}[\bar{u}_+(p_1)u_-(p_2)] &= S_+(p_1,p_2) \\ [\bar{u}_-(p_1)u_+(p_2)] &= S_-(p_1,p_2) = -[S_+(p_1,p_2)]^* \end{aligned} \qquad (27)$$

being

$$\begin{aligned} S_+(p_1,p_2) &= \frac{[\bar{w}_-(k_0)\not{p}_1\not{p}_2 w_+(k_0)]}{\eta_1 \eta_2} = \frac{[\bar{w}_-(k_0)\not{p}_1\not{p}_2\not{k}_1 w_-(k_0)]}{\eta_1 \eta_2} \\ &= \frac{Tr\{\frac{1-\gamma_5}{2}\not{k}_0\not{p}_1\not{p}_2\not{k}_1\}}{\eta_1 \eta_2} \\ &= 2\,\frac{(p_1.k_0)(p_2.k_1) - (p_1.k_1)(p_2.k_0) - i\epsilon(k_0,k_1,p_1,p_2)}{\eta_1 \eta_2} \end{aligned} \qquad (28)$$

As the values of k_0 and k_1 are arbitrary we can evaluate the previous expression taking a simple and suitable choice. For instance

$$\begin{aligned} k_0^\mu &= (1,1,0,0) \\ k_1^\mu &= (0,0,1,0) \end{aligned} \qquad (29)$$

turns out to be a good choice if the majority of the particles tend to go in the z direction. If we take $p_i^\mu = (E_i, p_i^x, p_i^y, p_i^z)$ then we can write

$$\begin{aligned} S_\lambda(p_1,p_2) &= \lambda\left((p_1^y + \lambda i p_1^z)\frac{\eta_2}{\eta_1} - (p_2^y + \lambda i p_2^z)\frac{\eta_1}{\eta_2}\right) \\ &= \lambda\left((p_1^y + \lambda i p_1^z)\sqrt{\frac{E_2 - p_2^x}{E_1 - p_1^x}} - (p_2^y + \lambda i p_2^z)\sqrt{\frac{E_1 - p_1^x}{E_2 - p_2^x}}\right) \end{aligned} \qquad (30)$$

where

$$\eta_i = \sqrt{2(p_i.k_0)} = \sqrt{2(E_i - p_i^x)} \qquad (31)$$

This expression is the basic building block of helicity amplitudes in the same way that scalar products build $|M|^2$. In fact

$$2(p_1.p_2) = |S(p_1,p_2)|^2 \qquad (32)$$

4.2 Massive case

For massive spinors the previous calculation can be carried out in the same way and leads to the same result, namely

$$[\bar{u}_+(p_1)u_-(p_2)] = S_+(p_1, p_2)$$
$$[\bar{u}_-(p_1)u_+(p_2)] = S_-(p_1, p_2) \tag{33}$$

being $S_\pm$ the same function obtained before, but now there are two additional non-vanishing spinor products:

$$[\bar{u}_+(p_1)u_+(p_2)] = Y_+(p_1, p_2)$$
$$[\bar{u}_-(p_1)u_-(p_2)] = Y_-(p_1, p_2) = Y_+(p_1, p_2) = Y(p_1, p_2) \tag{34}$$

being Y a function which can be computed as before using the definition properties of u_λ, namely

$$\begin{aligned} Y(p_1, p_2) &= \frac{[\bar{w}_+(k_0)(\not{p}_1 + m)(\not{p}_2 + m)w_+(k_0)]}{\eta_1 \eta_2} \\ &= m_1 \frac{Tr\{\frac{1+\gamma_5}{2}\not{k}_0\not{p}_2\}}{\eta_1 \eta_2} + m_2 \frac{Tr\{\frac{1+\gamma_5}{2}\not{k}_0\not{p}_1\}}{\eta_1 \eta_2} \\ &= m_1 \left(\frac{\eta_2}{\eta_1}\right) + m_2 \left(\frac{\eta_1}{\eta_2}\right) \end{aligned} \tag{35}$$

so that

$$[\bar{u}_\lambda(p_1)u_\lambda(p_1)] = Y(p_1, p_1) = 2 \ m_1 \tag{36}$$

as expected because of the spinor normalization used. In this case, the scalar product of the particle 4-momenta satisfies:

$$2(p_1.p_2) = |S(p_1, p_2)|^2 + m_1^2 \left(\frac{\eta_2}{\eta_1}\right)^2 + m_2^2 \left(\frac{\eta_1}{\eta_2}\right)^2 \tag{37}$$

5 Z functions

As said at the beginning, the links between fermionic lines shows up as functions of the form

$$Z(p_1, \lambda_1; p_2, \lambda_2; p_3, \lambda_3; p_4, \lambda_4) \equiv [\bar{u}_{\lambda_1}(p_1)\Gamma^\mu u_{\lambda_2}(p_2)][\bar{u}_{\lambda_3}(p_3)\Gamma'_\mu u_{\lambda_4}(p_4)] \tag{38}$$

being in the general case

$$\Gamma^\mu \equiv \gamma^\mu(C_L P_L + C_R P_R) \qquad P_L = \left(\frac{1-\gamma_5}{2}\right) \qquad P_R = \left(\frac{1+\gamma_5}{2}\right) \tag{39}$$

We will see now how these functions can be computed.

5.1 Massless case

Lets realize that for chiral spinors the effect of chiral projectors P_L and P_R can be trivially taken into account. It is also trivial to demonstrate that in this case

$$[\bar{u}_\lambda(p_1)\gamma^\mu u_\lambda(p_2)] \; \gamma_\mu = 2 \; u_\lambda(p_2)\bar{u}_\lambda(p_1) + 2 \; u_{-\lambda}(p_1)\bar{u}_{-\lambda}(p_2) \qquad (40)$$

which is the so-called Chisholm identity. Therefore, as the massless spinors we have are chiral the helicities should necessarily verify $\lambda_1 = \lambda_2 = \lambda$ and $\lambda_3 = \lambda_4 = \lambda'$ and therefore, we can trivially write

$$\begin{aligned}
Z(\lambda, p_1, p_2; \lambda', p_3, p_4) &\equiv [\bar{u}_\lambda(p_1)\gamma^\mu u_\lambda(p_2)][\bar{u}_{\lambda'}(p_3)\gamma_\mu u_{\lambda'}(p_4)] \\
&= 2 \; [\bar{u}_{\lambda'}(p_3)u_\lambda(p_2)][\bar{u}_\lambda(p_1)u_{\lambda'}(p_4)] \\
&+ 2 \; [\bar{u}_{\lambda'}(p_3)u_{-\lambda}(p_1)][\bar{u}_{-\lambda}(p_2)u_{\lambda'}(p_4)]
\end{aligned} \qquad (41)$$

so that we finally have

$$Z(\lambda, p_1, p_2; \lambda', p_3, p_4) = -2 \; \delta_{\lambda\lambda'} S_\lambda(p_1, p_3) S_{-\lambda}(p_2, p_4) - 2 \; \delta_{-\lambda\lambda'} S_\lambda(p_1, p_4) S_{-\lambda}(p_2, p_3) \qquad (42)$$

5.2 Massive case

In the massive case, spinors are not chiral and therefore we do not have any trivial simplification of the P_L and P_R projectors. Moreover Chisholm identity does not hold directly and therefore we need to proceed in a different manner.

One possible way of attacking the calculation of Z is decomposing the u spinors as follows:

$$\begin{aligned}
u_\lambda(p, m) = \frac{\not{p} + m}{\eta} w_{-\lambda}(k_0) &= \frac{\not{p}}{\eta} w_{-\lambda}(k_0) + \frac{m}{\eta} w_{-\lambda}(k_0) \\
&= w_\lambda(p) + \mu \; w_{-\lambda}(k_0)
\end{aligned} \qquad (43)$$

being

$$w_\lambda(p) \equiv \frac{\not{p}}{\eta} w_{-\lambda}(k_0) \qquad \mu \equiv \frac{m}{\eta} \qquad (44)$$

$w_\lambda(p)$ does not verify Dirac equation but is certainly a chiral object by definition. Additionally it is easy to demonstrate that

$$[\bar{w}_\lambda(p_1) w_\lambda(p_2)] = S_\lambda(p_1, p_2) \qquad (45)$$

Moreover, if we define

$$\tilde{Z}(\lambda, q_1, q_2; \lambda', q_3, q_4) \equiv F^\mu(\lambda, q_1, q_2) F_\mu(\lambda', q_3, q_4) \qquad (46)$$

where

$$F^\mu(\lambda, q_1, q_2) \equiv [\bar{w}_\lambda(q_1)\gamma_\mu w_\lambda(q_2)] \qquad (47)$$

being $q_i \equiv p_i$ or k_0 where p_i denotes a particle 4-momentum, it is a lengthy but straightforward task to prove that $\tilde{Z}$ can be computed easily if at the level of F^μ we make explicit the difference between $w_\lambda(p_i)$ and $w_\lambda(k_0)$ by using in all the occurrences the substitution rule:

$$\begin{aligned}
F^\mu(\lambda, k_0, q_j) &= -f^\mu(-\lambda, q_j, k_0) \\
F^\mu(\lambda, p_i, q_j) &= f^\mu(\lambda, p_i, q_j)
\end{aligned} \qquad (48)$$

and applying afterwards the index elimination property (Chisholm identity like)

$$f^\mu(\lambda, q_1, q_2) \; f_\mu(\lambda', q_3, q_4) =$$
$$= -2 \; \delta_{\lambda\lambda'} SM_\lambda(q_1, q_3) SM_{-\lambda}(q_2, q_4) - 2 \; \delta_{-\lambda\lambda'} SM_\lambda(q_1, q_4) SM_{-\lambda}(q_2, q_3) \quad (49)$$

being the quantities $SM_\lambda(q_i, q_j)$ a natural extension of the spinor products $S_\lambda(p_i, p_j)$ defined as:

$$\begin{aligned} SM_\lambda(p_i, p_j) &= -SM_\lambda(p_j, p_i) = [\bar{u}_\lambda(p_i) u_{-\lambda}(p_j)] = S_\lambda(p_i, p_j) \\ SM_\lambda(p_i, k_0) &= -SM_\lambda(k_0, p_i) = \eta_i \end{aligned} \quad (50)$$

that is, $SM_\lambda(q_i, q_j)$ is an antisymmetric function of q_i and q_j and therefore:

$$SM_\lambda(k_0, k_0) = SM_\lambda(p_i, p_i) = 0 \quad (51)$$

After applying these rules the calculation of all the possible values of the Z function is straightforward, and defining

$$Z(\lambda_i, \lambda_j, \lambda_k, \lambda_l) \equiv Z(p_i, \lambda_i; p_j, \lambda_j; p_k, \lambda_k; p_l, \lambda_l) \quad (52)$$

we find the following expressions for half of the possible helicity configurations:

$$\begin{aligned} Z(+,+,+,+) &= -2\{SM_+(p_3, p_1) SM_-(p_4, p_2) C'_R C_R - \mu_1 \mu_2 \eta_3 \eta_4 C'_R C_L \\ &\quad - \eta_1 \eta_2 \mu_3 \mu_4 C'_L C_R\} \\ Z(+,+,+,-) &= -2\eta_2 C_R \{SM_+(p_4, p_1) \mu_3 C'_L - SM_+(p_3, p_1) \mu_4 C'_R\} \\ Z(+,+,-,+) &= -2\eta_1 C_R \{SM_-(p_2, p_3) \mu_4 C'_L - SM_-(p_2, p_4) \mu_3 C'_R\} \\ Z(+,+,-,-) &= -2\{SM_+(p_1, p_4) SM_-(p_2, p_3) C'_L C_R - \mu_1 \mu_2 \eta_3 \eta_4 C'_L C_L \\ &\quad - \eta_1 \eta_2 \mu_3 \mu_4 C'_R C_R\} \\ Z(+,-,+,+) &= -2\eta_4 C'_R \{SM_+(p_3, p_1) \mu_2 C_R - SM_+(p_3, p_2) \mu_1 C_L\} \\ Z(+,-,+,-) &= 0 \\ Z(+,-,-,+) &= -2\{\mu_1 \mu_4 \eta_2 \eta_3 C'_L C_L + \mu_2 \mu_3 \eta_1 \eta_4 C'_R C_R - \mu_2 \mu_4 \eta_1 \eta_3 C'_L C_R \\ &\quad - \mu_1 \mu_3 \eta_2 \eta_4 C'_R C_L\} \\ Z(+,-,-,-) &= -2\eta_3 C'_L \{SM_+(p_2, p_4) \mu_1 C_L - SM_+(p_1, p_4) \mu_2 C_R\} \end{aligned}$$
$$(53)$$

where C_L and C_R are the left and right handed couplings of the first bilinear and C'_L and C'_R are the ones of the second bilinear. The rest can be obtained by exchanging $+ \leftrightarrow -$ and $L \leftrightarrow R$.

6 Calculational procedure

The actual steps needed to perform a calculation using the helicity amplitudes method are:

(1) draw all the Feynman diagrams you want to compute to figure out how many topologically different you have. Obviously gauge invariant subsets have to be complete.

(2) write down the amplitude for one representative of each topology. In order to allow an easy transformation of such amplitude to obtain the one from any diagram topologically equivalent is important to use standard Feynman rules plus the following prescriptions:

i.- take
$$m_i = \begin{cases} m & \text{for particles} \\ -m & \text{for antiparticles} \end{cases} \quad (54)$$

and then just substitute every
$$\begin{aligned} v_i &\to u_i \\ \bar{v}_i &\to \bar{u}_i \end{aligned} \quad (55)$$

ii.- write each 4-momentum showing up explicitly in the amplitude absorbing its sign in the way
$$b_i p_i \quad (56)$$

where
$$b_i = \begin{cases} 1 & \text{if } p_i \text{ is incoming} \\ -1 & \text{if } p_i \text{ is outgoing} \end{cases} \quad (57)$$

these two rules allow absorbing the characteristics of *particle* ↔ *antiparticle* and *incoming* ↔ *outgoing* of the specific amplitude studied, in the masses m_i and the direction parameters b_i in such a way that the obtained amplitude corresponds to a diagram in which the only information relevant in each external leg is just one label. Therefore, we have build up the GENERIC AMPLITUDE for the diagram topology studied. Any diagram topologically equivalent to the one studied can be obtained from the generic amplitude just by an overall permutation of particle labels and helicities (and occasionally conjugation). This would be more clearly seen in the examples discussed later.

(3) express the generic amplitude in terms of Z and Y functions (massive case) or directly in terms of *spinor products* (massless case) by using the two following substitutions:

i.- if you have external massless gauge bosons:
$$\epsilon_\lambda^{\mu *} = N \ [\bar{u}_\lambda(k)\gamma^\mu u_\lambda(p)] \quad (58)$$

being
$$\begin{aligned} N &= (2|SM_\lambda(k,p)|^2)^{-1/2} \text{ the normalization} \\ p &= \text{any 4-momentum but k} \end{aligned}$$

This substitution allows you to transform the vector boson polarization vector in a bilinear. The existing freedom in choosing p corresponds to fixing the gauge and therefore different p correspond to different gauges. As the final result of a calculation must be gauge-independent, we must obtain the same result regardless on the election of p. The fact that p is actually just a simple label, allows to easily check for the consistency of the calculation.

ii.- if you have propagators in a fermionic line:

$$\not{p}_i = \{\sum_\lambda u_\lambda(p_i)\bar{u}_\lambda(p_i)\} - m_i \tag{59}$$

regardless on whether p_i is a fermion or a boson 4-momentum

(4) In the case of massless particles you can obtain rather compact expressions by just adding analytically all the amplitudes for each helicity configuration λ_i, whereas in the case of massive particles the complexity of the Z and Y functions strongly suggests using directly a computer program in which the generic amplitudes are written down as callable functions of the particle labels and helicities to proceed to a numerical calculation.

7 Monte Carlo generators

When using this method to implement a Monte Carlo event generator the general way of proceeding is the following:

(1) Generate a phase space point with a suitable distribution density mapping as much as possible the relevant $|M|^2$ peaks

(2) Fill the arrays $SM(i,j)$ (and $Y(i,j)$ if needed) where the indices i and j run over all particle labels and also k_0

(3) Take a helicity configuration λ_i and add the contribution of all the diagrams by using the generic amplitude functions which in turn are written in terms of functions Z (this is just a numerical complex number calculation)

(4) For this helicity configuration $M(\lambda_i)$ is just a complex number and $|M(\lambda_i)|^2$ is proportional to the probability of this configuration in the generated phase space point.

(5) Add (or average) over the desired helicity configurations and then you will have your matrix element squared

8 Example: $e^-(p_1) + e^+(p_2) \to e^-(p_3) + e^+(p_4) + \gamma(k)$

For this process, at the tree level in the Electroweak theory there are 16 possible diagrams (fig. 1) and in some detection configurations it is very important keeping the electron mass, namely collinear photon or outgoing electron (positron) parallel to the incoming one. Using the standard method the calculation of this process shows the following problems:

(1) there are $16 \times 17/2 = 136$ terms in $|M|^2$ which have to be computed

(2) keeping masses produces very long expressions which in practice can be handled only by using an algebraic computer language such as REDUCE for instance

(3) during the calculation of $|M|^2$ in phase space points giving the bulk of the contribution in collinear configurations you have numerical unstability even using FORTRAN double precision variables due to strong cancellations among different diagram contributions

Figure 1: Feynman diagrams of the process $e^-(p_1) + e^+(p_2) \to e^-(p_3) + e^+(p_4) + \gamma(k)$ at the tree level

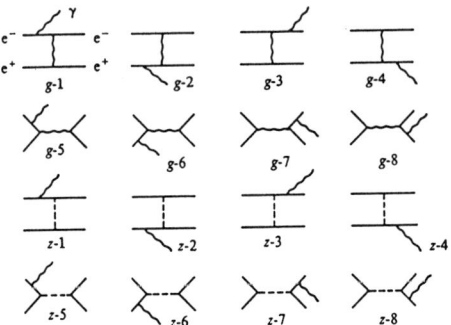

These problems vanish using the helicity amplitudes method as we will see now. Looking at the diagrams of figure 1 it is clear that all of them have exactly the same topology. Therefore it is enough to build one generic amplitude for the calculation of all the diagrams. Lets take for instance diagram $g-1$ (or $z-1$) and write down the amplitude using standard Feynman rules (fig.2):

$$iM_1 = \frac{ie^3 T_1(p_i, \lambda_i)}{\{(p_1-k)^2 - m_e^2\} F((p_2-p_4)^2)} \quad (60)$$

being

$$T_1(p_i, \lambda_i) = [\bar{u}(p_3, \lambda_3)\Gamma^\mu(\not{p}_1 - \not{k} + m_e)\not{f}^*(k, \lambda_k)u(p_1, \lambda_1)][\bar{v}(p_2, \lambda_2)\Gamma_\mu v(p_4, \lambda_4)] \quad (61)$$

where

Figure 2: g-1 diagram

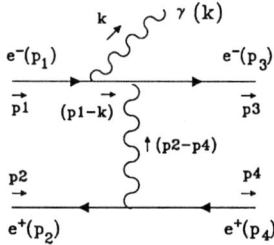

Figure 3: Generic diagram

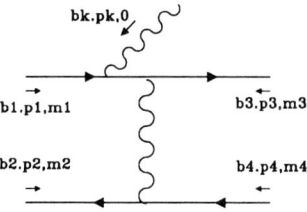

$$\Gamma^\mu = \begin{cases} \gamma^\mu & \text{photon} \\ \gamma^\mu(C_L P_L + C_R P_R) & Z^0 \end{cases} \quad (62)$$

and

$$F(y) = \begin{cases} y & \text{photon} \\ (y - M_Z)^2 + i M_Z \Gamma_Z & Z^0 \end{cases} \quad (63)$$

As we are going to use this diagram as the generic one, we have to re-write its amplitude by using the rules recommended:

$$iM_1 = \frac{ie^3 T_1(p_i, \lambda_i)}{\{(b_1 p_1 + b_k k)^2 - m_e^2\} F((b_2 p_2 + b_4 p_4)^2)} \quad (64)$$

being

$$T_1(p_i, \lambda_i) = [\bar{u}(p_3, \lambda_3)\Gamma^\mu(b_1(\not{p}_1 + m_1) + b_k\not{k})\not{\epsilon}^*(k, \lambda_k)u(p_1, \lambda_1)][\bar{u}(p_2, \lambda_2)\Gamma_\mu u(p_4, \lambda_4)] \quad (65)$$

taking

$$\begin{aligned} m_1 &= m_3 = m_e \\ m_2 &= m_4 = -m_e \end{aligned} \quad (66)$$

and

$$\begin{aligned} b_1 &= b_2 = 1 \\ b_3 &= b_4 = b_k = -1 \end{aligned} \quad (67)$$

What we get is an amplitude which in fact corresponds to the diagram of figure 3 where in b_i, p_i and m_i i is just a label which can be used for any particle in the process. Then it is trivial to see that for instance, for diagrams $g - 5$ ($z - 5$) we have (fig.4):

$$iM_5 = -iM_1(2 \leftrightarrow 3) \quad (68)$$

Figure 4: g-5 diagram

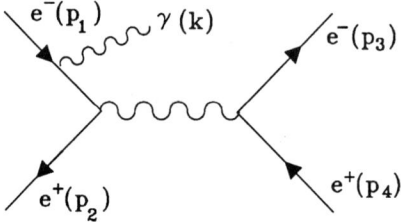

Figure 5: g-2 diagram

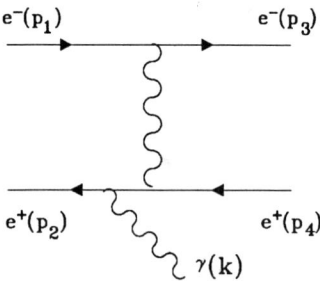

where the $(-)$ sign comes from the t channel to s channel relative sign. In the same way, for diagrams $g-2$ ($z-2$) we have (fig.5):

$$iM_2 = -(iM_1)^*(1 \leftrightarrow 2, 3 \leftrightarrow 4; \lambda_k \leftrightarrow -\lambda_k) \tag{69}$$

where the conjugation reverses the fermionic arrows and changes the photon helicity (besides producing an overall $(-)$ sign due to an odd number of couplings).

Therefore, as all the amplitudes can be computed using M_1 (and permutations of labels and conjugations) we just have to compute M_1. To do so, following the rules explained above, lets rewrite the photon polarization vector as

$$\epsilon^{\mu*}(k, \lambda_k) = N \ \bar{u}(k, \lambda_k)\gamma^\mu u(p, \lambda_k) \tag{70}$$

being

$$N = \frac{1}{\sqrt{2|SM_\lambda(k,p)|^2}}$$
$$p \quad \text{any 4-momentum different from } k \tag{71}$$

and lets substitute

$$\begin{aligned} \not{p}_1 + m_1 &= \sum_\lambda u(p_1, \lambda)\bar{u}(p_1, \lambda) \\ \not{k} &= \sum_\lambda u(k, \lambda)\bar{u}(k, \lambda) \end{aligned} \tag{72}$$

and then we can write

$$\begin{aligned} T_1(p_i, \lambda_i) = N \sum_\lambda (\ b_1 \ &[\bar{u}_{\lambda_3}(p_3)\Gamma^\mu u_\lambda(p_1)][\bar{u}_{\lambda_2}(p_2)\Gamma_\mu u_{\lambda_4}(p_4)] \\ &[\bar{u}_\lambda(p_1)\gamma^\mu u_{\lambda_1}(p_1)][\bar{u}_{\lambda_k}(k)\gamma_\mu u_{\lambda_k}(p)] \\ -b_k \ &[\bar{u}_{\lambda_3}(p_3)\Gamma^\mu u_\lambda(k)][\bar{u}_{\lambda_2}(p_2)\Gamma_\mu u_{\lambda_4}(p_4)] \\ &[\bar{u}_\lambda(k)\gamma^\mu u_{\lambda_1}(p_1)][\bar{u}_{\lambda_k}(k)\gamma_\mu u_{\lambda_k}(p)]) \end{aligned} \tag{73}$$

so that in terms of Z functions we simply have

$$\begin{aligned} T_1(p_i, \lambda_i) = N \sum_\lambda (b_1 \quad & Z(p_3, \lambda_3; p_1, \lambda; p_2, \lambda_2; p_4, \lambda_4; C_L, C_R, C_L, C_R) \\ & Z(p_1, \lambda; p_1, \lambda_1; k, \lambda_k; p_1, \lambda_k; 1, 1, 1, 1) \\ -b_k \quad & Z(p_3, \lambda_3; k, \lambda; p_2, \lambda_2; p_4, \lambda_4; C_L, C_R, C_L, C_R) \\ & Z(p_3, \lambda; p_1, \lambda_1; k, \lambda_k; p_1, \lambda_k; 1, 1, 1, 1)) \end{aligned} \tag{74}$$

being $C_L = C_R = 1$ for photonic diagrams

Given a particular set of 4-momenta and helicities p_i, λ_i one can evaluate the Z functions for the generic diagram and, with the appropriate permutations and conjugations the actual amplitudes for all the diagrams contributing to the process. Finally

$$|i\bar{M}|^2 = \frac{1}{4}\sum_\lambda |\sum_j (iM_j)|^2 \tag{75}$$

where j runs over diagrams and λ runs over helicities.

Choosing different possibilities for p (that is, choosing labels $1, .., 4$) the consistency of the calculation is trivially checked.

9 Conclusions

We have sketched a powerful method for computing tree level Feynman diagrams which we have already used in many processes [3-4]. For instance

$$\begin{aligned} e^+e^- &\to e^+e^-\gamma \\ e^+e^- &\to e^+e^-\gamma\gamma \\ e^+e^- &\to \nu\bar{\nu}\gamma \end{aligned} \tag{76}$$

The basic tool are just few Fortran functions computing spinor products and Z functions which are available from the authors on request. The only actual limitation on the complexity of the problem attacked are your available computer resources.

References

1. P. de Causmaker et al., Nucl. Phys. **B206**,53 (1982)
   G.R. Ferrar and F. Neri, Phys. Lett. **130B**,109 (1983)
   S. Jadach and Z. Was, Acta Phys. Pol **B15**,1151 (1984)
2. F.A. Berends, P.H. Daverveldt and R. Kleiss, Nucl. Phys. **B253**,441 (1985)
   R. Kleiss and W.J. Stirling, Nucl. Phys.**B262**,235 (1985)
   P.H. Daverveldt, Ph. D. Thesis, Leiden Univ. (1985)
3. C. Mana and M. Martinez, Nucl. Phys.**B287**,601 (1987)
4. C. Mana, M. Martinez and F. Cornet, DESY prep. 86/114 (1986)
   M. Martinez and R. Miquel, UAB prep. UAB-LFAE-87-01 (1987)
   F.A Berends, G.H.J.Burgers, C. Mana, M. Martinez and W.L.Van Neerven, Nucl. Phys. **B301**,538 (1988)
   R. Miquel, Ph. D. Thesis, Universitat Autonoma de Barcelona (1989)

Noodles and Stars allow a precise and efficient calculation of the Z–line shape and the polarization asymmetry

CARL JUNG-CHOON IM

Stanford Linear Accelerator Center
Stanford University, Stanford, California 94309

and

Department of Physics, Stanford University, Stanford, CA 94305

ABSTRACT

We give a pedagogical introduction to the star functions and the Noodle method. The Z-line shape and the polarization asymmetry at SLC/LEP can be evaluated elegantly and efficiently using the star functions and the Noodle method.

Presented at the Workshop on QED Structure Functions, Ann Arbor, Michigan, May 22-25, 1989.

© 1990 American Institute of Physics

1. Introduction

Measurements at SLC/LEP must be compared to accurate theoretical predictions of the standard Glashow-Weinberg-Salam theory before the electro-weak parameters can be determined accurately. Given the experimental goal of about 1% accuracy in the Z-line shape and the polarization asymmetry measurements, one has to include radiative corrections to tree-level predictions. The issue here is how to carry out this rather complicated calculation in a convenient and efficient way. This paper is divided into five sections. Section 1 introduces the *starred* functions and shows how these quantities greatly simplify expressions for renormalized matrix elements in the standard electro-weak theory. Section 2 discusses motivations for the Noodle method by critically examining traditional Monte Carlo methods, and section 3 introduces the Noodle method. Section 4 contains a detailed example of an application of the Noodle method, and, finally, section 5 compares the Noodle method to other Monte Carlo methods. Detailed presentation of our results have been presented elsewhere.[1]

2. The Star Functions

Let us begin by considering the tree-level neutral-current amplitude:

$$M_{NC} = \frac{e^2 Q Q'}{q^2} + \frac{e^2(I_3 - s_\theta^2 Q)(I_3' - s_\theta^2 Q')}{s_\theta^2 c_\theta^2} \frac{1}{q^2 - M_Z^2},$$

where the unprimed quantities refer to the initial legs of the annihilation diagram, and the primed quantities refer to the final legs. A precise prediction of the neutral current amplitude requires radiative corrections to the tree level graphs. Owing to the renormalizability of the Standard Model, these radiative corrections merely change the values of the bare parameters, converting each of these fixed coupling constants to a universal *starred* function that runs with q^2. Of course, the presence of M_{Z*} in the expression for M_{NC} does not mean that the observed mass of the Z runs with q^2; the physical mass M_Z of the Z is defined as the pole of the full Z-propagator and satisfies the equation $M_Z^2 - M_{Z*}^2(M_Z^2) = 0$. So far, the renormalization program using the *starred* quantities seems identical to any other renormalization procedure. The only crucial difference is that we do not make the point of renormalization for each running coupling constant explicit. For example, instead of writing the running electric charge as

$$\alpha(q^2) = \frac{\alpha(M^2)}{1 - \frac{\alpha(M^2)}{3\pi}\log(\frac{q^2}{M^2})},$$

we simply write it as

$$\alpha_*(q^2).$$

When this simple trick is applied to the $SU(2) \times U(1)$ theory, a considerable simplification occurs in the calculation of the 1-loop matrix elements. The result for the truncated diagram for $e^+e^- \to f^+f^-$ turns out to be

$$M_{NC,\text{1-loop}} = \frac{e_*^2 Q Q'}{q^2} + \frac{e_*^2(I_3 - s_{\theta*}^2 Q)(I_3' - s_{\theta*}^2 Q')}{s_{\theta*}^2 c_{\theta*}^2} \frac{1}{q^2 - M_{Z*}^2}.$$

Figure 1 shows how $s_{\theta*}^2$ runs with q^2. The absolute height of the graph has to be fixed by an experiment. Fixing the height of the graph is exactly the momentum subtraction procedure. For practitioners, however, the *star* scheme is clearly better than the momentum subtraction scheme for two reasons. First, renormalized matrix elements simplify considerably when they are expressed in terms of the *starred* quantities. Second, since these *starred* quantities are process-independent, as all renormalized quantities are, once these *starred* quantities are computed, the problem of computing a renormalized matrix element reduces to that of expressing the matrix element in terms of the *starred* quantities.

The only remaining class of radiative corrections are the initial and the final leg Bremsstrahlung diagrams. Since the observables that we are interested in are the Z-line shape and the polarization asymmetry, we are only interested in the final state fermions and the indirect effects of the initial and the final state Bremsstrahlung radiation on them. We therefore follow Trentadue and Nicrosini[2] and Kuraev and Fadin[3] and ignore any explicit consideration of photon dynamics and write

$$\sigma(S) = \int_0^1 dx_+ \int_0^1 dx_- D(S, x_+) D(S, x_-) \sigma_0(x_+ x_- S) .$$

This has the following physical interpretation: The e^+ beam loses some energy due to Bremsstrahlung, and its energy is reduced to a fraction x_+ of its original value with probability $D(S, x_+)$. Likewise, the e_- beam energy is reduced to a fraction x_- with probability $D(S, x_-)$. Then, at the interaction point, the beam energies are spread out according to the distributions $D(S, x_+)$ and $D(S, x_-)$, and the actual scattering occurs with an effective $q^2 = x_+ x_- S$. Although this physical picture is rather rough, the expression for $\sigma(S)$ include the same contribution to the total cross section as a complete α^2 calculation would and are adequate for evaluating scattering amplitudes without any phase-space cut.

We, however, would like to apply phase-space cuts. Consequently, we would like to find a similar expression for the differential cross section $\frac{d\sigma(S)}{d\Omega}$ that implicitly includes the details of the Bremsstrahlung photons. Here we made the following ansatz:

$$\frac{d\sigma(s)}{d\Omega_{lab}} = \int dx_+ dx_- D(S,x_+)D(S,x_-)\frac{d\sigma_0(S)}{d\Omega_{cm}}\frac{d\Omega_{cm}}{d\Omega_{lab}}.$$

The discrepancy between this ansatz and more complete treatments of Bremsstrahlung is found to be negligible.[1]

3. Integration of the Differential Cross Section

In order to obtain a physical prediction, one still has to integrate the differential cross section with respect to a given phase-space cut. Since a given phase-space cut may be quite complicated, a given cut may contain an acollinearity cut, for example, use of Monte Carlo integration method is essential. Traditional Monte Carlo methods[4], however, are needlessly inefficient for integrating physical differential cross sections and some must be supplied with various analytic formulae in addition to the expression for the integrand. The Noodle method eliminates these complications. Let us see in detail what these complications are.

Consider the integral

$$I = \int_0^1 dx \int_0^1 dy \int_0^1 dz\ f(x,y,z) \tag{1}$$

The Monte Carlo method is based on the following equation:

$$\frac{Vol(S)}{N}\sum_{i=1}^{i=N} f(x_i) = \int_S f \pm \sqrt{\frac{Var(f)}{N}} \tag{2}$$

where each x_i is a randomly and uniformly chosen point in the space S.

Consequently, to estimate I accurately for small values of N, I is first rewritten as

$$\int_0^1 dx \int_0^1 dy \int_0^{G(x,y)} dG \, \frac{f(x,y,z(G))}{g(x,y,z(G))}, \text{ where } G(x,y) = \int_0^1 dz \, g(x,y,z) \quad (3)$$

with the corresponding Monte Carlo sum

$$\frac{Vol(\tilde{S})}{N} \sum_{i=1}^{i=N} \frac{f(x_i, y_i, z(G_i))}{g(x_i, y_i, z(G_i))} \pm \sqrt{\frac{Var(\frac{f}{g})}{N}} \quad (4)$$

where each (x_i, y_i, G_i) is chosen randomly and uniformly from $\tilde{S}(figure\ 2)$. Then g is chosen so that $Var(\frac{f}{g})$ is small.

For this method to work, g has to satisfy two mutually antagonistic criteria. It is crucial that g has a rather complicated form so that it closely approximates f. Yet at the same time, g has to have a simple form so that it can be integrated analytically. Choosing such a g is an art form and is the first complication that one encounters when one tries to apply the method.

Now, suppose that such a g has somehow been constructed. The Monte Carlo sum is then performed by first generating a $(x_i, y_i, z(G_i))$ triplet, then by evaluating the summand. The usual algorithm for generating (x_i, y_i, G_i)s is to generate a random point inside the smallest box containing $\tilde{S}$ and reject the point if it lies outside $\tilde{S}$. This algorithm becomes inefficient when $\tilde{S}$ has $spikes(figure\ 3)$. Moreover, after a (x_i, y_i, G_i)-triplet has been generated, it is not easy to invert

$$G_i = \int_0^{z(G_i)} dz \, g(x, y, z) \quad (5)$$

to generate the corresponding (x_i, y_i, z_i)-triplet.

4. The Noodle Method

For pedagogical reasons, we apply the Noodle method to the simplest possible case, namely that of integration over a cubical region, and once again, consider the problem of evaluating the integral

$$I = \int_0^1 dx \int_0^1 dy \int_0^1 dz \, f(x,y,z) \, .$$

In the Noodle method, S is partitioned into regions in each of which f is roughly constant. Then, for each region, g is defined as constant and equals f at some point in the region. One such choice of g is

$$g(x,y,z) = f(\frac{i}{M}, \frac{j}{M}, 0); \frac{i}{M} < x < \frac{i+1}{M}, \frac{j}{M} < y < \frac{j+1}{M}, 0 < z < 1 \quad (6)$$

for some M and all $0 \leq i,j < M$. Again, for pedagogical reasons, we do not partition S along the z-direction; if, however, one needed to cut along the z-direction, one could easily do so. The corresponding $\tilde{S}$, without cuts along the z-direction is shown on *figure* 4a. The partition of S induces a partition of $\tilde{S}$ via the change of coordinates z into G:

$$\frac{i}{M} < x < \frac{i+1}{M}, \frac{j}{M} < y < \frac{j+1}{M}, 0 < G < G_{i,j}; G_{i,j} = \frac{1}{M^2} g(\frac{i}{M}, \frac{j}{M}).$$

Each of these sets is called a noodle.

By construction, g approximates f closely, and, since g is constant over each noodle, g can be integrated exactly. This solves the problem of finding a suitable g.

Moreover, defining g locally constant allows a fast and efficient generation of the $(x_i, y_i, z(G_i))$-triplets. In order to see this, notice that **uniform distribution in $\tilde{S}$ is equivalent to distribution g in S**. Consider $\rho(0 < x - x_0 < \delta x, 0 < y - y_0 < \delta y, 0 < G - G_0 < \delta G) = \delta x \delta y \delta G$, which is the uniform distribution in $\tilde{S}$. Since $G - G_0 = g(x, y, z_0)(z - z_0)$, we have $\rho(0 < x - x_0 < \delta x, 0 < y - y_0 < \delta y, 0 < z - z_0 < \frac{\delta G}{g}) = \delta x \delta y \delta G$. Since δG is small but otherwise arbitrary, replacing δG by $g \delta z$ yields the desired result. Consequently, we may generate (x_i, y_i, z_i)-triplets with distribution g by uniformly generating (x_i, y_i, G_i)-triplets, and then finding the corresponding (x_i, y_i, z_i)-triplets: First, we arbitrarily order noodles in a line to make a long noodle called the meganoodle:

NOODLE(0)= 0

NOODLE(i)= NOODLE($i-1$)+volume of the ith noodle

Length of the meganoodle = NOODLE(last noodle)

This is shown in *figure* 4b. Then, we generate a random number between 0 and the length of the meganoodle. This picks out a noodle in $\tilde{S}$. Let us say that this noodle is defined by

$$\frac{i}{M} < x < \frac{i+1}{M}, \frac{j}{M} < y < \frac{j+1}{M}, 0 < G < G_{i,j}; G_{i,j} = \frac{1}{M^2} g(\frac{i}{M}, \frac{j}{M})$$

for some i and j. But this is equivalent to choosing a box

$$\frac{i}{M} < x < \frac{i+1}{M}, \frac{j}{M} < y < \frac{j+1}{M}, 0 < z < 1$$

in S with probability g. Since the x and the y-coordinates of the noodle and the corresponding box are the same, once a noodle is chosen, finding the corresponding box in S is immediate. Finally, since g is constant within the box, any point within the set is equally likely to be chosen. Accordingly, we simply pick a random point (x, y, z) within the box. Hence we generate points in S with probability g with only a binary search algorithm and three random number generations. In this simple minded example, the method rejects no point once it is generated and, hence, is quite efficient.

For this method to work better than the traditional Monte Carlo method, it is crucial to partition S into regions so that f is indeed approximately constant over each region. One way to achieve this is to choose M large enough so that, over S, $\frac{f}{g}$ takes values in a small interval around 1, say $(1 - \epsilon, 1 + \epsilon)$. Consequently, $\frac{1}{M}$ in our example is the maximum scale in $x - y$ space over which the fractional change in f is less than ϵ. Thus, if the graph of f has a tall but narrow spike at a point but is otherwise flat, the necessary value of M dictated by the width of the spike is clearly an overestimate for the rest of the region of integration. It would be nice if the integration routine automatically partitioned the space of integration in an intelligent way.

Many integration routines such as VEGAS[5] and DIVONNE2[6] provide a solution by randomly sampling the range of the integrand and, based on this sampling, further partitioning the region in which the integrand has an unacceptable fluctuation. This solution is adequate for a general integrand that is known to be free of isolated sharp spikes. If, on the other hand, if the spikes are there, then even a large number of sampling does not guarantee that all of the spikes are detected. These "hidden" spikes manifest themselves in large fluctuations of the weights. Consequently, for a given accuracy, a relatively long time is spent on partitioning the region of integration. In fact, the general solution of finding an efficient partition of the space of integration seems quite hard.

Fortunately, for many integrations in high energy physics involve integration of functions whose singularities are quite well known. For this class of integrands, the criterion for partitioning a region can be given by our knowledge of general features of the integrand, rather than by a random and uniform sampling of the range of the integrand. The following is the algorithm that the Noodle method uses:

start with a small number of noodles;
procedure bakery;
    index:= 1;
    while (index $\leq$ length of the noodle) do
    begin
        while noodle_is_wild(indexth noodle) do
        begin
            partition the noodle into smaller noodles;
            let one of the noodles be the indexth noodle;
            put the rest of the new noodles at the end of the ordering;
            increase the number of noodles by the number of new noodles;
        end;
        noodle(index):= noodle(index-1)+ volume of the indexth noodle;
        index:= index+1;
    end;
end;
function noodle_is_wild(noodle):boolean;
    noodle_is_wild:= (f fluctuates unacceptably inside the noodle)

end;

User has to define the *wild* function. At first sight, it might seem that we have merely replaced the problem of coming up with the approximate integrand by the problem of supplying the wild function, but for a given integrand, the corresponding wild function is much easier to define than a good approximant of the integrand.

5. An Example

We illustrate the method by using the noodle method to compute the $e^+e^- \to \mu^+\mu^-$ cross section after the effects of the initial state bremsstrahlung have been taken into account. The cross section is given by[2]

$$\sigma = \int_{x_{cut}}^{1} dx_+ \int_{x_{cut}}^{1} dx_- \int d\Omega \; D(x_+)D(x_-)\frac{d\sigma_0}{d\Omega_{CM}}(\theta_{cm}, x_+x_-S)\Theta(p_{\mu^+}^{lab}, p_{\mu^-}^{lab}) \;, \quad (7)$$

where σ_0 is the uncorrected cross-section, and $\Theta(x) = 1$ if x is inside the phase space cut and 0, otherwise. For the sake of simplicity, we consider the case of no phase space cut.

In order to integrate this numerically, it is necessary to regulate the poles of D by a suitable change of variables. The resulting integral is

$$\int_0^{P_{cut}} dP_+ \int_0^{P_{cut}} dP_- \int_0^{2\pi} d\phi \int_{-1}^{1} dcos\theta \; w_d(P_+)w_d(P_-)\frac{d\sigma}{d\Omega}(cos\theta, x_+(P_+)x_-(P_-)S) \quad (8)$$

where $P_+ = (1 - x_+)^{\frac{\beta}{2}}$, the singular part of the integrated electron structure function. We shall evaluate this integral for $\sqrt{S} = 94 Gev$. The corresponding

$\hat{S}$ is shown on *figure* 5a. Since the Ds are always approximately 1, the only *wild* regions are when $\sqrt{x_+ x_-}S$ is near the Z-pole or the *photon*-pole. Moreover, the noodles should not be too big. Hence, it is natural to cut S evenly in the $x_+ x_-$-plane and to consider the two extreme values of $x_- x_+ S$ to decide whether a particular noodle is wild.

Following the Noodle method, at the beginning of the program, we make noodles:

procedure bakery;

arrays $x_-^{min}, x_-^{max}, x_+^{min}, x_+^{max}$, NOODLE

number of noodles:= 1;

index:= 1;

x_-^{min}(index):= x_{cut};

x_+^{min}(index):= x_{cut};

x_-^{max}(index):= 1;

x_+^{max}(index):= 1;

while (index $\leq$ number of the noodles) do

while is_wild(x_-^{min}(index)x_+^{min}(index)S, x_-^{max}(index)x_+^{max}(index)S) do

begin

x_-^{min}(number of noodles+1):= x_-^{min}(index);

x_+^{max}(number of noodles+1):= x_+^{max}(index);

x_-^{max}(number of noodles+1):= $\frac{1}{2}(x_-^{max}$(index)+x_-^{min}(index));

x_+^{min}(number of noodles+1):= $\frac{1}{2}(x_+^{max}$(index)+x_+^{min}(index));

x_-^{min}(number of noodles+2):= x_-^{max}(number of noodles+1);

x_+^{max}(number of noodles+2):= x_+^{max}(index);

x_-^{max}(number of noodles+2):= x_-^{max}(index);

x_+^{min}(number of noodles+2):= x_+^{min}(number of noodles+1);

x_-^{min}(number of noodles+3):= x_-^{min}(number of noodles+2);

x_+^{max}(number of noodles+3):= x_+^{min}(number of noodles+2);

x_-^{max}(number of noodles+3):= x_-^{max}(number of noodles+2);

x_+^{min}(number of noodles+3):= x_+^{max}(number of noodles+2);

x_+^{max}(index):= x_+^{max}(number of noodles+3);

x_-^{max}(index):= x_-^{max}(number of noodles+1);

number of noodles:= number of noodles + 3;

end;

$x_+ := \frac{1}{2}(x_+^{min} + x_{max}^+)$;

$x_- := \frac{1}{2}(x_-^{min} + x_{max}^-)$;

$P_+ := (1 - x_+)^{\frac{\beta}{2}}$;

$P_- := (1 - x_-)^{\frac{\beta}{2}}$;

$\Delta P_+ := (1 - x_{min}^+)^{\frac{\beta}{2}} - (1 - x_{max}^+)^{\frac{\beta}{2}}$;

$\Delta P_- := (1 - x_{min}^-)^{\frac{\beta}{2}} - (1 - x_{max}^-)^{\frac{\beta}{2}}$;

NOODLE(index):=

NOODLE(index-1)+$w_d(P_-)\dot{w}_d(P_+)\frac{d\sigma}{d\Omega}(cos\theta, x_+x_-S)\Delta P_+ \Delta P_- \Delta cos\theta$;

end;

function is_wild(S_{min}, S_{max});

wild if $S_{min} < M_Z^2 + 9\Gamma_Z M_Z$ and $M_Z^2 - 9\Gamma_Z M_Z < S_{max}$;

wild if $S_{min} < S_\gamma$;

wild if $\sqrt{S_{max}} - \sqrt{S_{min}} \geq \Delta E$;

otherwise not wild;

end;

The sampling of the integration region proceeds in two steps. First, we pick a noodle with the relative probability of NOODLE(index). Second, we pick a random point $(P_i^+, P_i^-, cos\theta_i)$ within the noodle:

procedure generate weights;

    pick a random number r between 0 and the length of the noodle;

    find l such that NOODLE($l-1$) $< r \leq$ NOODLE(l);

    $\Delta P^+ := (1 - x_{min}^+(l))^{\frac{\beta}{2}} - (1 - x_{max}^+(l))^{\frac{\beta}{2}};$

    $\Delta P^- := (1 - x_{min}^-(l))^{\frac{\beta}{2}} - (1 - x_{max}^-(l))^{\frac{\beta}{2}};$

    r_1 and r_2 are random numbers between 0 and 1;

    $P_+ := (1 - x_{max}^+)^{\frac{\beta}{2}} + r_1 \Delta P^+;$

    $P_- := (1 - x_{max}^-)^{\frac{\beta}{2}} + r_2 \Delta P^-;$

    $x_+ := 1 - P_+^{\frac{2}{\beta}};$

    $x_- := 1 - P_-^{\frac{2}{\beta}};$

    weight:= $\frac{w_d(P_-)w_d(P_+)\frac{d\sigma}{d\Omega}(cos\theta_l, x_+ x_- S)\Delta P^+ \Delta P^- \Delta cos\theta}{NOODLE(l) - NOODLE(l-1)};$

end;

6. Implementation and Comparison

We begin this section with an important remark on a reliable estimation of errors in Monte Carlo integration methods. As higher and higher accuracy is required of Monte Carlo calculations, **Monte Carlo** calculations should include reliable error estimates. The error formula in equation (2) comes from the central limit theorem. The central limit theorem says the following: Let $X_1, X_2, \ldots, X_n$ denote the items of a random sample from any distribution that has mean μ and variance σ^2. Then the random variable $\sum_1^n X_i$ has a limiting distribution that is normal with mean μ and variance $\frac{\sigma^2}{N}$. Naively, equation (2) is exactly the statement of the central limit theorem. But there is a problem. The central limit theorem assumes that we know the true mean and the true variance of the Xs. We do not. In fact, the true mean of the Xs is what we are trying to estimate! In our case, we would like to estimate the true mean of f over S by

$$\frac{1}{N}\sum_{1}^{N} f(x_i) .$$

Is this a good estimator of the true mean? If N is large enough, then the Monte Carlo summation will eventually sample over entire S, and the estimate should be very accurate. If, however, N is small enough that the generated values of x_is sample only over a small region of S, then the Monte Carlo sum would be a poor estimator of the true mean. Stated mathematically, the Monte Carlo sum actually estimates $<f\rho>$, not $<f>$, where ρ is the empirical distribution of the x_is for a given value of N. For large values of N, ρ approaches a uniform distribution over the entire region of integration, but for small values of N, depending on random number generators, ρ can be far away from being uniform. In order to remedy this, one must independently measure $<\rho>$ and σ_ρ^2 and solve for $<f>$ and σ_f^2 in terms of $<\rho>, <f\rho>, \sigma_{f\rho}$, and σ_ρ. But it turns out that measuring $<\rho>$ and σ_ρ is rather complicated and depends on the way the Monte Carlo method is implemented. The best solution out of this appears to be to ensure that ρ is very close to the uniform distribution by choosing a good random number generator and by choosing a large enough value for N. If ρ is significantly different from the uniform distribution, the results of Monte Carlo calculation would seem to fluctuate by more than what the naive error estimate in equation (2) would allow.

For 10^4 Monte Carlo samples, we reproduce the results obtained by Berends et al. at the Z-peak.[7] The resulting weight distribution is shown on *figure* 6. The results of a more complete calculation that contains the initial state radiation, the weak vertex functions, and the vector boson self-energies appears elsewhere.[1]

7. Acknowledgement

The author gratefully acknowledges valuable discussions with Bryan Lynn, Patricia Rankin, and Giovanni Bonvicini.

REFERENCES

1. D. Kennedy, B. Lynn, C. Im, R. Stuart, SLAC-PUB-4128, 1988 (Submitted to *Nuclear Physics B*)

2. O. Nicrosini and L. Trentadue, Phys. Lett. 196B,551(1987)

3. E.A.Kuraev and V.S.Fadin, Sov. J. Nucl. Phys. 41,466(1985)

4. F. James, Rept. Prog. Phys. 43,1145(1980)

5. G. P. LePage, CLNS-80/447(Unpublished)

6. J. H. Friedman, CGTM-188, 1977(Unpublished)

7. F. A. Berends, G. Burgers, W. L. van Neerman, Phys. Lett. 185B,395(1987)

8. Captions

Figure 1

An example of a star function, $s_\theta{}^2(q^2)$

Figure 2

The change of variables from z to G improve the convergence rate of the Monte Carlo sum.

Figure 3

The conventional Monte Carlo Method becomes inefficient for certain shapes of the space of integration, such as the one shown here.

Figure 4

In the Noodle method, the integrand is approximated by step functions. The resulting $\tilde{S}$, corresponding to the same integrand as in figure 2, is shown on the left. The blocks are then arranged into the shape of a noodle, as shown on the right.

Figure 5

The graphs of $\tilde{S}$ at $94 GeV$ (a), $110 GeV$ (b), $2\dot{}0 GeV$ (c) for integrand appearing in equation 8 are shown here. Each group of blocks of the same height, the group of blocks appearing at the $(P_+, P_-) = (1,1)$ corner on figure 5a, for example, corresponds to a single noodle. This correspondence is explicitly illustrated on figure 5c. In order to show detail, only the portion of P_+ P_- in which both coordinates are greater than 0.8 is shown. The photon pole, appears at $(P_+, P_-) = (1,1)$ and the other protruding region corresponds to the Z-pole.

Figure 6

The distribution of the ratio between the integrand in equation 8 and the approximate integrand constructed using noodles at $\sqrt{S} = 94 GeV$.

Figure 7

The distribution of the ratio between the integrand in equation 8 and th approximate integrand constructed using noodles at $\sqrt{S} = 1 TeV$.

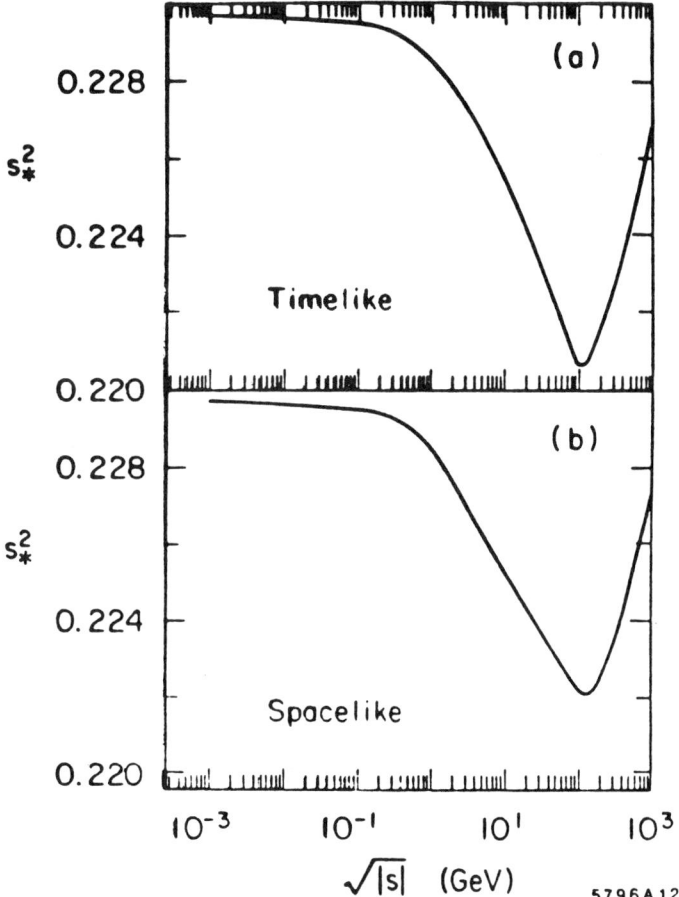

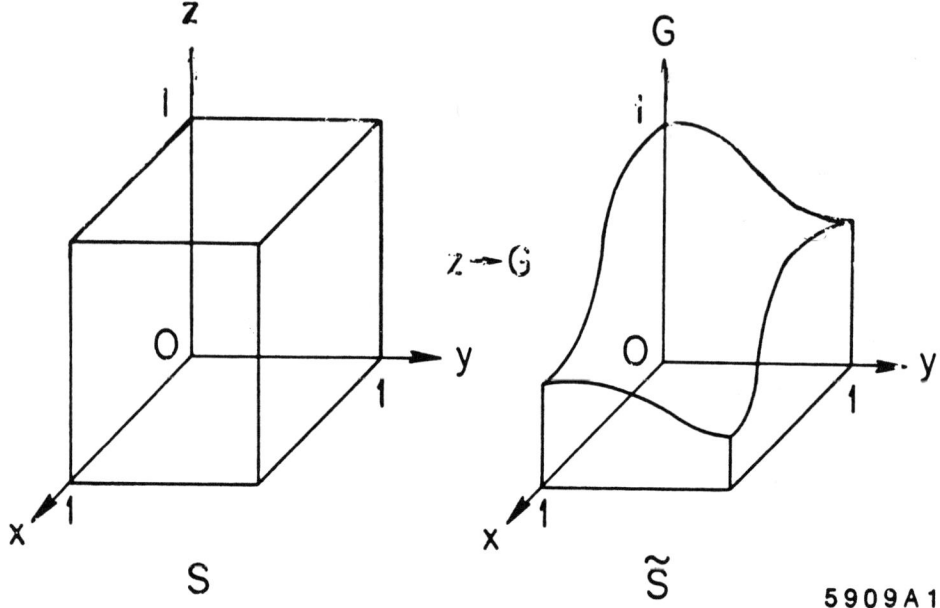

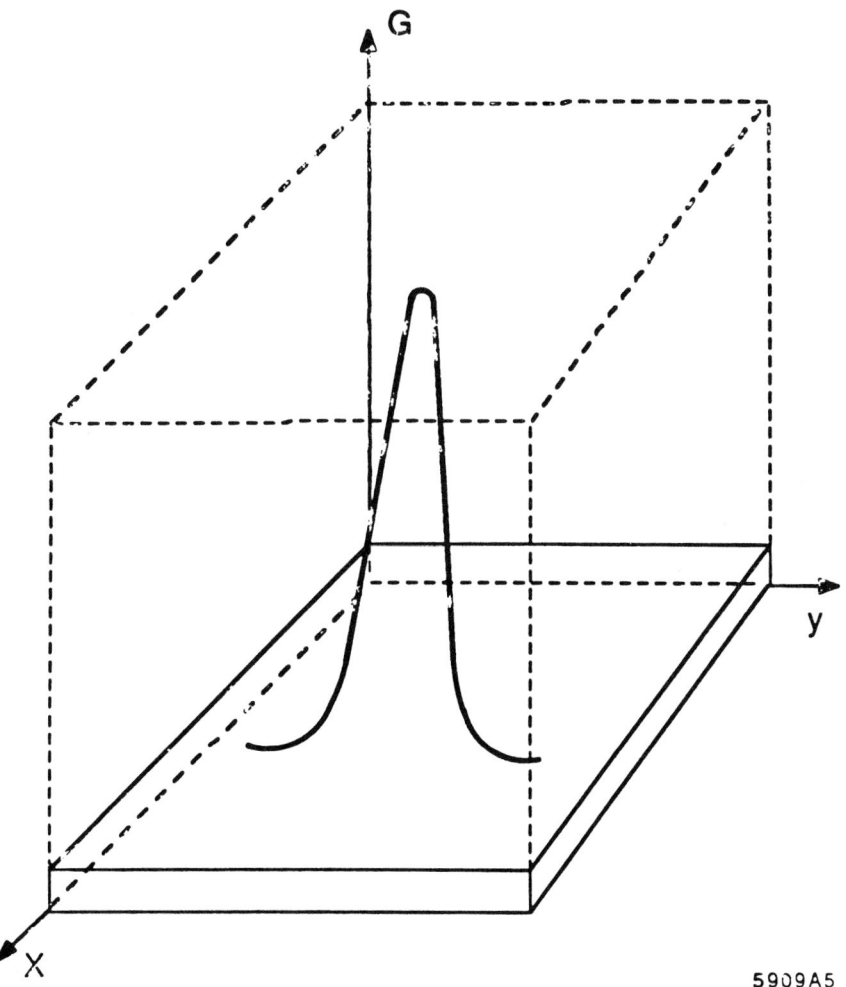

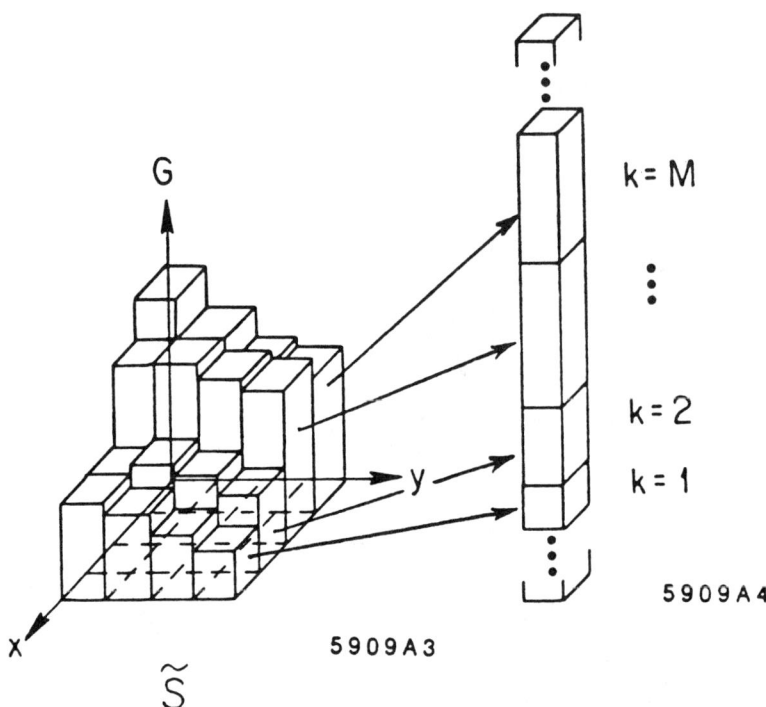

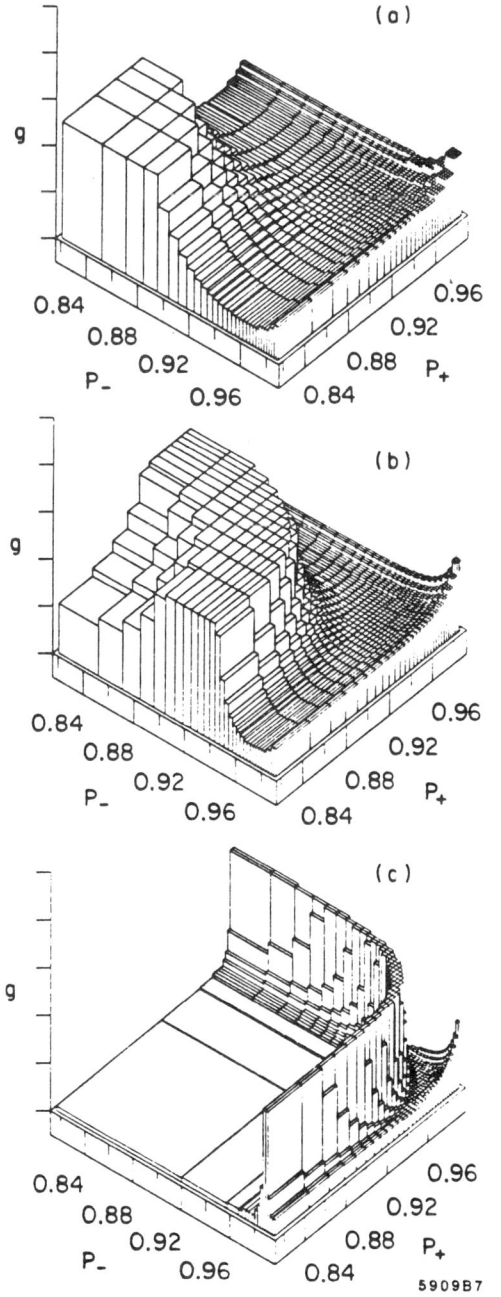

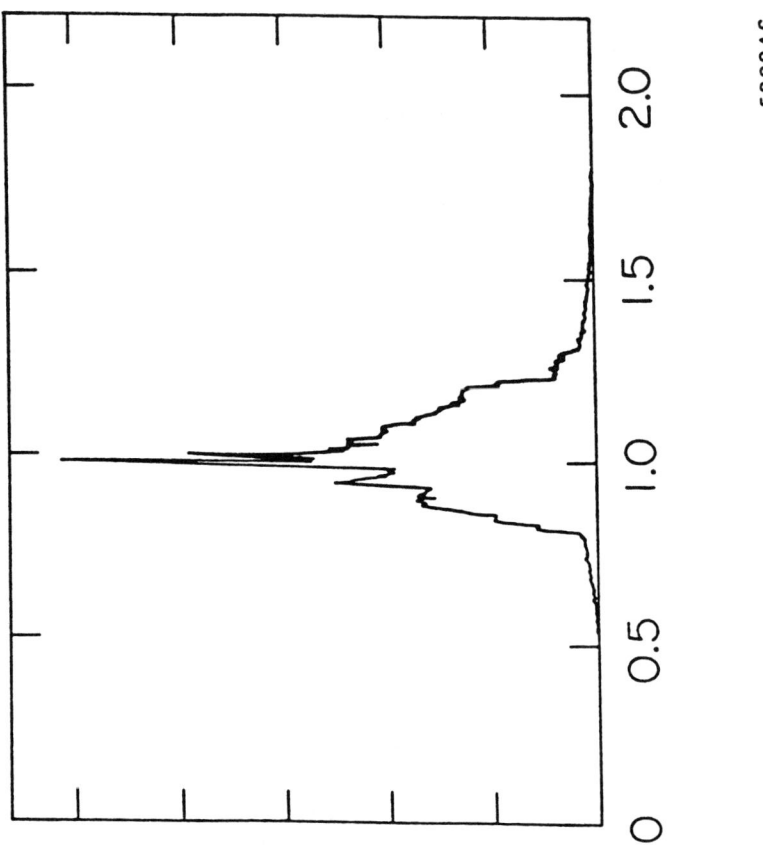

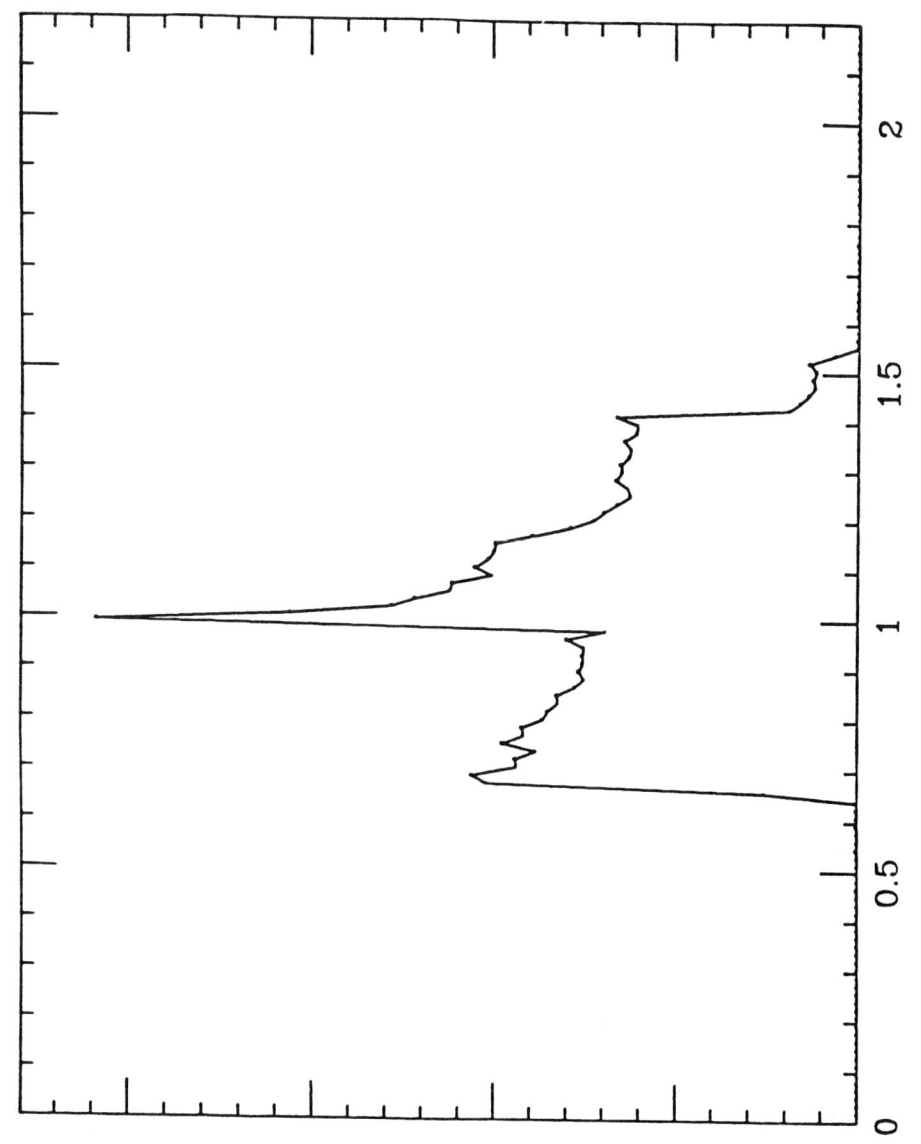

Theory and Experimental Data

RADIATIVE CORRECTIONS AT SLC AND LEP*

PATRICIA RANKIN

Department of Physics, University of Colorado, Boulder, Colorado, 80309-390.

Abstract

This paper discusses the extraction of electroweak parameters from measurements of the Z^0 line shape and asymmetries at the Z^0 pole. The likely limits to the experimentally attainable accuracy are compared to the limitations due to the theoretical implementation of radiative corrections.

*Talk presented at the Michigan Workshop on " QED Structure functions", Ann Arbor, Michigan, May 22-25, 1989. Work supported in part by the Department of Energy, contract DE-0286ER-40253.

Introduction

The start of data taking at the Stanford Linear Collider in April 1989 began a new era in the precision testing of the Electroweak Standard Model.(Figure 1 shows an xy view of a Z^0 decay to hadrons as seen in the Mark II detector). The high precision attainable has brought with it the need for extreme care in extracting the parameters of the theory from the experimental data. An understanding of the effects of radiative corrections and of how to include them in an analysis is needed by experimentalists. In the past errors in the application of radiative corrections have led to mistakes in the extraction of the masses of the J/Ψ and Υ particles.[1] However, errors in the extraction of the mass of the Z^0 will have much more serious consequences since they will directly affect our understanding of the Standard Model.

The simplest version of the Standard Model requires three experimental measurements to renormalize the bare parameters of the theory (the SU(2) coupling strength, the U(1) coupling strength, and the vacuum expectation value of the Higgs field). If the particle content of the theory is not fixed however, predictions based on the theory will have an inherent uncertainty. The gauge boson self energies depend on the masses of particles such as the top quark, and on the number, type, and masses of any Higg's particles which exist. Contributions to the self energies may also come from currently unknown particles. The introduction of a fourth parameter into the model, usually termed the ρ parameter, essentially allows the removal of this uncertainty by relating the particle content to a direct measurement.[2] Once the model is fully constrained by four measurements (α_{cm}, G_F, the Z^0 mass and perhaps the W mass), any further measurement of any of the parameters of the model tests the theory. The better the measurements made, the greater the significance of the check of the models self consistency.

This paper begins with a discussion of how well the Z^0 line shape can be determined, the experimental problems which must be considered, and related theoretical issues such as the effects of initial state radiation. The paper continues with a discussion of the asymmetry measurements which can be made by experiments at the SLC and LEP and which can be used to test the predictions of the Standard Model. The theoretical sensitivity of these measurements to indirect effects, such as those due to a heavy top quark will be compared to the likely sensitivity attainable in practice given the sources of experimental errors and the difficulty of controlling or unfolding the effects of initial state radiation.

The Z^0 Line shape

The measurement of the Z^0 line shape is very simple in principle. The cross-section is measured at several different center-of-mass energies and then the data is fit to give the overall line shape. The absolute and relative (point to point) errors on the measurement of beam energies and luminosity will determine how close the measured line shape is to the true line shape.

The accuracy with which the absolute energy at the center-of-mass is known is currently around 40 MeV at the SLC (similar initial accuracies have been achieved at LEP).[3] The point to point reproducibility at LEP is limited to about 40 MeV by temperature fluctuations from fill to fill, the relative error at the SLC is 30 MeV. The errors on measurements at LEP will decrease significantly if the beams can be transversely polarized and this transverse polarization measured. In this case the error on energy measurements at LEP will be about 2 MeV. It is also expected that the error on beam energy measurements made at the SLC will decrease with time as the effects due to the details of how the beams collide are understood and corrected for.

The situation as regards luminosity measurements is less clear cut. It is generally felt by experimentalists that it is tough to make an absolute measurement of luminosity to better than 2%. However the incentive to try to do so at the Z^0 peak is very strong, since one method of measuring the total Z^0 width depends on measuring the muon cross-section to high precision.

$$\sigma_{\mu\mu}(s = M_{Z^0}^2) \propto \frac{\Gamma_{ee}\Gamma_{\mu\mu}}{\Gamma_{tot}^2}.$$

Assuming the theoretical values for the leptonic partial widths (these are completely dominated by the axial couplings which do not depend on the Z^0 mass) the error on the measurement of the total width is given by

$$\Delta\Gamma_{tot} = \frac{\Gamma_{tot}}{2} \times \frac{\Delta\sigma_{\mu\mu}}{\sigma_{\mu\mu}}.$$

So a 2% measurement corresponds to a contribution of about 25 MeV/c^2 to the error on the total width measurement.[4]

There are two basic types of luminosity monitors used by SLC/LEP experiments. Both operate at sufficiently small angles for the contribution from s-channel Z^0 exchange to the Bhabha rate to be either negligible or calculable with negligible error. The first kind of monitors, usually termed precision monitors, typically cover an angular range between about 50 to 150 milli-radians (for example, the Mark II small angle monitor) and see a Bhabha rate similar to the peak Z^0 decay rate. The second type of monitors are primarily intended to provide rapid, relative estimates of luminosity. These detectors are placed at very small angles to the beam axis (the Mark II monitor at very small angles (the mini-SAM) covers about 15 to 25 milli-radians, the OPAL monitor 5 to 10 milli-radians). Since the t-channel exchange rate is so high at these angles, the Bhabha cross-section going as

$$\sigma_0 \propto \frac{16\pi\alpha^2}{s}\left(\frac{1}{\Theta_{\min}} - \frac{1}{\Theta_{\max}}\right),$$

these detectors see a cross-section several times that of the peak Z^0 cross-section.

While the close-in, high rate monitors may not be primarily intended for use in making absolute luminosity measurements they can provide a useful cross check on the values obtained by the precision monitors. Furthermore, solving the theoretical problems associated with their use will also benefit the evaluation of data from the precise monitors. The Bhabha rate measured by these inner detectors is very sensitive to exactly how the detectors are aligned relative to the interaction point. To decrease sensitivity to longitudinal motion of the interaction point the acceptances of each side of the detector can be made slightly different. This asymmetric acceptance makes the detectors more sensitive to changes in the angle between the electron and positron caused by the radiation of photons by the initial state. When an energetic photon is radiated it tends to align with the scattered lepton. If the detector has symmetric acceptance the correction due to photon radiation to the number of events which have both leptons inside the cuts has been shown to be small. Effects due to multiple photons being radiated by the initial state mostly cancel. However in the asymmetric case there is an added correction which is due to events were the radiation pushes a track, and thus an event, which would have been rejected into the tighter cut region. Gains at small angles are not balanced by losses at larger angles. The exact size of this effect depends on the details of the Bhabha detector. The closer in the detector, and the more asymmetric the acceptance cuts made, the bigger the problem may be. The need for comparatively better Bhabha code to analyse data from Bhabha detectors which make asymmetric cuts compared to those which make symmetric ones illustrates a general rule - the less inclusive a measurement is the more accurately initial state photon radiation must be modelled by a Monte Carlo of the process.

Although arguments exist which imply that first order (single photon radiated by the initial state) Monte Carloes are adequate for physics at the SLC/LEP,[5] the only convincing proof is to explicitly calculate the higher order corrections and show them to be small. After all, first order Bhabha Monte Carloes needed exponentiation before they could be used to fit the observed acollinearity distributions at PEP/PETRA. Programs under development include BHLUMI,[6] and an extension of MOE.[7] These programs can also avoid the need to divide the radiated photons up into two classes; when a "soft" photon is radiated it has a low enough energy for the final state to be treated as two-body, a "hard" radiated photon requires the final state to be treated as three body. First order Monte

Carloes must make this division. The soft photon contribution is summed with the genuine two-body contribution to avoid problems with the infra-red singularity associated with the two-body final state, and must be large enough to make the combined cross-section positive. The value found for the total cross-section can depend on the photon energy which is used to classify photons as soft (below this energy) or hard. Higher order Monte Carloes which avoid this division therefore eliminate a potential normalization error.

The work which has recently gone into studying Bhabha Monte Carloes has resulted in a steady improvement in the accuracy with which absolute cross-sections can be predicted,[8] and highlighted a potential problem. The conversion of the Bhabha rate to an absolute luminosity measurement requires a calculation of the vacuum polarization (Δr) of the photon. The use of different routines to calculate this can lead to apparent discrepancies in the results of Bhabha Monte Carloes. The accuracy with which the vacuum polarization can be calculated is limited by the "hadronic uncertainty". This uncertainty arises from the use of low energy data to evaluate the contribution of light quarks to the photon's vacuum polarization. This contribution cannot be accurately calculated but must be estimated using dispersion relations. The estimate is limited in accuracy by the errors on the low energy measurements.[9] A very careful analysis of the existing data was made recently by Burkhart, Penso, Jegerlehner and Verzegnassi,[10] who find the contribution to Δr due to the known quarks (u,d,s,c,b) to be 0.0288$\pm$0.0009. It has recently been pointed out[11] that the calculations also need to include a correction due to the hadronic interactions of heavy quarks. These effects, while small (they change Δr by about 2×10^{-3}) are of about the same order as the light quark uncertainty. The hadronic uncertainty, and the strong interaction contribution can be put into perspective by considering that if α_{em}, G_F, and M_z are kept fixed, the change in Δr corresponds to about a 25 MeV/c^2 uncertainty in predictions of the W mass.

Once the raw measurements of cross-section have been made at several energy points they must be fitted to a line shape to allow the mass, width and Z^0 cross-section to be extracted. The mass of the Z^0 is defined to correspond to the center of mass energy at which the real part of the Z^0 self energy is zero. The resonance line shape is distorted by initial state photon radiation. Analytic calculations which only include the effects of initial state radiation to first order (single photon radiated by the initial state) overestimate the separation of the Z^0 pole and the peak in the resonance cross-section by about 100 MeV (compared to the true separation which is also around 100 MeV !). These calculations also underestimate the peak cross-section (by a few percent). Analytic calculations which include higher order effects either by explicit calculation of second order contributions or by exponentiation are in very good agreement with each other. If the energy dependence of the Z^0 width (the imaginary part of the Z^0 self energy) is included in the calculations, the resonance peak occurs about 94 MeV above the Z^0 pole for a Z^0 mass of 91 GeV/c^2 and a width of 2.5 GeV/c^2.[12]

As well as predicting the energy separation of the peak of the Z^0 resonance from the Z^0 pole, the analytic calculations can also relate the observed peak cross-section to the cross-section at the pole (before the inclusion of the effects of initial state radiation) and the resonance width to the energies at which the observed cross-section has fallen off to half the peak cross-section . For hadrons, the observed peak cross-section is about 0.74 of the cross-section before initial state radiation and the energies at which the cross-section has half of its peak value are raised by about 60 MeV (below) and 430 MeV (above) relative to their original positions. This means that the separation of the half maxima positions is increased by a factor of about 1.14.

The inclusion of the effects of photon-Z^0 interference in the calculations slightly modifies the derived Z^0 line-shape for muons. The half maxima are raised in energy by about 50 MeV (below) and 450 MeV (above) increasing their separation by a factor of 1.15 (the muon line shape appears to be about 25 MeV wider than the hadronic line shape). The peak cross-section (which occurs about 1 MeV lower than it does for hadrons) is a slightly larger fraction (about 0.745) of the uncorrected cross-section. This enhancement of the muon cross-section relative to the hadron cross section must be allowed for when measuring branching fractions.[13]

Final state radiation effects are small compared to the effects of initial state radiation, but not when compared to the experimental accuracy with which the width can be measured. Although final state photon radiation increases the Z^0 width by less than 5 MeV/c^2, gluon radiation by quark final states has a significant effect. The observed partial width to hadrons ($\Gamma_{h,obs}$) is given by

$$\Gamma_{h,obs} = \Gamma_h(1 + \frac{\alpha_s}{\pi} + 1.41(\frac{\alpha_s}{\pi})^2 + 64.8(\frac{\alpha_s}{\pi})^3......)$$
$$= \Gamma_h(1 + 0.04 + 0.0024 + 0.0046 + ...)$$

where Γ_h is the predicted width before the inclusion of final state effects. The analysis of D'Agostini, de Boer, and Grindhammer gives the overall factor for the increase in hadronic widths due to gluon radiation to be 1.046 ± 0.005.[14] The expansion in powers of α_s ceases to be well behaved at the fourth term (the coefficient of the fourth term is larger than that of the third). Technically, this can be treated by truncating the expansion at the third term. It has been argued (for example by Kuhn[11] that it is in any case inconsistent to include the fourth term in the expansion without also adding in the effects of lower order diagrams, neglected earlier, which are sensitive to the mass splitting between the top and bottom quark. These latter effects are gaining in importance as the lower limit on the mass of the top quark keeps increasing. The b-quark partial width decreases by about 1 MeV/c^2 as the top quark mass is increased from 80 GeV/c^2 to 180 GeV/c^2. This effect is small, larger effects are seen if the b-quark partial width is compared to the down quark partial width. The b-quark partial width is about 1 MeV/c^2 less than the down-quark partial width if the top quark mass is 80 GeV/c^2, but is about 7 MeV/c^2 less if the top quark mass is raised to 180 GeV/c^2. Raising the top quark mass from 80 to 180 GeV/c^2 increases the total Z^0 width by about 25MeV/c^2.

The extremely good agreement between higher order analytic calculations allows them to act as a standard of comparison for assessing line shape Monte Carloes. If the experiments are assumed to have perfect acceptance and no cuts are applied to the generated events the predicted Z^0 line shapes agree very well with the standards as long as the Monte Carloes include higher order effects. The five Monte Carloes which satisfy this criterion are DYMU2,[15] EXPOSTAR,[16] YFS2/KORALZ3,[17] MOE10,[18] and ZBATCH(ZSHAPE).[19] The reader interested in a more detailed description of these Monte Carloes, and in their current status is urged to refer to the summary reports of Ronald Kleiss.[5] The ZBATCH and EXPOSTAR[20] Monte Carloes should be considered to be "semi-analytic", they do not explicitly generate events, but are useful for studying effects such as those due to changes in the input parameters. These programs are in good agreement with each other and the standard calculations except, perhaps, in overall normalization.

The differences of about 1% between the results of the various programs can again be traced mostly to the routines used to calculate the self energies of the vector bosons. The corrections to the self energies (usually called electroweak corrections or oblique radiative corrections to distinguish them from initial state radiative corrections) vary slightly from author to author.[21] The problem is not due to differences between calculations of the same diagrams but is a result of terms which are not common to all calculations. One set of such terms, sensitive to the mass difference between members of an iso-doublet is only starting to become interesting numerically as the experimental limits on the top quark mass continue to increase. A consensus is being reached over which terms to include in calculations. However, the possibility that some numerically important terms may not yet have been calculated should not be completely discounted.[22]

To summarize, experimentally the Z^0 mass can be measured to better than 50 MeV/c^2 fairly easily (and to within 20 MeV/c^2 with care). The Z^0 width is expected to be known to within 30 MeV/c^2 from direct measurements of the line shape, and the aim is to measure the Z^0 cross-section to one percent. The effects of initial state radiative corrections are understood and can be controlled at this level of accuracy. The overall normalization (which depends among other things on the electroweak self energy corrections) can be agreed on within this level of precision.

Standard Model Parameters

Once the Z^0 line shape has been measured, the next step is to make increasingly accurate measurements of Standard Model parameters such as the couplings of particles to the Z^0. A comparison of the measured couplings to predictions will check the internal consistency of the model and probe for contributions to the self energies of the vector bosons from particles too heavy to be produced directly in Z^0 decays.

The coupling of particles to the Z^0 is determined by the sum of the squares of the axial(a) and vector(v) couplings. The vector couplings are of particular interest since they are sensitive to the ratio of the electromagnetic coupling to the weak coupling at the Z^0 pole. The ratio of the bare weak and electromagnetic couplings gives the tangent of the Weinberg angle (θ_w). This angle can be defined in physical terms using the masses of the W and Z^0 gauge bosons as,

$$\sin^2\theta_w = 1 - \frac{M_w^2}{M_z^2}.$$

This definition, sometimes referred to as the on-mass-shell, or the Marciano-Sirlin definition is very convenient. Since the physical values of the masses are used, the effects of all diagrams (of any order) contributing to this quantity are automatically included. Implicit in this definition is the assumption that the Higgs structure of the model is the simplest possible (that is, that there are only two Higgs doublets, and only one physical Higgs particle remains after spontaneous symmetry breaking gives mass to the W and Z^0 gauge bosons). Currently, the value of $\sin^2\theta_w$ found by direct measurements at $p\bar{p}$ colliders is 0.226 ± 0.009.[23] A much more accurate value of $\sin^2\theta_W$ can, in principle, be derived in terms of the Z^0 mass, α_{em}, and G_F. The only problem with this method is the unknown value of the top quark mass which introduces an uncertainty into the calculation.[24]

The on-mass-shell definition of $\sin^2\theta_W$ only gives the approximate value of the quantity relevant to discussions of particle couplings to the Z^0 The bare parameters of the theory are renormalized by relating them to physical quantities measured at particular energy scales. This renormalization only fixes the values of these parameters to lowest order when they are used to predict the results of measurements which involve different momentum transfers. The closer the renormalization conditions are to the conditions the new measurements are made at, the better the predictions will be. The apparent dependence of the predictions on the choice of renormalization measurements (or of "renormalization scheme") is removed by including higher order diagrams into the calculations. It is clearly preferable to work with quantities that include a more complete set of the diagrams instead of an approximation if one wants to test the theory (otherwise discrepancies between prediction and experiment lack significance).

The desire to be able to make accurate predictions (for physics at the Z^0 pole, and for other experiments) has lead to the development of "running" coupling constants which give the effective value of $\sin^2\theta_W$ ($\sin^2\theta_W$ corrected for the conditions under which the measurement is made). One such example is the quantity $\sin^2_*\theta$ which was proposed by Bryan Lynn.[25] The calculation of these quantities depends on the particle content of the model and on the masses of those particles. It is therefore not possible to make a unique prediction for the value of $\sin^2_*\theta$ at the Z^0 pole. However, once we make one measurement of this quantity the particle content needed to give the measured value must be consistent with that needed by other measurements. Once we can no longer adjust the particle content to make all measurements consistent with each other, we have disproved the model.

The effective value of $\sin^2\theta_W$ at the Z^0 pole ($\sin^2_*\theta(M_z)$) is less sensitive to the exact value of the top quark mass than $\sin^2\theta_W$ is. The differences between $\sin^2_*\theta$ and $\sin^2\theta_W$ are shown graphically in Figure 2. The importance of using $\sin^2_*\theta$ in calculations at the Z^0 pole increases with increasing top quark mass. The sensitivity needed for measurements of $\sin^2_*\theta$ in different processes is set by the size of the effects induced by changing the input parameters. If everything else is kept fixed; a 45 MeV/c^2 increase in the Z^0 mass decreases $\sin^2_*\theta$ by 0.0003,[26] increasing the top quark mass from 60 to 180 GeV/c^2 decreases $\sin^2_*\theta$ by 0.0035, and while decreasing the Higgs mass from 100 to 10 GeV/c^2 decreases $\sin^2_*\theta$ by 0.001, raising the Higgs mass to 400 GeV/c^2 increases $\sin^2_*\theta$ by only 0.0007.

Asymmetry measurements with unpolarized beams

The most direct way to measure the vector coupling constant for a particular fermion final state is to measure the forward-backward asymmetry for Z^0 decays into that state. If the beams are unpolarized, the forward-backward asymmetry (A_{FB}) at the Z^0 pole is given by

$$A_{FB}^{f\bar{f}}(M_{Z^0}) = \frac{3a_e v_e a_f v_f}{(a_e^2 + v_e^2)(a_f^2 + v_f^2)},$$

where a_e, v_e, a_f and v_f represent the axial and vector couplings of the electron and the fermion of interest to the Z^0.

The sensitivity (S) of the forward-backward asymmetry to $\sin^2_*\theta$ (the effective value of $\sin^2\theta_W$ at the Z^0 pole) depends both on the value of $\sin^2_*\theta$ and on the value of the asymmetry (which depends on $\sin_*^2 \theta$ and the fermion type).

$$S_f = A_{FB}^{f\bar{f}} * s_f$$

$$s_f = \frac{4}{v_e} - \frac{4Q_f}{v_f} - \frac{(8v_e - 8v_f Q_f)}{(1+v_e^2)(1+v_f^2)}$$

Q_f is the electric charge carried by the final state fermion. Table 1 compares the sensitivity of measurements of the muon forward-backward asymmetry to measurements of the b-quark asymmetry for selected values of $\sin_*^2 \theta$. A reasonable upper limit for $\sin_*^2 \theta$ is about 0.235. This corresponds to a Higgs mass of 500 GeV/c^2, a Z^0 mass of 91 GeV/c^2 and a top quark mass of 80 GeV/c^2. Raising the Z^0 mass or the top quark mass lowers $\sin_*^2 \theta$, decreasing the Higgs mass also lowers $\sin_*^2 \theta$.[27] The error on $\sin_*^2 \theta$ is given by

$$\delta \sin_*^2 \theta = \delta A_{FB}^{f\bar{f}}/S_f$$

As can be seen the muon asymmetry rapidly looses sensitivity as the value of $\sin_*^2 \theta$ approaches 0.25.

Table 1.
A comparison of the sensitivity of the unpolarized
forward-backward asymmetry for muons and for b quarks
for various values of $\sin_*^2 \theta$

$\sin_*^2 \theta$	Muons $A_{FB}^{\mu\bar{\mu}} * s_\mu = S_\mu$	b-quark $A_{FB}^{b\bar{b}} * s_b = S_b$
0.225	0.029× 78 = 2.31	0.139× 39=5.50
0.23	0.019× 99 = 1.87	0.112 ×50=5.61
0.235	0.011× 132=1.42	0.084 × 67=5.61
0.24	0.005 × 199=0.95	0.006 × 100=5.60

Figures 3 and 4 show the muon and b-quark asymmetries as a function of center of mass energy for a Z^0 mass of 91 GeV/c^2, a Higgs mass of 100 GeV/c^2 and top quark masses of 60 and 180 GeV/c^2. As one moves away from the pole these asymmetries change rapidly. Away from the Z^0 pole the dominant contribution to the asymmetry comes from the interference between photon exchange and Z^0 exchange, which depends on the axial couplings. However, at the Z^0 pole there is no interference and the asymmetry depends on the vector couplings of particles to the Z^0. The vector couplings are functions of $\sin^2_* \theta\, (M_z)$ so it is these which one really wants to measure. The strong energy dependence requires close monitoring of the machine energy at which data is taken. A 50 MeV change in beam energy changes the muon forward-backward asymmetry by 0.004 (similar to the statistical error on a 200 pb^{-1} measurement), the b-quark asymmetry is much less sensitive and changes by less than 0.0005. The strong energy dependence makes these asymmetries (particularly the muon) sensitive to the details of initial state radiative corrections which effectively change the energy of the collision (and therefore the value of the asymmetry which is measured). The need to get angular correlations correct complicates the evaluation of initial state effects since not only the energy loss but also the momentum change due to radiated photons must be calculated correctly.

The Monte Carloes[15-18] which agree so well with each other and with analytic calculations when predicting the Z^0 line shape do not agree so well over predictions of the forward-backward asymmetry.[15] The muon forward-backward asymmetry is a particularly stringent test of Monte Carloes. Until recently no standard was available to check these Monte Carloes against. However, the recent development of CALASY,[28] a semi-analytic Monte Carlo (which plays the same role in forward-backward asymmetry measurements that ZBATCH and EXPOSTAR do in line shape measurements) will help.[29] Initial-state radiative corrections also cause the flattening out of the energy dependence of the asymmetry above the Z^0 pole, due to the tendency for the electrons and positrons to radiate onto the pole.

Both the lack of sensitivity of the muon asymmetry to $\sin^2_* \theta$ and the particularly strong energy dependence of this asymmetry make the b quark asymmetry appear to be a better choice to use when extracting the value of $\sin^2_* \theta$. There is also an apparent statistical advantage since the Z^0 decays more frequently to b-quarks than to muons by about a factor of 4.6. However, the statistical advantage in using b-quarks is diluted by the efficiency for detecting b-quarks (and assigning them the correct charge). The other advantages are partially diminished by the fact that the interpretation of the b quark measurement is complicated by the occurrence of b-quark mixing, and by strong interaction effects. It has been estimated[30] that an integrated luminosity of 200 pb^{-1} will lead to a measurement of the muon asymmetry with an error of 0.0035 (changing the top quark mass from 80 GeV/c^2 to 180 GeV/c^2 changes this asymmetry by 0.004) and the b-quark asymmetry with an error of 0.005. The error on the muon asymmetry measurement converts to an error on the measurement of $\sin^2_* \theta$ of 0.0025, while the b-quark measurement yields an error of 0.0009 (for $\sin^2_* \theta$ 0.235).

Polarization

The cross-section at the Z^0 pole is relatively insensitive to changes in (for example) the top quark mass. This is because changes in the partial widths are largely compensated for by the change in the total width of the resonance. (This cancellation would be perfect if effects due to photon-Z^0 interference could be ignored.) This insensitivity is shown in Figure 5 which gives the change in the muon cross-section at the pole as a function of the Z^0 mass when the top quark mass is changed from 60 to 180 GeV/c^2. The effects on the right and left-handed cross-sections considered individually are not so small but since the changes have opposite sign they compensate for each other. However, the almost perfect cancellation seen for a 92 GeV/c^2 Z^0 (the standard used in many studies) is not quite so good for lighter or heavier Z^0 masses. For a 91.1 GeV/c^2 Z^0, the change in the visible cross-section (hadrons, taus and muons) corresponding to this change in top quark masses is about 0.07% (which is experimentally unobservable). The right-handed and left-handed cross-sections change by -3% and 2.5% respectively. Polarizing the beams (or one of the beams) producing the Z^0 is desirable because it allows the isolation of the left and right-handed components of the cross-section.

The sensitivity to electroweak effects of the right and left-handed components of the total cross-section can be exploited by studying the left-right asymmetry, A_{LR}, which measures the difference between them. On the Z^0 pole, this asymmetry depends only on the couplings of electrons to the Z^0 and is given by,

$$A_{LR}(M_{Z^0}) = \frac{2Pa_e v_e}{a_e^2 + v_e^2},$$

where P is the magnitude of the polarization. This asymmetry is more sensitive to M_{Z^0}, consequently requiring a higher level of accuracy in the direct measurement. It also has the advantage that it requires a lower level of stability in the energy measurement.

The sensitivity of this asymmetry to $\sin_*^2 \theta$ is not dependent on the value of $\sin_*^2 \theta$ (it is always 8). Figure 6 shows clearly how sensitive A_{LR} is to changes in the top quark mass. The energy dependence of this asymmetry is fairly weak so that it is comparatively insensitive to the effects of initial state photon radiation. The asymmetry at the Z^0 pole does not depend on the final state (Figure 7) which means that all events can be used to measure it. A discussion of the insensitivity of this measurement to final state effects and detector acceptances can be found in the papers by Lynn.[31] Bryan deserves much credit for persuading many experimentalists to work at making this measurement possible.

The prospects for polarizing the electron beam at the SLC are very good. The minimum requirement is the installation of two spin-rotator magnets, the installation of a third magnet allows the maximum beam polarization at any beam energy. The installation of a polarized electron source is already in progress. Plans are to measure the beam polarization to at least 3 % initially and to improve this to 1 % over time.[32] Polarizing the beams longitudnally at LEP will be more difficult but the problems involved are being studied.[33]

It is expected that interesting measurements of the left-right asymmetry will be available in 1991. Combined with all the other measurements of electroweak parameters which will then be available it should be possible to test the Standard Model to the limits of our current understanding.

Summary and Acknowledgements

The next few years will provide many precise tests of the Standard Model. The expected accuracy of experimental measurements is well matched by the current accuracy of theoretical predictions and no doubt both will improve with time given the impetus of data from the SLC and LEP. The best prospect for a measurement which challenges our understanding of the Standard Model is the left right asymmetry measurement.

I would like to thank Giovanni Bonvicinni for organizing a very successful workshop and the conference staff at the University of Michigan for all their help. The author is grateful for support as an Alfred P. Sloan Research Fellow, and for support by the U.S. Department of Energy as an Outstanding Junior Investigator.

FIGURES

1) An xy view of a Z^0 decay to hadrons as viewed by the Mark II detector at the SLC.
2) A comparison of the values of $\sin^2\theta_W$ (solid line) and $\sin^2_*\theta$ (M_z) (dashed line) as a function of top quark mass. For this figure, and for figures 3,4,6, and 7, M_z is assumed to be 91 GeV/c^2 , and the Higgs mass is assumed to be 100 GeV/c^2 .
3) The muon forward-backward asymmetry as a function of center-of-mass energy for two top quark masses . The lower set of points corresponds to a 60 GeV/c^2 top quark mass, the upper to a 180 GeV/c^2 top quark mass. The vertical lines correspond to the error on the Monte Carlo simulation (the Monte Carlo EXPOSTAR was used).
4) The b-quark forward-backward asymmetry as a function of center-of-mass energy for two top quark masses. The lower set of points corresponds to a 60 GeV/c^2 top quark mass, the upper to a 180 GeV/c^2 top quark mass. The vertical lines correspond to the error on the Monte Carlo simulation (the Monte Carlo EXPOSTAR was used).
5) The difference (in nanobarns) in the unpolarized(solid line), left(dotted line), and right(dashed line) handed cross-sections for Z^0 decays to muons resulting from changing the top quark mass from 60 to 180 GeV/c^2 .
6) The left-right polarization asymmetry for muons as a function of center of mass energy for two top quark masses. The lower set of points corresponds to a 60 GeV/c^2 top quark mass, the upper to a 180 GeV/c^2 top quark mass. The vertical lines correspond to the error on the Monte Carlo simulation (the Monte Carlo EXPOSTAR was used).
7) The left-right polarization asymmetry for the muon final state (crosses) and the b-quark final state (diamonds), as a function of center of mass energy. The top quark mass is assumed to be 60 GeV/c^2 .

REFERENCES

1. See the contributions to this conference by G. Bonvicinni and by E.A. Kuraev.
2. See the contribution of G. Passarino to this conference.
3. For details of beam energy measurements at the SLC see J.Kent et al., Precision Measurements of the SLC Beam Energy, SLAC-PUB-4922,LBL-26976. For information on the LEP measurements see for example, D.Schailes report in the proceedings of the Rinberg Workshop on Electroweak Radiative Corrections.
4. This should be compared to the accuracy with which neutrino counting experiments on or close to the Z^0 pole plan to be able to measure the invisible width; see the report of B.Wilson in these proceedings.
5. See discussions by R. Kleiss on the status of Monte Carloes for Lep, for example in the proceedings of the Rinberg Workshop on Electroweak Radiative Corrections, and the report of the Monte Carlo working group in the second CERN yellow report on LEP physics.
6. This approach, due to Jadach and Ward is based on YFS theory, (see article to appear in Phys.Rev.D40, Dec. 1st issue,1989) the program philosophy is the same as that of KORALZ3.
7. G. Bonvicinni and L.Trentadue, preprint UM-HE-88-36. See also contributions to this conference.
8. The cross-sections predicted by current Bhabha Monte Carloes are now believed to be accurate to a percent, B.F.L. Ward, private communication.
9. Further experiments at low energies could reduce this error significantly.
10. H.Burkhart et al, in Polarization at LEP, vol 1 (p 145-157).
11. J. Kuhn, proceedings of the Rinberg Conference on Electroweak Radiative Corrections and the proceedings of the Nato Advanced Study Workshop on Radiative Corrections, Brighton, England.
12. Detailed formulae describing these effects can be found in R.N. Cahn, "Analytic forms for the e^+e^- annihilation cross-section near the Z^0 including initial state radiation", Phys.Rev. D36:2666,1987. Another useful treatment can be found in D.Y. Bardin et al," Energy dependent width effects in Z^0 line shape", Phys.Lett.B206:539,1988. An apparent disagreement as to the size of the energy dependent width effect between these two treatments can be traced to a difference in the form they assume for the Breit-Wigner when the width is constant. See also the "Z line Shape" by D.Y. Bardin et al. in the

second CERN yellow report on LEP Z^0 physics, this contains an exhaustive list of references.

13. For more details see W.Beenakker et al, "Rules of thumb for the Z^0 line shape", University of Leiden preprint, October 1989.
14. G. D'Agostini et al, " Determination of α_s and the Z Mass from measurements of the Total Hadronic Cross-Section", Desy 89-057.
15. J.E. Campagne and R.Zitoun, University of Paris preprint LPNHE-88.08
16. D.C. Kennedy at al, SLAC-PUB-4128(1988)
17. S.Jadach and B.F.L. Ward, Comp.Phys.Commun. ,1989,in press
18. G. Bonvicinni and L.Trentadue, preprint UM-HE-88-36
19. See G. Burgers contribution to Polarization at LEP, CERN 88-06.
20. EXPOSTAR is particularly suited to studying effects on the left right polarization asymmetry. It allows one to select the final state to be studied. However, although the partial widths for quark final states include QCD corrections, the cross-sections for these final states do not. Users should be aware of this when making line shape studies with this code.
21. The two main libraries are provided by Hollik et al and Stuart et al.
22. R. Stuart, private communication.
23. P. Langacker," The implications of recent M_z, M_w and neutral current measurements for the top quark mass", UPR-0400T, September 1989.
24. When the W mass is much better measured we can use it to predict the top mass, or vice versa. The current uncertainty in the W mass is too large for interesting limits on the top mass to be obtained.
25. There has been some discussion about the gauge invariance of some of the running coupling schemes, in particular the $\sin^2_*\theta$ scheme. The problems raised with the original formulation of this scheme (D. Kennedy and B. W. Lynn, Nucl. Phys. B322:1,1989.) have been addressed by Lynn in SU-ITP-867 (Aug 1989) which has been submitted to Phys.Lett. Similar calculations have been developed by W. Hollik, the interested reader is recommended to read his DESY report 88-188, for a complete discussion of this issue.
26. $\sin^2\theta_w$ is decreased by a similar amount. The direction of the change is different than one might expect because increasing the Z^0 mass by 45 MeV/c^2 corresponds to an increase of 50 MeV/c^2 in the W mass if everything else is kept fixed.
27. Increasing the Higgs mass to 1000 GeV would raise $\sin^2_*\theta$. Calculations give 0.0236, but they are unreliable for such a high Higgs mass.

28. D. Y. Bardin et al; contribution to the second CERN yellow report on LEP100 physics.

29. For the current (September 1989) status of these programs see D. Bardin et al," On Some New Analytic Calculations for the Process $e^+e^- \to f\bar{f} + (n\gamma)$",CERN-TH.5434/89.

30. J. Drees, Proceedings of the Rinberg Conference on Electroweak radiative corrections.

31. See B.W. Lynn,"High Precision Tests of Electroweak Physics on the Z^0 resonance", contribution to report on CERN yellow report on Polarization at LEP, and references therein.

32. For details refer to: D. Blockus et al; " Proposal for polarization at the SLC",SLAC-PROP-1, Stanford, April 1986.

33. G. Alexander et al (editors), "Polarization at LEP", CERN 86-06.

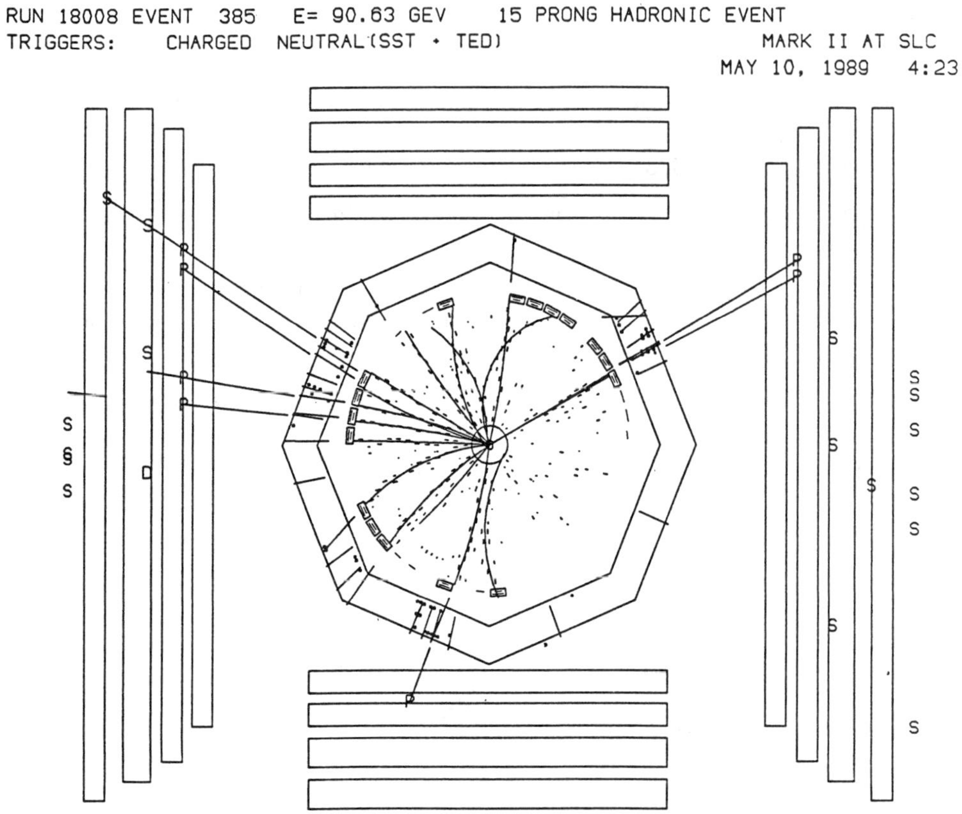

RUN 18008 EVENT 385   E= 90.63 GEV   15 PRONG HADRONIC EVENT
TRIGGERS: CHARGED NEUTRAL(SST + TED)   MARK II AT SLC
                                                       MAY 10, 1989   4:23

FIGURE 1

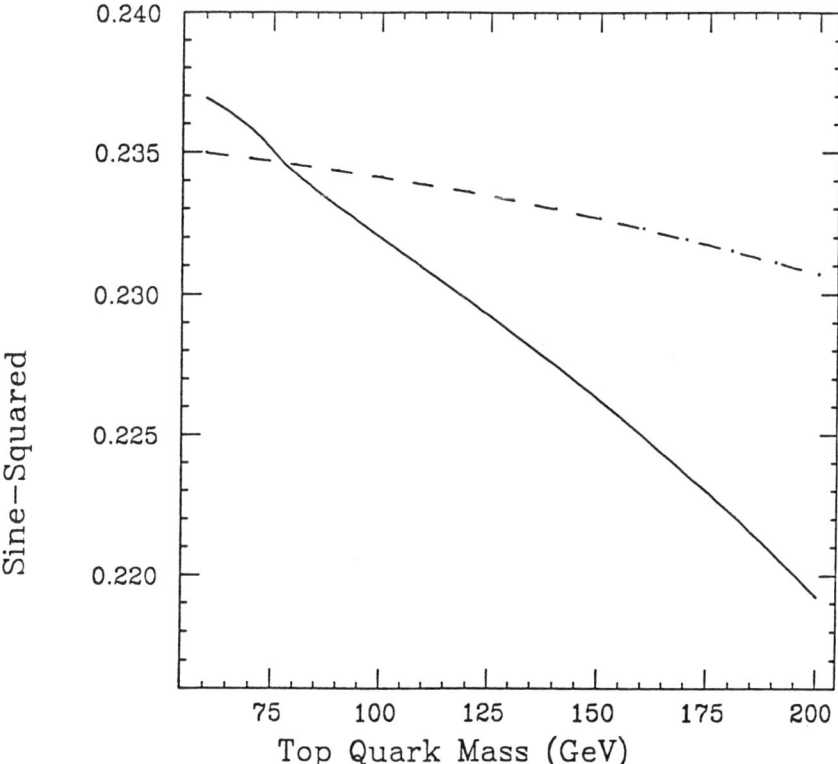

FIGURE 2

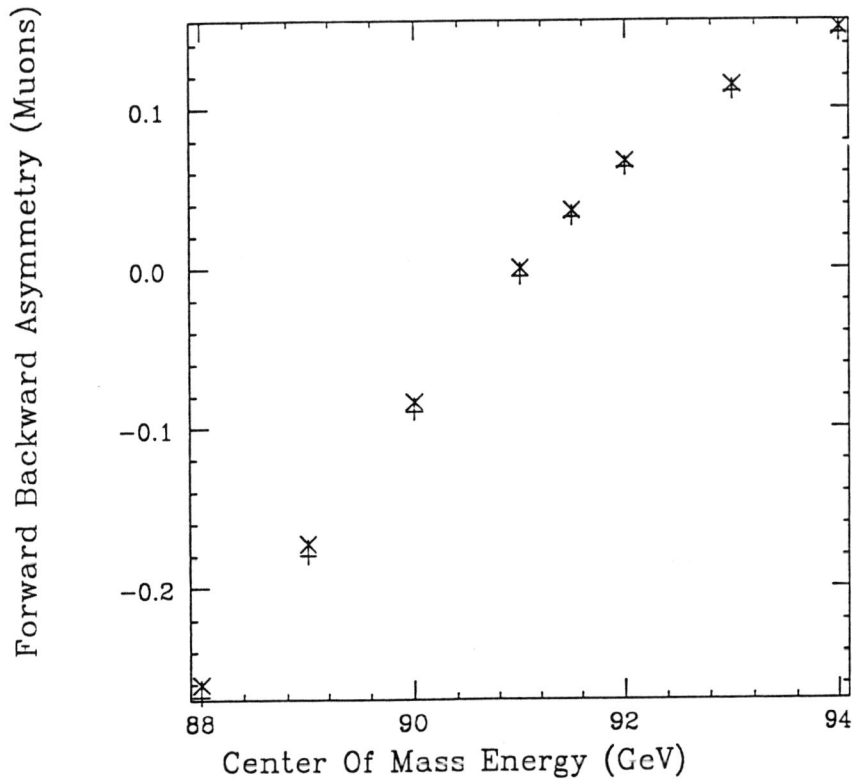

FIGURE 3

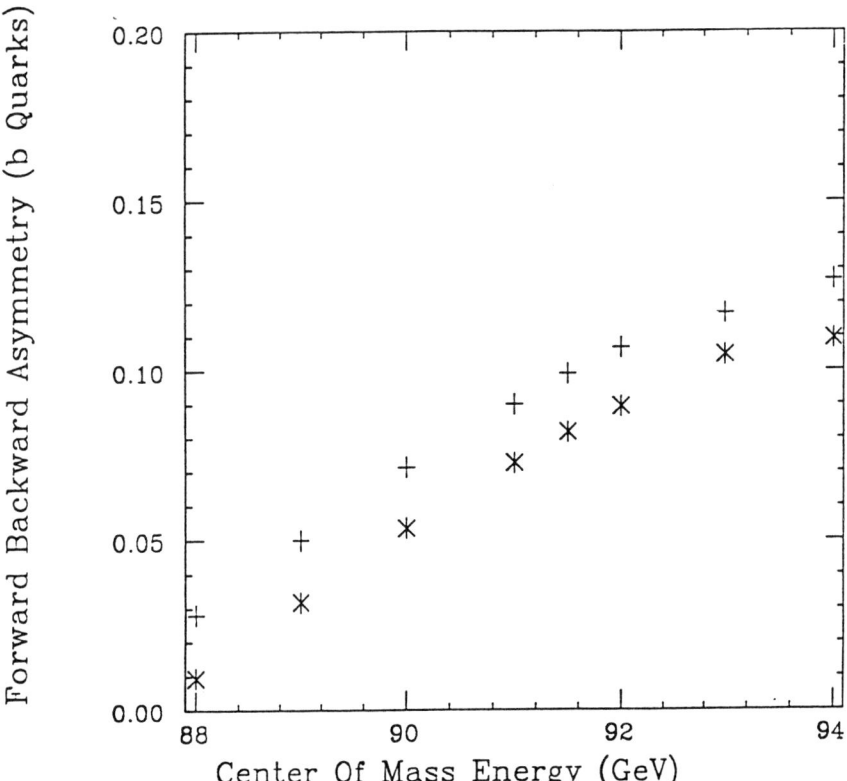

FIGURE 4

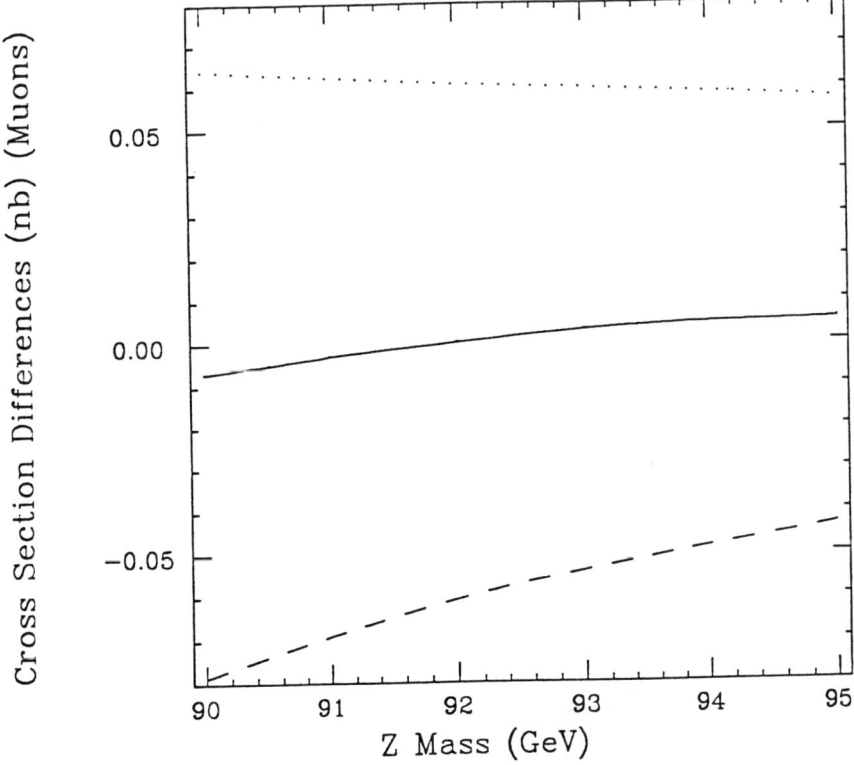

FIGURE 5

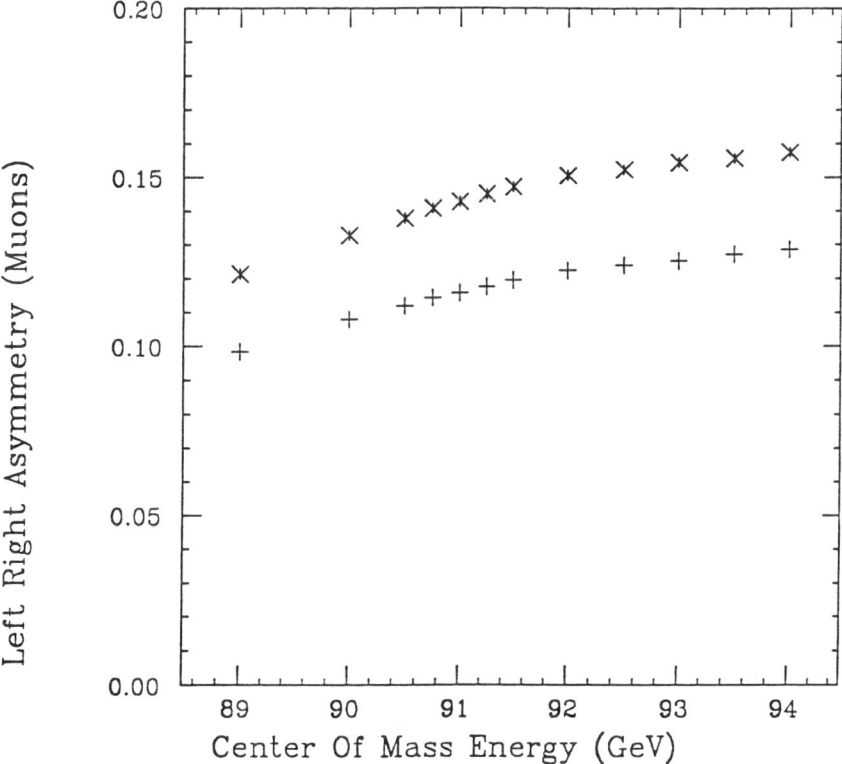

FIGURE 6

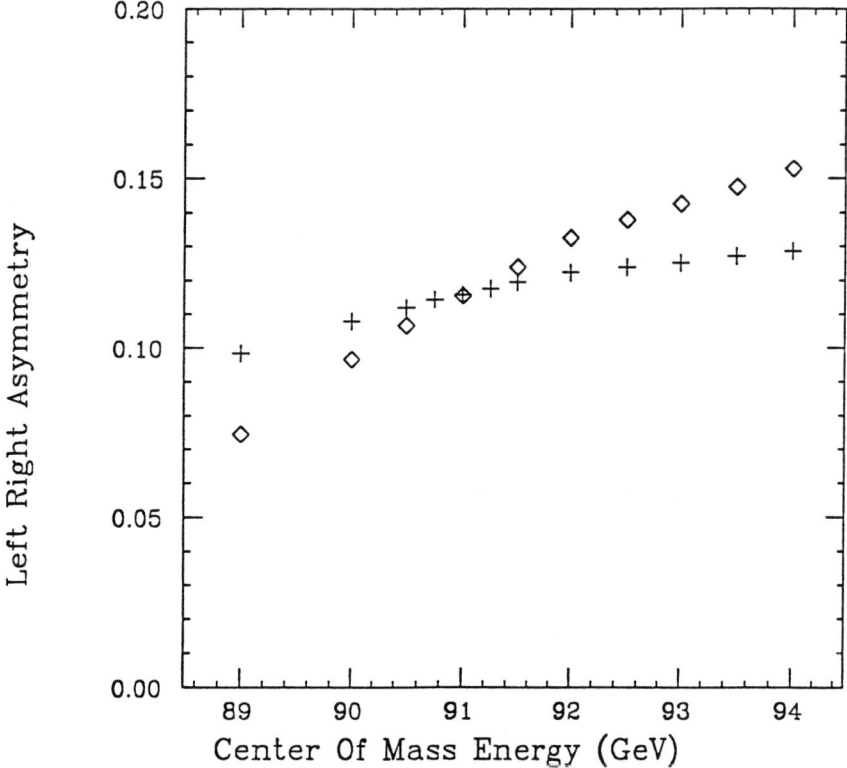

FIGURE 7

Status Report on Radiative Correction at KEK

J.Fujimoto

Department of Physics, Nagoya University

ABSTRACT

I describe the radiative correction to R-ratio at the TRISTAN. It is obtained with the one-loop $SU(2) \times U(1)$ radiative correction in the on-shell renormalization scheme within the framework of the standard electroweak theory. In the TRISTAN energy region, the pure weak correction reaches about 3%. The dependence of the radiative correction factor on the top-quark and Higgs masses is discussed.

1. Introduction

In the TRISTAN experiments we are now taking data around 60 GeV. In this energy region, the weak interaction is sizeable. For example the increase of R-ratio, the total hadronic cross section normalized to the point-like QED cross section, due to Z pole reaches more than 30%. Therefore the systematic treatment of the radiative correction is needed to derive the fundamental parameters of the electroweak theory from the data. In this paper the radiative correction to R at TRISTAN is brought into focus. The R ratio is expressed in the standard model by [1] :

$$R = 3\sum_q [\frac{\beta_q}{2}(3-\beta_q^2)\left(1 + C_1^V \frac{\alpha_s}{\pi} + C_2^V \left(\frac{\alpha_s}{\pi}\right)^2 + C_3^V \left(\frac{\alpha_s}{\pi}\right)^3\right) C_{VV}^q$$
$$+ \beta_q^3 \left(1 + C_1^A \frac{\alpha_s}{\pi} + C_2^A \left(\frac{\alpha_s}{\pi}\right)^2 + C_3^A \left(\frac{\alpha_s}{\pi}\right)^3\right) C_{AA}^q] \quad (1.1)$$
$$- 1.679 \left(\sum_q e_q\right)^2 \left(\frac{\alpha_s}{\pi}\right)^3,$$

with

$$C_{VV}^q = e_q^2 - 2e_q g_V^e g_V^q Re(\chi) + ((g_V^e)^2 + (g_A^e)^2)(g_V^q)^2 |\chi|^2,$$
$$C_{AA}^q = ((g_V^e)^2 + (g_V^e)^2)(g_A^q)^2 |\chi|^2,$$

$$\beta_q = \sqrt{1 - \frac{4m_q^2}{s}},$$

$$\chi = \frac{1}{16\sin^2\theta_W (1-\sin^2\theta_W)} \frac{s}{s - M_Z^2 + iM_Z\Gamma_Z},$$

where Γ_Z is the width of the Z boson and fixed to be 2.5 GeV because the Brite-Wigner formula is very good approximation in the TRISTAN energy region. The quantities $g_{V(A)}^e$ and $g_{V(A)}^q$ stand for the vector and axial vector coupling of electron and quarks, respectively:

$$g_V^e = -\frac{1}{2} - 2e\sin^2\theta_W, \quad g_A^e = -\frac{1}{2},$$
$$g_V^q = \pm\frac{1}{2} - 2e_q \sin^2\theta_W, \quad g_A^q = \pm\frac{1}{2} \quad \begin{cases} +, & \text{for up type quark} \\ -, & \text{for down type quark} \end{cases},$$

Here QCD correction coefficients ($C_i^{V(A)}$) up to the third order of the strong coupling constant, α_s, in the modified minimal subtraction scheme are given by:

$$C_1^V = \frac{4}{3}\pi\left(\frac{\pi}{2\beta_q} - \frac{3+\beta_q}{4}\left(\frac{\pi}{2} - \frac{3}{4\pi}\right)\right),$$

$$C_1^A = \frac{4}{3}\pi\left(\frac{\pi}{2\beta_q} - \left(\frac{19}{10} - 22\frac{\beta_q}{5} + \frac{7}{2}\beta_q^2\right)\left(\frac{\pi}{2} - \frac{3}{4\pi}\right)\right),$$

$$C_2^V = C_2^A = C_2 = 1.986 - 0.115 N_f, \quad C_3^V = C_3^A = C_3 = 70.985 - 1.2 N_f - 0.005 N_f^2$$

$$\frac{\alpha_s(s)}{4\pi} = \alpha_{s0}(s) - \frac{\beta_1 L}{\beta_0}\alpha_{s0}^2(s) + \left(\frac{\beta_1^2}{\beta_0^2}(L^2 - L - 1) + \frac{\beta_2}{\beta_0}\right)\alpha_{s0}^3(s),$$

$$\alpha_{s0} = \frac{1}{\beta_0 \ln(\frac{s}{\Lambda_{\overline{MS}}^2})}, \quad L = \ln(\ln(\frac{s}{\Lambda_{\overline{MS}}^2})),$$

$$\beta_0 = \frac{33 - 2 N_f}{3}, \quad \beta_1 = \frac{2(153 - 19 N_f)}{3}, \quad \beta_2 = \frac{1}{2}(2857 - \frac{5033}{9} N_f + \frac{325}{27} N_f^2),$$

(1.2)

where N_f is the number of actual flavors. Coefficients in higher order axial couplings, C_2^A and C_3^A, are approximated to be equal to corresponding vector ones, C_2^V and C_3^V, respectively. C_3^V was recently calculated by S. G. Gorishny et al.[2].

In this paper, we first present the renormalization scheme we used and its feature. Next we discuss the radiative correction factor to R-ratio and give some comments to obtain R-ratio from data. Then the summary will be given.

2. Renormalization scheme

We adopt the on-shell renormalization scheme developed by Kyoto group [3]. In this scheme the divergence are renormalized into the following set;

$$e, M_Z, M_W, m_f, M_{Higgs},$$

where e is the electromagnetic coupling constant, and M_W, M_Z, m_f and M_{Higgs} are masses of the W- and Z-boson, the fermion and the Higgs boson, respectively. And the renormalization is performed on the mass-shell of particles just like QED after symmetry breaking. Thus in this scheme we do not need extra factor for external lines. The ultraviolet divergence is regularized by dimensional regularization, and the infrared divergence is done by introduction of the fictitious photon mass. In the following calculation we take the 't Hooft-Feynman gauge.

In the energy region near Z_0 boson mass, the correction is large and it comes from the fermion vertex correction but not from Z_0 self energy in this scheme. To show this feature, we first introduce the renormalization constants of neutral gauge bosons. They are given by the following formula;

$$\begin{pmatrix} Z_0 \\ A_0 \end{pmatrix} = \begin{pmatrix} Z_{ZZ}^{1/2} & Z_{ZA}^{1/2} \\ Z_{AZ}^{1/2} & Z_{AA}^{1/2} \end{pmatrix} \begin{pmatrix} Z \\ A \end{pmatrix}, \qquad (2.1)$$

where the fields Z and A with (without) suffix 0 denote the bare (renormalized) physical fields of Z-boson and photon, respectively. These renormalization constants are obtained from the transverse parts of the proper self-energies, Π_T, through the on-shell conditions,

$$\begin{aligned} \Pi_T^A(0) &= 0, \Pi_T^{A'}(0) = 0, \\ \Pi_T^Z(M_Z^2) &= 0, \Pi_T^{Z'}(M_Z^2) = 0, \\ \Pi_T^{ZA}(M_Z^2) &= 0, \Pi_T^{ZA}(0) = 0, \end{aligned} \qquad (2.2)$$

where the suffix T denotes to take the transverse part. For 1-loop level, Z factors are obtained as follows;

$$\begin{aligned} Z_{AA}^{1/2} &= 1 - \frac{e^2}{8\pi^2}[\frac{1}{3}\sum_i Q_i^2(C_{UV} - \ln m_i^2) - \frac{3}{4}(C_{UV} - \ln M_W^2) - \frac{1}{6}], \\ Z_{ZZ}^{1/2} &= 1 - \frac{e^2}{24\pi^2}\frac{C_{UV}}{M_W^2(M_Z^2 - M_W^2)}[\frac{N}{3}(8M_W^4 - 10M_W^2 M_Z^2 + 5M_Z^4) \\ &\quad - \frac{1}{8}(18M_W^4 + 2M_W^2 M_Z^2 - M_Z^4)], \\ Z_{AZ}^{1/2} &= \frac{e^2}{12\pi^2}\frac{C_{UV}}{M_W^2\sqrt{M_Z^2 - M_W^2}}[-\frac{N}{3}(8M_W^2 - 5M_Z^2) + \frac{1}{8}(30M_W^2 + M_Z^2)], \\ Z_{ZA}^{1/2} &= -\frac{e^2}{8\pi^2}\frac{M_W}{\sqrt{M_Z^2 - M_W^2}}(C_{UV} - \ln M_W^2). \end{aligned} \qquad (2.3)$$

where

$$C_{UV} = 1/\epsilon - \gamma + \ln 4\pi,$$

and γ is the Euler constant. m_i and Q_i are the mass and the charge of fermion and M_W and M_Z are the W- and Z-boson masses, respectively. N is the number of generations. We can see that in $Z_{AA}^{1/2}$ there are terms $\ln m_i^2$ and $\ln M_W^2$, therefore it has a large value.

Next we consider the charge renormalization constant Y;

$$e_0 = Y e. \tag{2.4}$$

We determine it through eeA vertex(fig. 1) under the following condition;

$$\bar{u}(m_e)\Gamma^\mu u(m_e)|_{k^\mu=0} = 0, \tag{2.5}$$

where Γ^μ is the renormalized vertex. This condition corresponds to the Thomson limit. At the 1-loop level, the counter term which is needed to calculate Γ^μ is expressed as follows;

$$\begin{aligned}\Gamma_C^\mu = &- e\gamma^\mu(Y + Z_{AA}^{1/2} + \frac{Z_L^e + Z_R^e}{2} - \frac{Z_L^e - Z_R^e}{2}\gamma_5) \\ &+ \frac{e}{4M_W\sqrt{M_Z^2 - M_W^2}} Z_{ZA}^{1/2}\gamma^\mu[3M_Z^2 - 4M_W^2 + M_Z^2\gamma_5],\end{aligned} \tag{2.6}$$

where Z_L^e and Z_R^e are the renormalization constants of the electron two-point function obtained from the on-shell condition. The constant Y can be obtained from (2.4) as the function of $Z_{AA}^{1/2}$, $Z_{ZA}^{1/2}$, Z_L^e and Z_R^e;

$$\begin{aligned}Y &= Y(Z_{ZZ}^{1/2}, Z_{ZA}^{1/2}, Z_L^e, Z_R^e), \\ &= 1 - \frac{e^2}{16\pi^2}[\frac{7}{2}(C_{UV} - \ln M_W^2) + \frac{1}{3} - \frac{2}{3}\sum_i Q_i^2(C_{UV} - \ln m_i^2)].\end{aligned} \tag{2.7}$$

Since the terms $\ln m_i^2$ and $\ln M_W^2$ come from $Z_{AA}^{1/2}$ and $Z_{ZA}^{1/2}$, Y has a large value.

Before going into the Z-boson vertex, we consider the W-boson− vertex in Fig. 2. Its coupling is as follows;

$$\Gamma_\mu^{tree}(W, f_I, f_i) = \frac{eM_Z}{2\sqrt{2(M_Z^2 - M_W^2)}} U_{Ii}^\dagger \gamma_\mu(1 - \gamma_5), \tag{2.8}$$

where U_{Ii} is the Kobayashi-Maskawa matrix element. From this tree coupling we can construct the counter term,

$$\Gamma_\mu^C(W, f_I, f_i) = Y G_2 (Z_L^{1/2})_{IJ} \Gamma_\mu^{tree}(W, f_J, f_j)(Z_L^{1/2\dagger})_{ji} Z_W^{1/2}, \quad (2.9)$$

where

$$G_2 = \frac{G_Z}{H}, \quad (2.10)$$

with

$$G_Z = \sqrt{1 + \frac{\delta M_Z^2}{M_Z^2}},$$
$$H = \sqrt{1 + \frac{\delta M_Z^2 - \delta M_W^2}{M_Z^2 - M_W^2}}, \quad (2.11)$$

and $\delta M_Z^2, \delta M_W^2$ and $Z_W^{1/2}$ are the Z- and W-bosons mass renormalization constants and the wave function renormalization constant of W-boson, respectively. Consequently, the vertex correction is large due to the counter term including the large terms originated from $Z_{AA}^{1/2}$, i.e. running α effects. The same discussion is valid for Z-vertex. On the other hand, the contribution from Z-boson self-energy is very small at $q^2 \sim M_Z^2$, because Z-boson is renormalized at $q^2 = M_Z^2$ as expressed in eq.(2.2).

We take the on-mass shell renormalization scheme just like QED but there are other on-mass shell renormalization schemes. For example the scheme developed by Böhm and Hollik [4] does not require the following condition;

$$A_T^{Z'}(M_Z^2)|_{phys.} = 0. \quad (2.12)$$

In their scheme due to the condition used instead of eq.(2.12), the contribution from Z-boson self energy is large, while the vertex correction is small. But because of the renormalization scheme independence, the summation of Z-boson self energy and the vertex correction is the same between each other.

3. The radiative correction factor

The R ratio is experimentally obtained by

$$R = \frac{N_{obs} - N_{BG}}{\epsilon(1+\delta)L} \cdot \frac{1}{\sigma^0_{\mu\mu}},$$

where N_{obs} and N_{BG} are the number of observed hadronic events and that of estimated background contamination, respectively. ϵ is the detection efficiency, and $1+\delta$ is the radiative correction factor. L is the integrated luminosity, and $\sigma^0_{\mu\mu}$ is the lowest order point-like QED cross section of μ pair production.

Here we consider the following three types of the correction;

1. Berends-Kleiss-Jadach's initial state correction [5] .

2. Initial state correction in electroweak theory.

3. Full correction in electroweak theory.

The correction of type 1, expressed in fig. 3, is commonly used in PEP/PETRA experiments. It is essentially the QED correction applied irrespective of the Z-boson vertices, that is, the renormalization is carried out in QED assuming the QED vertex counter term both for the vector and axial-vector vertices. It is not obtained in electroweak theory. Therefore this correction have ambiguity in determination of axial vector coupling between Z boson and the fermion. TRISTAN used this correction in the early stage of its analysis. [6] . It is installed in the LUND Monte Carlo(M.C.) [7] and the EPOCS M.C. [8] .

The Feynman diagrams for type 2 are depicted in fig. 4. It corresponds to the electroweak version of type 1. As the renormalization is carried out in the frame-work of the on-shell scheme of electroweak theory, there is no ambiguity to determine the coupling. But as mentioned above, this correction depends on the renormalization scheme, because the final vertex is not corrected.

Type 3 is the full $O(\alpha)$ correction in electroweak theory. For this correction, the Feynman diagrams in fig. 5 are added to ones of type 2. [9] . For the b-quark pair production, we kept the mass of top- and bottom-quark in the vertex correction and in the box diagrams [10] .

When we choose M_Z =91.9 GeV/c^2, m_t = 40GeV/c^2 and m_ϕ =100GeV/c^2, we obtain M_W =80.63 GeV/c^2 by M_Z-M_W relation in 1-loop level [11] . At $\sqrt{s}$=60 GeV, the calculated R-ratios are as follows;

$$R_0 = 4.901,$$
$$R_1 = 6.109,$$
$$R_2 = 6.211, \quad (3.1)$$
$$R_3 = 6.300,$$

where R_0, R_1, R_2 and R_3 are the calculated R-ratio of tree level, corrected by type 1, by type 2 and by type 3, respectively. These values do not include QCD correction and calculated in the maximum fractional energy k_{max} for the radiated photon which is set 0.99 except that for b-quark it is 0.97. The radiative correction factor of type 1 is;

$$\frac{R_1}{R_0} = 1 + \delta_I = 1.245. \quad (3.2)$$

The ratio between R-ratio corrected by type 2 or 3 and one corrected by type 1 is as follows;

$$\frac{R_2}{R_1} = 1.017,$$
$$\frac{R_3}{R_1} = 1.031. \quad (3.3)$$

The above difference, 3.1%−1.7%=1.4%, mainly comes from pure weak correction to the final vertex, while the QED correction to the final state is very small, order of $1 + \alpha/\pi \sim 0.2\%$, and the contributions from the pure weak boxed diagrams are less than 0.1%. Consequently, in TRISTAN energy region, the pure weak correction to R-ratio reaches about 3%. This value is consistent with Burgers-Hollik's result [12] . Fig. 6 shows the value of $1 + \delta$ calculated by type 3 as a function of $\sqrt{s}$. We can see that at $\sqrt{s}$=70GeV the pure weak correction reach 7.4%. Fig. 7 shows the dependence of $1+\delta$ on M_Z, M_{top} and M_{Higgs} at $\sqrt{s}$=60GeV. For M_{top} <100GeV/c^2, the dependence of $1 + \delta$ on M_{top} is very small, but at M_{top}=200GeV/c^2 the deviation reaches −1.3%. While the deviation caused by changing M_{Higgs} does not exceed 0.5%. in the region 10GeV/c^2< M_{Higgs} <1000GeV/c^2.

The detection efficiency ϵ is simply defined to be the fraction of the number of simulated events which pass through the selection cuts in the number of generated events. We simulate the events using the LUND M.C. with only initial state correction since the other radiative correction such as the final state correction does not change the detection efficiency.

4. Summary

In analysis of R-ratio at TRISTAN, the full radiative correction obtained in the on-shell renormalization scheme in electroweak theory are applied. In TRISTAN energy region, the pure weak correction reaches about 3%. Through the loop effects, the heavy top-quark makes the radiative correction factor reduced.

This work is the collaboration with TRISTAN theory group organized by Prof. Shimizu. I would like to thank the TOPAZ collaboration and particularly T. Tauchi for discussions, and I. Adachi for valuable help in preparation of my talk.

REFERENCES

1. K. G. Chetyrkin *et al.*, *Phys. Lett.* **85B**, 277 (1979);
   W. Celmaster and R. J. Gonsalves, *Phys. Rev. Lett.* **44**, 560 (1980);
   T. Appelquist and H. D. Politzer, *Phys. Rev. D* **12**, 1404 (1975);
   J. Jersak *et al.*, *Phys. Lett.* **98B**, 363 (1981).

2. S. G. Gorishny, A. L. Kataev, and S. A. Larin, a talk presented at the Hadron Structure'87 Conference, Bratislava, Nov, 1987; *Phys. Lett.* **212B**, 238 (1988).

3. K. Aoki, Z. Hioki, R. Kawabe, M. Konuma and T. Muta, *Prog. Theor. Phys.* **73**(1982)1.

4. M. Böhm, W. Hollik and H. Spiesberger, *Fortschr. Phys.* **34**(1987)687.

5. F. A. Berends and R. Kleiss, *Nucl. Phys.* **B178**, 141 (1981);
   F. A. Berends, R. Kleiss and S. Jadach, *Nucl. Phys.* **B202**, 63 (1982).

6. TOPAZ Collaboration, I. Adachi *et al.*, *Phys. Rev. Lett.* **60**, 97 (1988)

7. T. Sjöstrand and M. Bengtsson, Computer Physics Communication **43**(1987)367.

8. K. Kato and T. Munehisa, KEK-Report 87-5

9. M. Igarashi, N. Nakazawa, T. Shimada, Y Shimizu, *Nucl. Phys.* **B263**, 347 (1986)

10. J. Fujimoto and Y. Shimizu, *Mod. Phys. Lett.* A3(1988)581;
    J. Fujimoto, Ph.D. Thesis, Nagoya University, 1987

11. Z. Hioki, *Prog. Theor. Phys.* **68**(1982)2134, **71**(1984)663;
    W. J. Marciano and A. Sirlin, *Phys. Rev. D* **29**, 945 (1984).

12. W. de Bore, DESY 89-067

FIGURE CAPTIONS

1) eeA vertex.

2) W-boson–fermions vertex

3) The Feynman diagrams for the radiative correction to $e^+e^- \to q\bar{q}$ included in the BKJ.

4) The Feynman diagrams for the initial state radiative correction to $e^+e^- \to q\bar{q}$ in electroweak theory.

5) The added Feynman diagrams for the full $O(\alpha)$ correction to $e^+e^- \to q\bar{q}$ in electroweak theory.

6) The value of $1 + \delta$ calculated by full electroweak theory as a function of $\sqrt{s}$ with M_Z=91.9GeV/c², M_{top}=40GeV/c² and M_{Higgs}=100GeV/c². The value of the BKJ is also shown by a dotted curve.

7) The dependence of $1 + \delta$ on (a) M_Z, (b) M_{top}, and (c) M_{Higgs} at $\sqrt{s}$=60GeV, where vertical axis is the deviation of $1 + \delta$ from the standard constants.

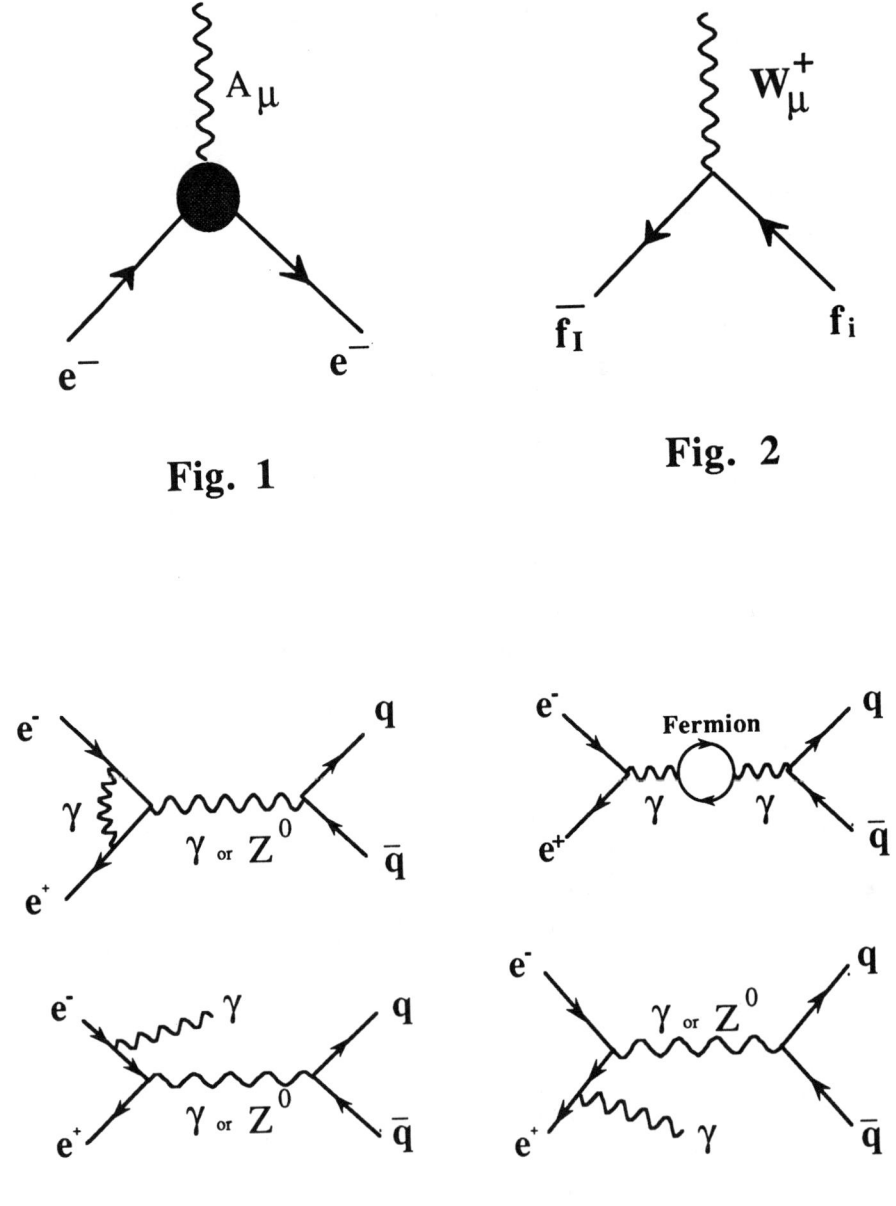

Fig. 1

Fig. 2

Fig. 3

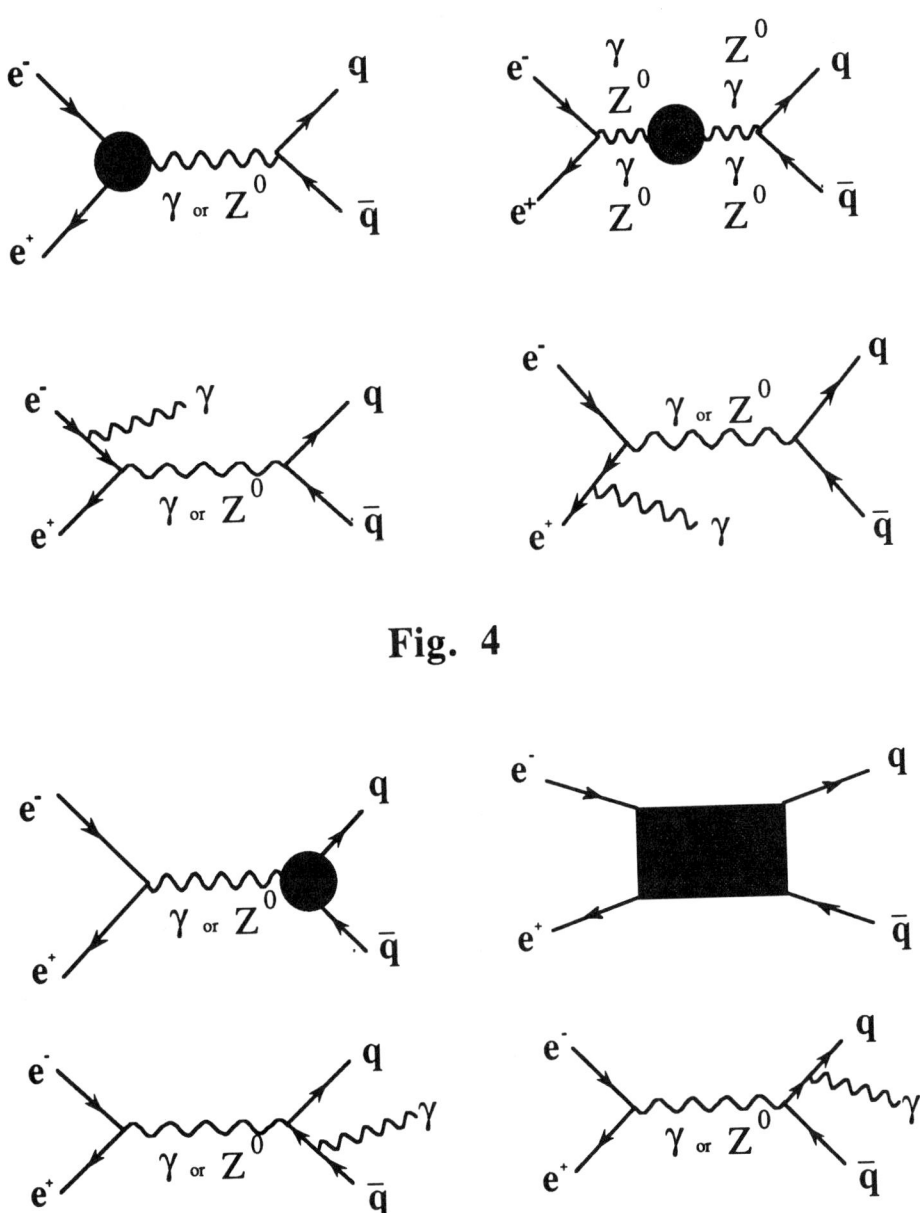

Fig. 4

Fig. 5

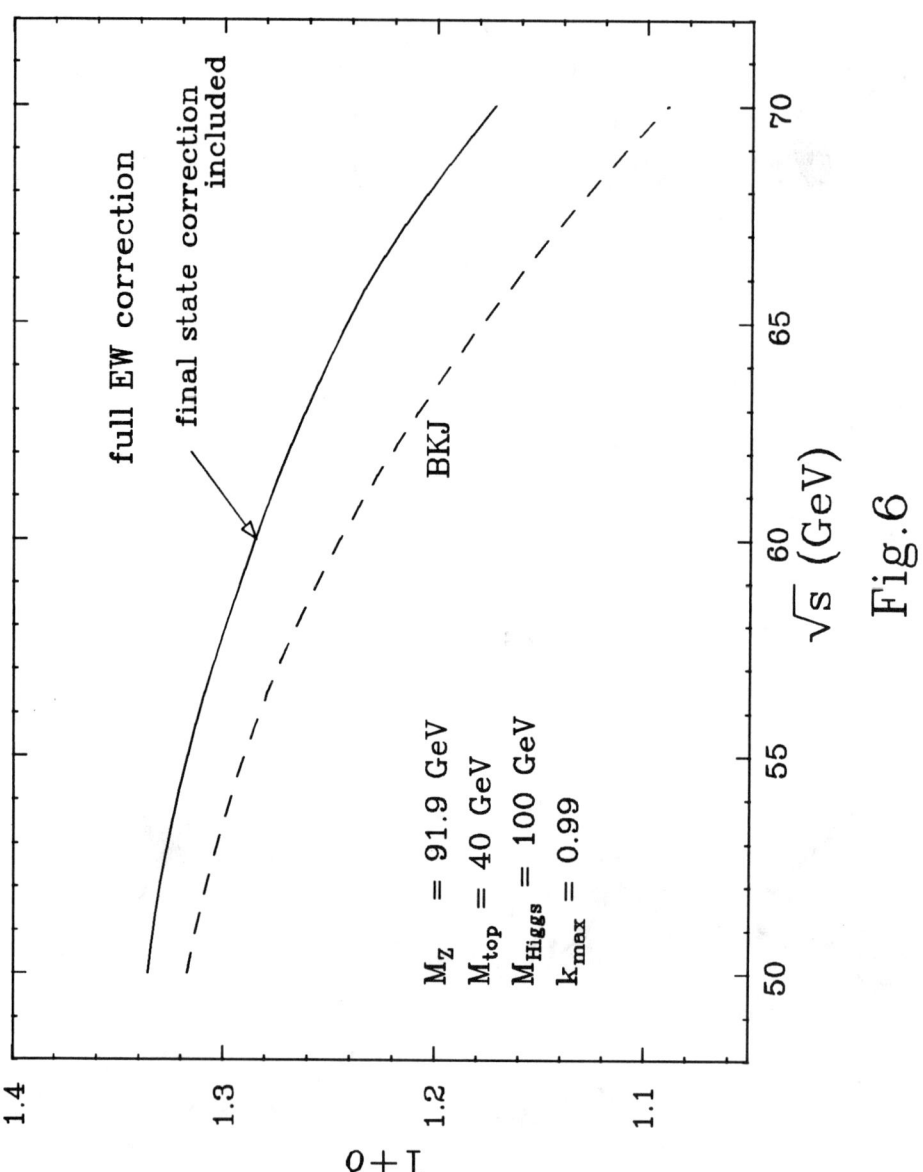

Fig.6

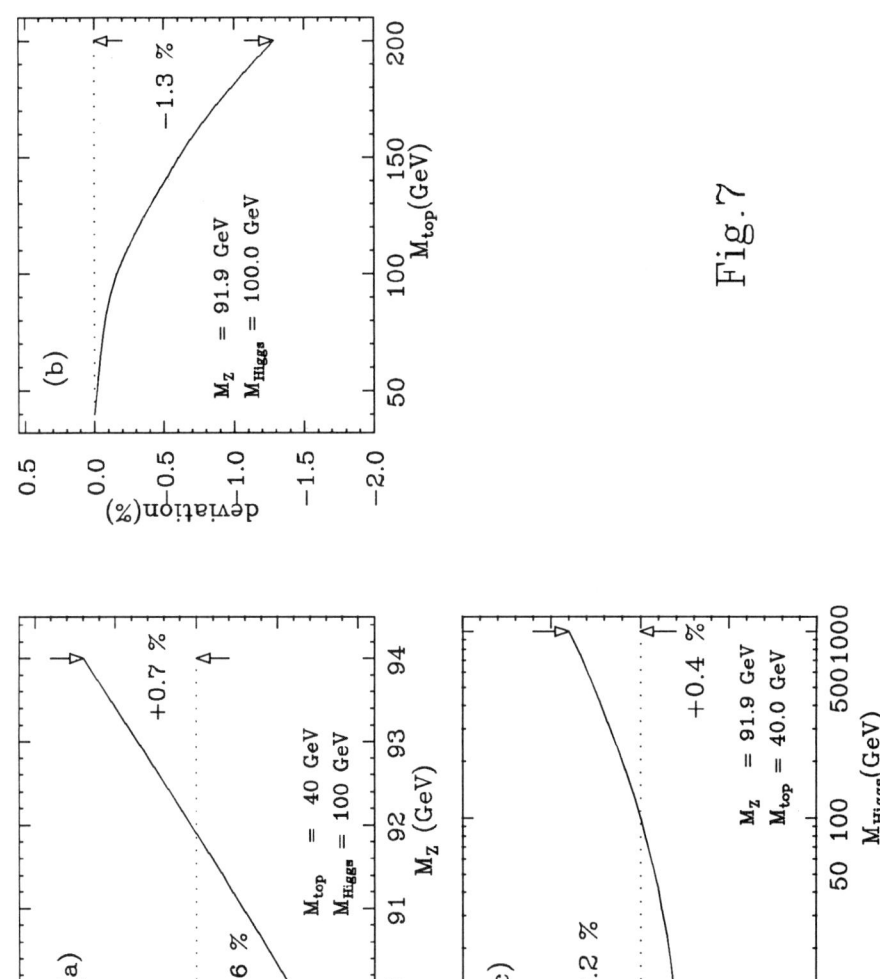

Fig. 7

HEAVY FLAVOR RESONANCES AND QED RADIATIVE CORRECTIONS*

J. ALEXANDER, G. BONVICINI, P. DRELL, R. FREY AND V. LUTH

Lawrence Berkeley Laboratory, University of California, Berkeley CA 94720

University of Michigan, Ann Arbor, MI 48109

Stanford Linear Accelerator Center, Stanford University, Stanford CA 94309

ABSTRACT

An application of high precision QED against experimental data is presented. When the corrections to ψ and Υ families are improved according to the method described below, the masses and widths of the resonances below open flavor trheshold change by up to three standard deviations from presently accepted experimental values.

Presented by G. Bonvicini,
Ann Arbor Workshop on QED Structure functions, May 22-25, 1989.

---

*Work supported by the Department of Energy, contracts DE–AC03–76SF00515 (SLAC), DE–AC03–76SF00098 (LBL), and DE–AC02–76ER01112 (U. of Mich.).

In this paper we examine in detail the QED corrections to very narrow resonances such as the ψ and Υ particles. We focus on the determination of their mass M, their total width Γ, and their partial width to electrons Γ_e^0. Our objective is to review experimental results that were obtained in analyses with incorrect radiative corrections. Our analysis shows in fact that the errors incurred are sometimes bigger than the uncertainties quoted for the current world averages.[1]

In the presence of a very narrow resonance, the radiative corrections depend critically on the exact treatment of the infrared region, and differences in the formulae used to fit the data have an impact on many resonance parameters. A basic understanding of the infrared divergences associated with the vanishing photon mass was first achieved by Bloch and Nordsieck in 1937.[2] They stated that in charged particle scattering the number of photons emitted is undetermined, and that the cross section for the emission of zero energy and no photons is exactly zero. Many treatments of radiative corrections that exist in the literature violate this theorem by containing terms that correspond to a finite, non-zero probability for the emission of no photons. In particular, a result from truncated perturbation expansion alone produces elastic (*i.e.*, with no photon emitted) terms which violate the Bloch-Nordsieck theorem. All the results which were shown to produce errors in the extraction of the resonance mass and width at the Z^0 include such terms.[3]

1. Initial State Radiative Corrections to Narrow Resonances

In e^+e^- collisions, the nominal collision energy, $\sqrt{s} = 2E$, is set by E, the energy of the incident beams. The actual c.m. energy available for the annihilation is reduced by Bremsstrahlung to $\sqrt{s(1-k)}$, where kE is the total energy of the emitted photons. The observed cross section, $\sigma_{obs}(s)$ at the nominal energy $\sqrt{s}$, can be written as a convolution of the cross section $\sigma_0(s(1-k))$ and a dimensionless sampling function $f(k,s)$,[3]

$$\sigma_{obs}(s) = \int f(k,s)\sigma_0(s(1-k))\,dk. \tag{1.1}$$

and

$$\sigma_0 = \sigma_{nonres} + \sigma_{peak}\frac{s\,\Gamma^2}{(s-M^2)^2 + s\,\Gamma^2}, \tag{1.2}$$

where M is the mass and Γ is the total width of the resonance.

Expressions for $f(k,s)$ which attain the required precision of 1% have been obtained by several authors in the literature.[4-7] It is well known that $f(k,s)$ is dominated by inital state effects.[8] Effects of final state radiation on the cross section are usually ignored at the fraction of a percent level.

We employ the following expression for $f(k,s)$, truncated to first order in the hard photon terms, and to second order in the vertex terms:

$$f(k,s) = (1+\delta_{vp})(1+K)[\beta k^{\beta-1}(1+\delta_1+\delta_2) - \beta(1-\frac{k}{2})]. \tag{1.3}$$

β is the electron equivalent radiator thickness,

$$\beta = \frac{2\alpha}{\pi}(\log\frac{s}{m_e^2} - 1). \tag{1.4}$$

The δ_n terms arise from the leading parts of the vertex correction diagrams of order n, δ_{vp} is the vacuum polarization contribution and K is the K−factor. All these terms are reported in the Appendix.

In the past, most experimenters have fit the narrow resonances of the ψ and Υ families using a different expression for $f(k,s)$, based on the classic work of Jackson and Scharre,[9]

$$f'(k,s) = \delta_{tot}\delta(k) + \beta k^{\beta-1} - \beta(1 - \frac{k}{2}), \qquad \delta_{tot} = \delta_1 + \delta'_{vp} + K. \qquad (1.5)$$

$\delta(k)$ refers to the Dirac function. This expression has been obtained from a first order perturbative calculation with the inclusion of exponentiation.

There are essential differences between the distribution functions $f(k,s)$ and $f'(k,s)$ and the way they get convoluted according to Eq. (1.1). The differences occur in second order in α. First, in the formulation by Jackson and Scharre, the photon vacuum polarization δ'_{vp} is approximated by the electron loop δ_e only, excluding contributions from hadrons, muons, and τ leptons, δ_h, δ_μ, and δ_τ. Secondly, the vertex correction, $(1 + \delta_1)$, should multiply the Bremsstrahlung term $k^{\beta-1}$, at least to first order, and hence should enter as an overall multiplicative factor to the soft term, as in our definition of $f(k,s)$. The factorization of the virtual terms arises naturally from those semi-classical formalisms which are based on factorization principles.[10][11] This factorization of the virtual corrections can be checked to first order by doing an explicit second order calculation. A second order calculation does not, however, determine unambiguously that the δ_2 term factorizes, though it is a natural choice and it agrees with the Bloch-Nordsieck theorem. In the definition of $f'(k,s)$, the virtual corrections were not properly separated and the $\delta(k)$ term gives a finite probability for the electron and positron to annihilate without soft photon emission, in direct disagreement with the Bloch-Nordsieck theorem. This locally distorts the cross section by a fraction $\delta_{tot} \approx 14\%$ at 10 GeV and $\delta_{tot} \approx 10\%$ at 3 GeV.

The convolution integral of a Breit-Wigner resonance cross section with $f(k,s)$ can be solved analytically. We use the expression given in the Appendix, which was derived by Cahn[5] for the Z^0 resonance. We have added the photon vacuum polarization and δ_2 terms. We also account for the energy spread of the incident beams (which can be two orders of magnitude bigger than the resonance width, therefore forbidding direct observation of the resonance structure) by further convoluting the cross section with a Gaussian resolution function of the appropriate width σ_E. The error associated with Eq.(1.3) is of order 1%, mostly in the normalization due to vacuum polarization uncertainties.

2. Distortion of the Resonance Shape and Analysis Method

A resonance is described by its mass, M, and two of the following three parameters: the total width, Γ, the cross section integral A, and the cross section at the peak, σ_{peak}. These three parameters are related by the equation

$$A = \frac{\pi}{2} \Gamma \sigma_{peak}. \tag{2.1}$$

A can be related to the measured partial width to electrons, Γ_e^{exp}, and the branching ratio for this process, in our case B_{had}, by

$$A = \frac{\pi}{2} \Gamma \sigma_{peak} = \frac{6\pi^2}{M^2} \Gamma_e^{exp} B_{had} \quad \text{with} \quad B_{had} = \frac{\Gamma_{had}}{\Gamma}. \tag{2.2}$$

Under the assumption that the total width is the sum of the partial width to hadrons and charged lepton pairs and that the leptonic widths are all equal, we have

$$\Gamma = \Gamma_{had}^{exp} + m\Gamma_e^{exp}, \quad \text{and} \quad mB_e + B_{had} = 1. \qquad (2.3)$$

Here m stands for the number of partial widths into lepton pairs, $m = 2$ for charmonium and $m = 3$ for bottomonium states. The leptonic branching ratios are determined experimentally, and therefore the relations above can be used to measure the quantities Γ and Γ_e^{exp}.

We note explicitly the nature of Γ_e^{exp}, defined in Eq. (2.3), and draw the distinction with the quantity of theoretical interest,[12] Γ_e^0. Γ_e^{exp} is the physical coupling of the resonance to leptons through one photon, and is obtained from the data by making all radiative corrections *except* vacuum polarization corrections. This is the quantity which, divided by the measured branching ratio, gives the total width. The value of Γ_e^0, on the other hand, is drawn from the data by making all radiative corrections *including* vacuum polarization. Thus Γ_e^0 reflects the coupling strength at tree level only. The quantity Γ_{had}, which couples to the resonance mostly through three gluons, does not have QED vacuum polarization corrections, and in this case $\Gamma_{had}^{exp} = \Gamma_{had}^0$.

Historically, experimenters have generally included some level of vacuum polarization in their corrections, and have therefore implicitly extracted Γ_e^0. For the remainder of our discussion we follow this precedent, though at the end we include values for Γ_e^{exp} in summary tables. The relationship between the two quantities is

$$\Gamma_e^{exp} = (1 + \delta_{vp})\Gamma_e^0. \qquad (2.4)$$

Since radiative effects in the final states are negligible, the branching ratios do not depend on radiative corrections. Thus differences in the formulation of the radiative corrections will cause changes in two parameters, the integral A and the partial width Γ_e^0. They will scale proportionally, with a factor that depends on the branching ratio for the particular channel under study. If one studies simultaneously the resonance cross sections into hadrons, muon pairs, and electron pairs, the three integrals will change by the same fraction, giving approximately the same change to Γ_e^0, while the ratio between the three integrals (which determines the branching ratios) remains unchanged.

The difference between our treatment of the radiative corrections and the formulation by Jackson and Scharre is illustrated in Figure 1. We plot the difference between the cross section for the $\Upsilon(9460)$ calculated with $f(k,s)$ and $f'(k,s)$ using the same input parameters. $f'(k,s)$ overestimates the cross section on the resonance and below it, and underestimates the cross section above the resonance. We illustrate in this figure both the case where the vacuum polarization in $f'(k,s)$ includes all terms, $\delta'_{vp} = \delta_e + \delta_\mu + \delta_\tau + \delta_h$, and where it is reduced to the electron loop, $\delta'_{vp} = \delta_e$. This latter case is, in fact, the formulation that most previous experiments had used to fit narrow resonances. It is evident that the use of the electron loop alone in the vacuum polarization reduces the difference in the predicted cross section at the peak resulting from the incorrect treatment of the virtual terms in $f'(k,s)$.

The magnitudes of the shifts in the parameters obtained by the fit to the resonance enhancement will depend on details that will vary from experiment to experiment, such as the ratio of resonant to non-resonant cross section, the amount of integrated luminosity taken on the peak, and the energy spread of the machine.

To correctly reproduce the complicated interplay of the fit parameters and to study the dependence and correlations among them, we resort to a technique of simulating the data obtained by various experiments to measure the ψ and Υ resonances. We generate data points by calculating the cross section at a given energy $\sqrt{s}$ using our definition $f(k,s)$ and errors proportional to $\sqrt{\sigma_{obs}}$. Subsequently, the generated data points are fit by functions based on both $f(k,s)$ and $f'(k,s)$. We study the changes to the fitted resonance parameters using the hadronic cross sections only. The four free parameters of the fit are M, Γ, σ_E and σ_{nonres}. B_e, the branching ratio into electrons, is fixed at the world average value.[1]

For a compact presentation of the results in the following section we find it convenient to introduce the ratio

$$C = \frac{\delta_{tot}}{\delta_1 + \delta_{vp} + \delta_2 + K}, \qquad (2.5)$$

with the term δ_{tot} as defined in $f'(k,s)$. The denominator is the factor multiplying the soft term in $f(k,s)$ when we expand the product with σ_0 and assume the virtual terms are small. In the denominator, we take $\delta_{vp} = \delta_e + \delta_\mu + \delta_\tau + \delta_h$, while the value of δ'_{vp} implicitly contained in δ_{tot} may be reduced to δ_e as in the Jackson and Scharre ansatz. Using this ansatz we obtain $C = 0.85$ and $C = 0.70$ at 3.1 GeV and 10 GeV, respectively. Using full vacuum polarization in δ_{tot} we obtain $C = 1.03$.

3. Analysis of Simulated Data

In this section, we show how we apply corrections to published experimental results on the parameters of narrow resonances based on fits to our simulated data. We deliberately consider only experiments listed in the 1986 Review of Particle Properties.[1] In changing values of the resonance parameters derived from previously applied radiative corrections to new values derived with our definition of the sampling function $f(k, s)$, we strictly use information contained in the original experimental[13] and theoretical[4][9][14] papers.

In correcting published values of the resonance parameters, we take account of the fact that experiments differ from one another in several significant ways. First, e^+e^- storage rings differ in their energy resolution. Second, in different experiments, the percentage of the total luminosity collected on the resonance peak, as compared to below or above the peak, can vary substantially. These effects introduce small ($\approx .5 - 1\%$) differences in the corrections to different experiments which have been taken into account. Finally, most of the measurements have been radiatively corrected based on the prescription by Jackson and Scharre.[9] For those, we typically derive changes in Γ_e^0 of 2% at the $\Upsilon(9460)$ by fits to simulated data. Adding the full and correct vacuum polarization to $f'(k, s)$, results in a large correction to Γ_e^0 of $\approx 9\%$. Other experiments derived resonance parameters using algorithms[4][14] which are identical to ours, except for the vacuum polarization terms, and the δ_2 term.

In summary, the fact that the changes to the resonance parameters vary from experiment to experiment is almost completely due to the differences in the radiative corrections (which in turn differ either in normalization or in the value of C).

The dependence of the correction to M and Γ_e^0 versus C will generate corrections to the leptonic width and mass as displayed in Figures 2 and 3. The shift in the mass ΔM is normalized to the energy resolution σ_E, because we find empirically that for a fixed ratio C the mass shift is proportional to σ_E. This behaviour can be understood, because the equivalent radiator thickness β is the same at the ψ and Υ to within 10%. These curves can be used to correct experimental results which are not listed here.

Table 1 lists the values previously measured along with our refitted ones for the experiments that determine the mass and widths of the ψ and Υ resonances and that are referenced in the 1986 Review of Particle Properties.[1] We would like to point out that our method is one of simulation; it shows fluctuations of typically 2-3% in the fitted parameters when cross sections are assigned errors that are comparable to those in published experiments. The overall error of our method, based on much smaller point to point errors, is conservatively estimated to be 1%.

Using the corrections to Γ_e^0, we have derived the corrections to Γ taking into account the error on the branching ratio. We decouple the measurement of Γ_e^0 and Γ by consistently using the world average branching ratio,[15] and not the particular value as measured by a given experiment.

Table 2 contains the summary of our results, presented in the form of new world averages for the resonance parameters that change significantly with our new analysis. Quantities which do not change the world average by at least 50% of a standard deviation are not listed here. The corrections to resonances above open flavor threshold resemble the corrections discussed for the Z^0,[3] and they are small.

4. Conclusions and Discussion

In conclusion, we have reviewed prescriptions for QED radiative corrections to resonance production by e^+e^- annihilation, and present a formula for QED corrections to narrow resonances (convoluted with a Gaussian resolution function to account for the spread in the beam energy) which has an estimated uncertainty of 1.0% in the 3-10 GeV energy region.

Recently two other papers[16][17] have dealt with the subject of radiative corrections to narrow resonances. Both use a formulation that is consistent with $f(k,s)$. Buchmüller and Cooper[17] rescale the results for the Υ states using only the peaks of the resonances, thereby obtaining changes to world averages which are slightly larger than ours. The correction method of Königsmann[16] gives results for the Υ resonances which are nearly identical to those of Ref. 17. However, his results for the ψ states differ substantially from ours, and we believe that this is because our method of simulating cross section data correctly accounts for the various nontrivial effects arising from a resonance fit. As discussed in Section 2, the J/ψ and ψ' resonance data are more sensitive to these effects than are the Υ data.

We have applied our prescription to correct existing measurements of the mass, total width, and electron partial width for the ψ and Υ resonances. The observed shifts are small, but when we combine the new values for all experiments and form new world averages, the changes are significant. The values of several quantities change as a result of our reevaluation of the radiative corrections by up to one standard deviation. The implications of our analysis for quarkonium potential models have been discussed elsewhere.[16][17]

Authors of this analysis are J. Alexander, G. Bonvicini, P.Drell, R. Frey and V. Lüth. We would like to thank L. Trentadue for useful discussions, and S. Cooper and K. Königsmann for helpful suggestions.

APPENDIX

We use in our analysis the form of the distribution function $f(k,s)$ convoluted with a Gaussian beam energy spread according to Cahn[5]. We add, however, the δ_2 term:

$$g(s) = \sigma_{peak}(1+\delta_1+\delta_2)\frac{\Gamma^2}{\Gamma^2+M^2}\left[\frac{s}{M^2}a^{\beta-2}\Phi(\cos\theta,\beta) - a^{\beta-1}\frac{\beta}{1+\beta}\Phi(\cos\theta,1+\beta)\right]$$

$$-\sigma_{peak}\beta\frac{\Gamma}{\sqrt{s}}\left[\tan^{-1}\frac{2M}{\Gamma} - \tan^{-1}\frac{2(M-\sqrt{s})}{\Gamma}\right].$$

The quantities a, $\cos\theta$, and $\Phi(\cos\theta,\beta)$ are defined as follows:

$$a^2 = \frac{b^2+c^2}{d}, \quad \cos\theta = -\frac{Mb+\Gamma c}{ad}, \quad \Phi(\cos\theta,\beta) = \frac{\pi\beta\sin((1-\beta)\theta)}{\sin\pi\beta\sin\theta}$$

where $b = M(s/M^2 - 1)$, $c = \Gamma s/M$ and $d = \Gamma^2 + M^2$. The terms δ_1, δ_2 are

$$\delta_1 = \frac{3}{4}\beta, \quad \delta_2 = -\frac{\beta^2}{24}(\frac{1}{3}\log\frac{s}{m_e^2} + 2\pi^2 - \frac{37}{4}).$$

The K−factor and δ_{vp} terms are

$$K = \frac{\alpha}{\pi}(\frac{\pi^2}{3} - \frac{1}{2}), \delta_{vp} = \delta_l + \delta_h \quad \text{with} \quad \delta_l = \delta_e + \delta_\mu + \delta_\tau.$$

The vacuum polarization contribution of charged leptons of mass m_i is

$$\delta_l = -\sum_i \frac{2\alpha}{\pi}(\frac{5}{9} + \frac{1}{3}\log\frac{m_i^2}{s}).$$

The hadronic part of the vacuum polarization, δ_h, is calculated numerically[18] and is $\delta_h = 1.1 \pm 0.5\%$ at 3 GeV and $3.4 \pm 1.0\%$ at 10 GeV. The quoted uncertainties are our estimates.

REFERENCES

1. Particle Data Group, Phys. Lett. 170B, (1986).

2. F. Bloch and A. Nordsieck, Phys. Rev. 52 (1937) 54.

3. J. Alexander et al., Phys. Rev. 37D (1988) 56.

4. M.Greco et al., Phys. Lett. 56B (1975) 367. Their result contains vertex pieces to first order and only the δ_e part of the vacuum polarization.

5. R.N.Cahn, Phys. Rev. 36D (1987) 2666. This result is not based on a coherent states calculation but is equivalent to M. Greco et al.

6. E.A.Kuraev and V.S.Fadin, Sov. J. Nucl. Phys. 41(3) (1985) 466.

7. F.A.Berends, G.J.H.Burgers and W.L.Van Neerven, Phys. Lett. 185B (1987) 395.

8. T. Kinoshita, J. Math. Phys. 3 (1962) 650; T.D. Lee and M. Nauenberg, Phys. Rev. 133B (1964) 1549.

9. J.D.Jackson and D.L.Scharre, Nucl.Instr.128 (1975) 13. See also F.A. Berends and G.J. Komen, Nucl. Phys. B115(1976) 114.

10. V.N. Gribov and L.N. Lipatov, Sov. J. Nucl. Phys. 15 (1972) 438.

11. E. Etim et al., Nuovo Cim. 51 (1967) 276.

12. V.N. Baier et al Phys. Rep. 78 (1981) 294; Ya. I. Azimov et al., JETP Lett. 21 (1975) 378; P. Tsai, SLAC-PUB-3129, unpublished.

13. See Ref. 1, Meson Full Listing, for a complete list of experimental references used.

14. Y.S.Tsai, SLAC-PUB-1515(December 1974). Here the virtual terms are treated correctly but because the nature of the J/ψ was not established at the time the vacuum polarization contributions are not explicitly included.

15. We use branching ratios from Ref. 1. Ref. 16 and Ref. 17 also use a single average branching ratio to obtain Γ, but more recent measurements are included there.

16. K. Königsmann, DESY 87-046(May 1987). We do not support the statement that the treatment by Kuraev and Fadin does not include vacuum polarization (see Eq. 6, Ref. 6). However, this is not relevant for the analysis of this paper.

17. W. Buchmüller and S. Cooper, MIT-LNS-159(March 1987).

18. F.A.Berends and G.J.Komen, Phys. Lett. 63B, (1976) 432.

TABLE CAPTIONS

1: Summary of the corrections to the parameters of ψ and Υ resonances, by experiment, as listed in Ref. 1, Meson Full Listing.

2: New world averages for those resonance parameters which change by more than 50% of a standard deviation. Also given are the percentage change in the experimental quantities, and the statistical significance of the change in units of overall experimental error.

FIGURE CAPTIONS

1) The difference between the cross section for the Υ (9460) calculated with $f(k,s)$ from Eq. (1.3) and with $f'(k,s)$ from Eq. (1.5) using the same input parameters. The solid line represents the difference for the full vacuum polarization terms in $f'(k,s)$, while the broken line gives the difference for only the electron contribution the vacuum polarization. C is defined later in the text.

2) Corrections to Γ_e^0 versus C for the five narrow resonances of the ψ and Υ families. The corrections to the $\Upsilon(10023)$, are roughly equal for $\sigma_E = 8$ MeV and for $\sigma_E = 4$ MeV.

3) Corrections to the mass M as a function of the ratio C. ΔM is given in units of the machine resolution σ_E.

Table 1

Quantity	Reference	New value Γ_e^0	New value Γ_e^{exp}	Old value
$\Gamma_e, J/\psi(3097)$	Boyarski	4.6 keV	4.8 keV	4.8 keV
$\Gamma_e, J/\psi(3097)$	Baldini	4.5 keV	4.7 keV	4.6 keV
$\Gamma_e, J/\psi(3097)$	Esposito	4.5 keV	4.7 keV	4.6 keV
$\Gamma_e, J/\psi(3097)$	Brandelik	4.5 keV	4.6 keV	4.4 keV
$\Gamma_e, \psi(3685)$	Lüth	2.0 keV	2.1 keV	2.1 keV
$\Gamma_e, \psi(3685)$	Brandelik	2.1 keV	2.2 keV	2.0 keV
$\Gamma_e, \Upsilon(9460)$	Berger	1.36 keV	1.46 keV	1.33 keV
$\Gamma_c, \Upsilon(9460)$	Bock	1.10 keV	1.18 keV	1.08 keV
$\Gamma_e, \Upsilon(9460)$	Albrecht	1.25 keV	1.34 keV	1.23 keV
$\Gamma_e, \Upsilon(9460)$	Niczyporuk	1.15 keV	1.24 keV	1.13 keV
$\Gamma_e, \Upsilon(9460)$	Tuts	1.18 keV	1.27 keV	1.15 keV
$\Gamma_e, \Upsilon(9460)$	Giles	1.42 keV	1.53 keV	1.30 keV
$\Gamma_e, \Upsilon(10023)$	Bock	0.40 keV	.43 keV	0.39 keV
$\Gamma_e, \Upsilon(10023)$	Niczyporuk	0.58 keV	.62 keV	0.56 keV
$\Gamma_e, \Upsilon(10023)$	Albrecht	0.60 keV	.65 keV	0.58 keV
$\Gamma_e, \Upsilon(10023)$	Tuts	0.58 keV	.62 keV	0.56 keV
$\Gamma_e, \Upsilon(10023)$	Giles	0.57 keV	.61 keV	0.52 keV
$\Gamma_e, \Upsilon(10355)$	Tuts	0.40 keV	.43 keV	0.39 keV
$\Gamma_e, \Upsilon(10355)$	Giles	0.46 keV	.49 keV	0.42 keV
$M, \Upsilon(9460)$	Artamonov	9460.5 MeV	-	9460.6 MeV
$M, \Upsilon(9460)$	Mac Kay	9459.87 MeV	-	9459.97 MeV

Table 2

Quantity	New world average	New world average, $\Gamma_e = \Gamma_\mu$	Fractional change	Statistical change
Γ_e^0, $J/\psi(3097)$	4.57± 0.51 keV	4.53± 0.35 keV	-4.0 %	0.5 σ
Γ_e^0, $\psi(3685)$	-	2.05 ± 0.21 keV	0	0
Γ_e^0, $\Upsilon(9460)$	-	1.279 ± 0.050 keV	4.5 %	1.1 σ
Γ_e^0, $\Upsilon(10023)$	-	0.569 ± 0.033 keV	6.0 %	1.0 σ
Γ_e^0, $\Upsilon(10355)$	-	0.423 ± 0.031 keV	5.2 %	0.7 σ
Γ_e^{exp}, $J/\psi(3097)$	4.77± 0.51 keV	4.72± 0.35 keV	+0.4 %	0.1 σ
Γ_e^{exp}, $\psi(3685)$	-	2.14 ± 0.21 keV	4.4 %	.4 σ
Γ_e^{exp}, $\Upsilon(9460)$	-	1.376 ± 0.050 keV	12.4 %	3.0 σ
Γ_e^{exp}, $\Upsilon(10023)$	-	0.612 ± 0.033 keV	14.0 %	2.3 σ
Γ_e^{exp}, $\Upsilon(10355)$	-	0.455 ± 0.031 keV	13.2 %	1.7 σ
Γ, $\Upsilon(9460)$	-	48.5 ± 3.2 keV	12.6 %	1.7 σ
Γ, $\Upsilon(10023)$	-	34.2 ± 7.3 keV	14.0 %	0.6 σ
M, $\Upsilon(9460)$	9459.93±0.19 MeV	-	0.001 %	0.5 σ

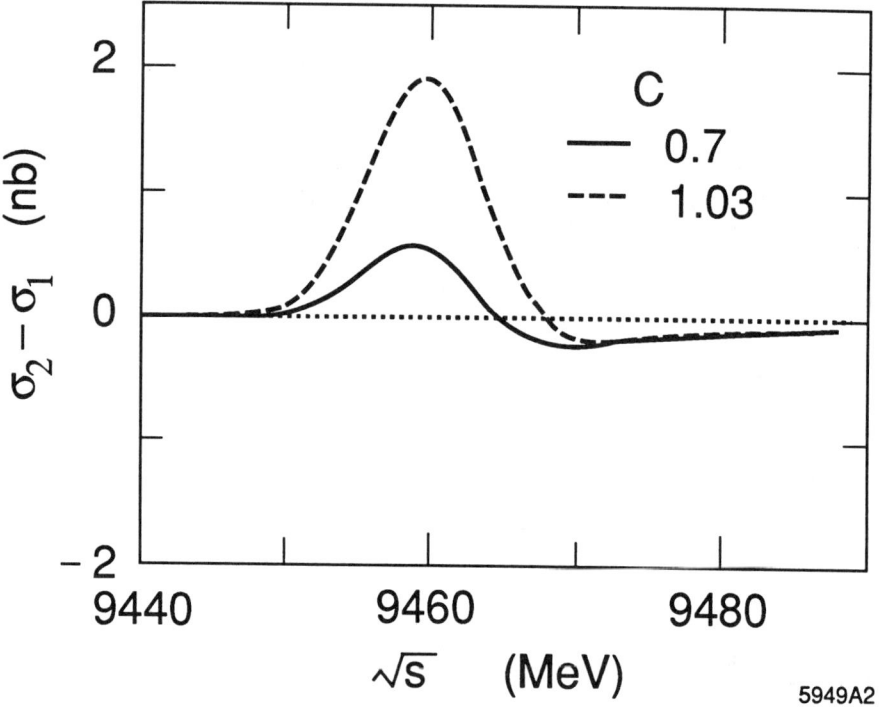

Fig. 1

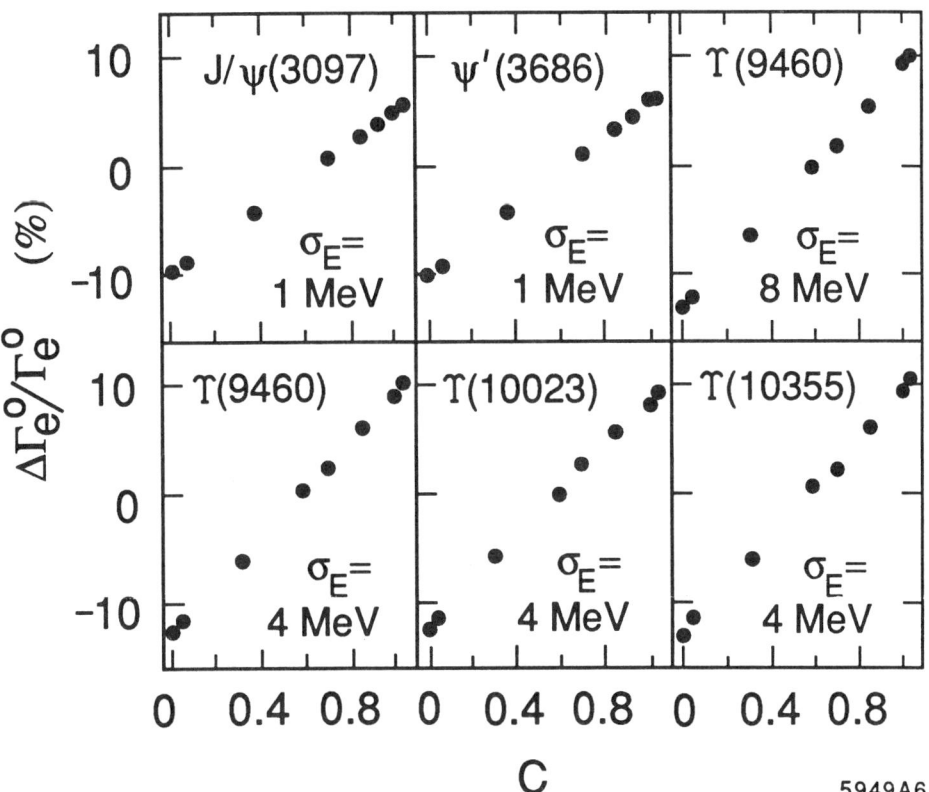

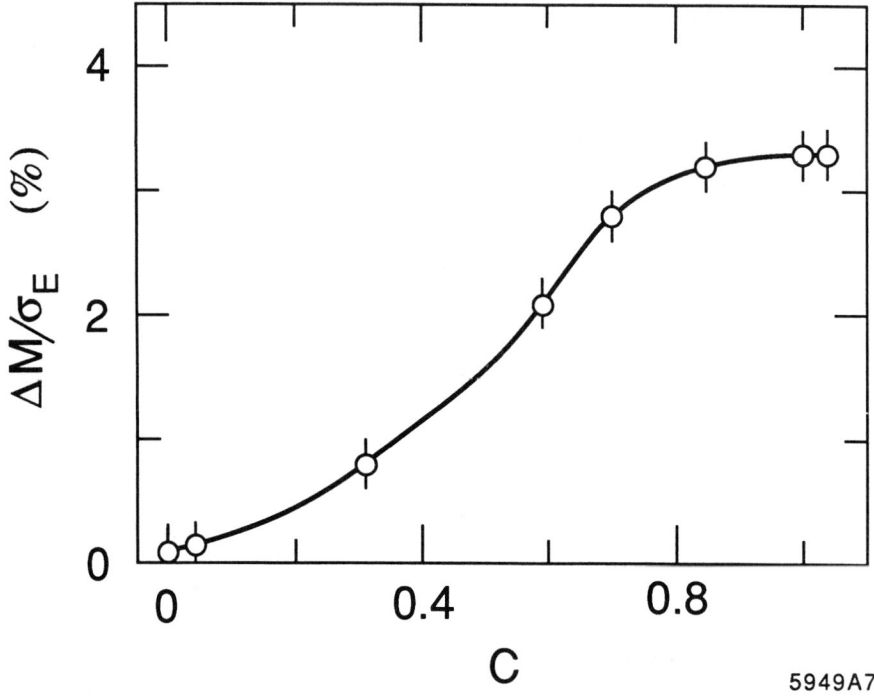

Fig. 3

AN INTRODUCTION ABOUT PRECISE MEASUREMENTS OF QED γ STRUCTURE FUNCTIONS

by

A. COURAU

Laboratoire de l'Accélérateur Linéaire

91405 ORSAY FRANCE

Pure QED processes are theoretically exactly computable. However precise measurements and theoretical expectations of QED γ structure functions within a given experimental acceptance are not so trivial. Yet such a study is quite interesting. It supplies on the one hand a good QED test and, on the other hand, a good exercise for testing the procedure used for the determination of the hadronic γ structure functions.

Let us note that, EXPERIMENTALLY, we never measure structure functions but only their contribution in some acceptance. Let us also note that measurements at 1% level imply statistical and systematical errors of the order of, or smaller than, 1%. This is obviously not within easy reach. But, actually, looking for some theoretical deviation or unexpected effects, it is often more efficient to select and analyse only a few well defined events.

I Main features

The classical way to measure the QED γ structure functions is to look at the deep inelastic $e\gamma \to e \ell^+ \ell^-$ process trough an $ee \to (e) \, e\ell^+\ell^-$ single tagged experiment. The target is the quasi-real γ corresponding to the unseen scattered electron, and the probe is the γ^* corresponding to the electron observed at small angle. Then in the $\gamma\gamma^*$ frame, one has

$$\frac{d\sigma^{\gamma^*\gamma \to \ell\ell}}{d\Omega^*} = \frac{d\sigma_T}{d\Omega^*} + \varepsilon \frac{d\sigma_L}{d\Omega^*} + \sqrt{2\varepsilon(\varepsilon+1)} \frac{d\sigma_{TL}}{d\Omega^*} \cos\varphi^* + \varepsilon \frac{d\sigma_{TT}}{d\Omega^*} \cos 2\varphi^*$$

where φ^* is the azimutal angle between the leptons and the scattered electron, and ε the polarisation of the virtual photon.

Up to now, all published results[1] were obtained by averaging on azimutal angles in the Laboratory. Assuming that this corresponds to integrating over φ^* there remain only 2 terms $\frac{d\sigma_T}{d\Omega^*} + \varepsilon \frac{d\sigma_L}{d\Omega^*}$; using the classical invariant scaling parameter $y = 1 - E'/E \cos^2 \theta/2$, where E is the beam energy, E' and θ the energy and the scattering angle of the tagged electron, all in the the lab. frame, one gets :

$$\frac{d\sigma}{d\Omega^*} = \frac{4\pi\alpha^2}{Q^4} \cdot (1-y) \frac{dF_2}{d\Omega^*} + xy^2 \frac{dF_1}{d\Omega^*}$$

so that, since y experimentally remains small with respect to 1, one practically only measures F_2.

Actually it was shown[2] that, all 4 terms, or at least 3 of then, can be determined by looking at the azimutal correlations within some experimental limits.

An other way to study the QED γ structure function in more detail has been recently proposed[3] using azimutal correlations in $\gamma^*\gamma^* \to \ell^+\ell^-$ through double tag measurement of, $ee \to ee\ell^+\ell^-$. Both "scattered electrons" are detected at small angle, both "produced leptons" at large angle.

The cross section contains 20 helicity terms ($\sum M_{\lambda\lambda'} M^*_{\bar{\lambda}\bar{\lambda}'}$). Assuming $q^2/W^2 \ll 1$ for both photons (i.e. for double tag at small angles), one can neglect 15 longitudinal terms (i.e. terms with at least one zero helicity subscript). There only remain 5 transverse terms which respectively depend on the azimutal angles φ and φ' between either scattered electron and one of the produced leptons through the factor 1, $\cos 2\varphi$, $\cos 2\varphi'$, $\cos 2(\varphi+\varphi')$, $\cos 2(\varphi-\varphi')$.

In the following, I shall only give the main features of those studies, describing first the general way and later the main difficulties and limitations mainly due to the event selection (and remaining background) and the dynamical and kinematical effects of the γ^* virtuality. I shall not discuss radiative corrections.

II SINGLE TAG MEASUREMENT

a) General procedure

Let us consider the single-tagged experiment ee→ (e) e $\ell^+\ell^-$ corresponding to the following graph, where $\underline{k}^2$ is very close to zero and $\underline{q}^2$ small :

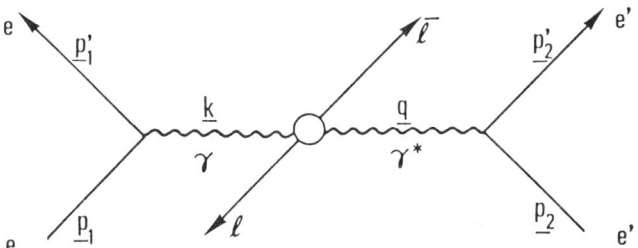

Fig. 1 Feynman diagram for ee→ ee$\ell^+\ell^-$ involving 2 photon exchange

We define :
$$s^2 = (\underline{p}_1+\underline{p}_2)^2 \quad ; \quad W^2 = (\underline{k}+\underline{q})^2 \quad ; \quad \underline{k} = z\underline{p}_1 \quad \underline{q} = (\underline{p}_2 - \underline{p}'_2)$$

and the classical Bjorken invariant parameters :
$$y = \frac{\underline{q}\,\underline{k}}{\underline{p}_2\,\underline{k}} \quad \text{and} \quad x = -\frac{(\underline{p}_2 - \underline{p}'_2)^2}{2\underline{k}\,\underline{p}_2} \simeq \frac{Q^2}{Q^2 + W^2} \quad (Q^2 = -\underline{q}^2)$$

wherefrom one derives the relations :
$$Q^2 = xy\,zs \quad \text{and} \quad W^2 = (1-x)y\,zs$$

so that :
$$W^2 = \frac{1-x}{x} Q^2 \quad \text{and} \quad y = \frac{1}{zs}\frac{Q^2}{x} = \frac{1}{zs}\frac{W^2}{1-x}$$

Then one writes the ee→ ee$\ell^+\ell^-$ cross section as follows :

$$\frac{d\sigma^{ee\rightarrow ee\ell^+\ell^-}}{dZ^*\,dy\,dQ^2\,d\Omega} = f_{\gamma/e}(z) \cdot f_{\gamma^*/e}(yQ^2) \cdot \frac{d\sigma^{\gamma\gamma^*\rightarrow\ell^+\ell^-}}{d\Omega}$$

with :
$$f_{\gamma/e}(z) = \frac{\alpha}{\pi z}\left(\ell n \frac{s}{4m_e^2} - 1\right) f(z)$$

$$f_{\gamma^*/e}(y.Q^2) = \frac{\alpha}{\pi y} \cdot \frac{1}{Q^2} f(y)$$

where : $f(X) = 1 - X + X^2/2$

Using as parameters x and W^2 rather than x, and Q^2, one gets :

$$\frac{d\sigma^{ee \to ee\ell\ell}}{dW^2 \, dx \, d\Omega} = \frac{\alpha^4}{32\pi^4} \left(\ell n \frac{s}{4m_e^2} - 1\right) \frac{\sqrt{1 - 4m_e^2/W^2}}{W^2} \frac{1-x}{x} \int_{z_{min}}^{1} f(z) f(y) \, \mathbf{I}\left(W^2, x, \theta, \varphi\right) \frac{dz}{z}$$

with : $\mathbf{I}(W^2, x, \theta, \varphi) = I_1 + \varepsilon I_2 - \sqrt{2\varepsilon(\varepsilon+1)} \, I_3 \cos\varphi + \varepsilon I_4 \cos 2\varphi$

where setting $I_{m\bar{m},n\bar{n}} = \sum M_{\lambda\bar{\lambda}} \cdot M^*_{\lambda'\bar{\lambda}'}$

$I_1(W^2, x, \theta) = I_{++,++}(W^2, x, \theta) + I_{++,--}(W^2, x, \theta)$
$I_2(W^2, x, \theta) = 2 I_{++,oo}(W^2, x, \theta)$
$I_3(W^2, x, \theta) = \sqrt{2} \left[I_{++,+o}(W^2, x, \theta) - I_{++,o-}(W^2, x, \theta)\right]$
$I_4(W^2, x, \theta) = -2 I_{++,+-}(W^2, x, \theta)$

$\varepsilon = (1-y)/(1-y+y^2/2)$
$z_{min}(W^2, x) = \frac{W^2}{s(1-X)}$

Assuming (for simplicity) $W \gg 2m_\ell$ and $u \equiv \cos\theta$ definetely smaller than one, one gets :

$I_1 = 8 (4\pi\alpha)^2 \cdot (1 - 2x + 2x^2) \frac{1+u^2}{1-u^2}$

$I_2 = 32 (4\pi\alpha)^2 \cdot x (1-x)$

$I_3 = 16 (4\pi\alpha)^2 \sqrt{x(1-x)} \cdot (1-2x) \frac{u}{\sqrt{1-u^2}}$

$I_4 = 16 (4\pi\alpha)^2 \cdot x (1-x)$

Let us note that, within our assumptions, the helicity terms scale, i.e. become independant of W.

Now changing the parameters from (x, W^2) to (x, Q^2) one has :

$$\frac{d\sigma}{dQ^2\, dx\, d\Omega} = \frac{2\alpha^2}{(4\pi)^4} \frac{\ln(s/4m_e^2)-1}{Q^4} \int_{z_{min}}^{1} \frac{dz}{z} f(z) f(y)\, \mathbf{I}\left(Q^2, x\, u\, \varphi\right)$$

Using the classical Structure Function F_T and F_L so that :

$$2x \frac{\partial F_T(x,u)}{\partial u} \propto 2x\, I_1\ ; \quad \frac{\partial F_L(x,u)}{\partial u} \propto 2x\, I_2$$

and defining :

$$\frac{\partial F_{TL}(x,u)}{\partial u} = 2x\, I_3\ ; \quad \frac{\partial F_{TT}(x,u)}{\partial u} = 2x\, I_4$$

one gets :

$$I \propto 2x \frac{dF_T(x,u)}{du} + \varepsilon \frac{dF_L(x,u)}{du} - \sqrt{2\varepsilon(\varepsilon+1)} \frac{dF_{TL}(x,u)}{du} \cos\varphi + \varepsilon \frac{dF_{TT}(x,u)}{du} \cos 2\varphi$$

with :

$$\frac{dF_T(x,u)}{du} \propto \left[(1-x)^2 + x^2\right] \frac{1+u^2}{1-u^2}\ ; \quad \frac{dF_L(x,u)}{du} \propto 8x^2(1-x)$$

$$\frac{dF_{TL}(x,u)}{du} \propto 4x(1-2x)\sqrt{x(1-x)}\, \frac{u}{\sqrt{1-u^2}}\ ; \quad \frac{dF_{TT}(x,u)}{du} \propto 4x^2(1-x)$$

b) Difficulties and Limitations

First of all, let us note that, as far as the unobserved electron and the corresponding quasi-real γ are concerned, $\Theta_e \sim 0$ is better defined than the measured Θ of the observed particles, but a somewhat large amount of events has $k^2 \gtrsim m_e^2$ (even $k^2 > m_\mu^2$)

i) Dynamical effects

For real $\gamma\gamma \to \ell^+\ell^-$ one has

$$\frac{d\sigma^{\gamma\gamma \to \ell^+\ell^-}}{d\Omega} = \frac{\alpha^2}{W^2}\beta\left[1 + 2\frac{\beta^4 \sin^4\theta \cos^2\theta + \beta^2(1-\beta^2)}{1-\beta^2\cos^2\theta}\right]$$

with $\beta = \sqrt{1 - \frac{4m_\ell^2}{W^2}}$

The denominator $(1-\beta^2 \cos^2)$ originates from the propagator of the exchanged lepton. When integrating over the full phase space of the lepton produced, it leads to

$$\sigma\gamma\gamma \to \ell^+\ell^- = \frac{4\pi\alpha^2}{W^2}\left[\ln\frac{W^2}{m_\ell^2} - 1\right]$$

For $\gamma\gamma^* \to \ell\ell$, the quasi pole of the propagator at $\cos\theta = 1$ is modified, surprisingly, by the virtuality of the quasi-real γ (and not of the more virtual one). Setting $X_2 = Q^2/W^2$ and $X_1 = k^2/W^2$, the expression of the Propagator is given by

$$P = \frac{W^2}{2}(1+X_1+X_2)\sqrt{1-\beta_{\gamma\gamma}\beta\cos\theta}$$

with $\beta_{\gamma\gamma} = \sqrt{1 - \frac{4X_1X_2}{(1+X_1+X_2)^2}}$

which for $X_2 \gg X_1$ becomes

$$\beta_{\gamma\gamma} \approx \sqrt{1 - \frac{4\mu^2}{W^2}} \quad \text{with } \mu^2 = x(1-x)k^2.$$

When integrating over the full phase space one gets:

$$\sigma\gamma\gamma^* \to \ell^+\ell^- = \frac{4\pi\alpha^2}{W^2}\left(\ln\frac{W^2}{\mu^2+m_\ell^2} + 1\right)$$

where μ^2 is not necessary negligible with respect to m_ℓ^2.

However, as long as $W \gg m_\ell$ and $\cos\theta < u_0 < 1$, our above-derived expressions remain unchanged.

ii) <u>Kinematical effects</u>

$\frac{d\sigma}{dW^2\, dx\, d\Omega}$ has been derived in the C-of-M frame, whereas the angular acceptance is defined in the Lab with respect to the beam direction. For single-tag experiment, where the virtual γ^* corresponds to an electron scattered at small but finite angle, the velocity of the C-of-M does not remain parallel to the beam direction, so that the approximation $\Theta = f(w^2, z, \theta)$ and $\phi = \varphi$ which is good for quasi real $\gamma\gamma$ collisions, is no longer valid.

Here, the Lorentz transformation implies that both φ and θ depend on W^2, x, z, Θ, ϕ. It results that the integration over Ω_{LAB} cannot be obtained from independant integrations over φ and θ and the acceptance in the C-of-M, Ω^*, is a quite complicated function of Ω_{LAB}, W^2, x and z. Therefore, averaging over

$\phi = 2\pi$, in the Laboratory, does not necessary mean integrating over $\varphi = 2\pi$ in the C-of-M frame.

Actually one can select events such that $\Delta\phi = 2\pi$ implies $\Delta\varphi = 2\pi$. This can be done by imposing certain limits on W^2, x and Θ. Nevertheless, even in this case, since θ is a function of w^2, x, z, Θ and ϕ, we are not allowed to average over ϕ and afterwords integrate over θ. Moreover, because of the z dependance of θ and φ, we are not allowed to integrate independenty over Ω_{LAB} and z as we did before.

There is no real problem when using Monte Carlo generation, provided that we don't neglect, a priori, the contributions of F_{TT} and F_{TL}.

When using analytical calculations, it can be shown that one can avoid the difficulties just mentioned by neglecting the transverse component of the velocity of the $\gamma\gamma$ system in the Lab. Frame. This can be done by requiring a more stringent constraint on W^2, Θ and Q^2 namely : $Q \ll \frac{W}{2}\sin\Theta$.

<u>Event selection and "background" corrections</u>

Actually we have assumed that, looking at single tag ee→(e) e$\ell^+\ell^-$ events,
(1) one selects the multiperiphical graph involving "$\gamma^*\gamma$" collisions
(2) one determines Q^2 from the electron observed at small angle and W^2 from the lepton pair observed at large angle.

Those assumptions are not well justified by just requiring one electron at small angle and a lepton pair at large ones. There still remains a somewhat large contamination mainly due to the dynamical ranges where the assumption (2) is not satisfied, to the contributions of other graphs (conversion, exchange, annihilation ...) and to experimental bias from other processes (as ee→$\gamma\gamma\ell\ell$ for example) or randoom events (cosmics, converted γ ...).

Let us note that we only observe 3 of the 4 particles, and generally don't identify the charge of the particle observed at small angle. Moreover it is obvious that the contamination is more important for ee production than for $\mu\mu$ production, since talking about "scattered" and "produced" electrons might be meaningless and there are additional exchange graphs and interferences.

Therefore, we obviously need to apply a lot of various additional cuts and/or constraints which reflects on the acceptance and counting rates of the selected events : the more drastic are the cuts, the cleaner are the events but, the lower is the counting rate. On the other hand, precise measurements need high statistics. One is thus led to compromise, applying less severe cuts and more corrections.

Let me just give two examples, of dangerous configurations.

We want to select the folowing diagram and corresponding kinematic configuration, for k^2 almost zero and q^2 small :

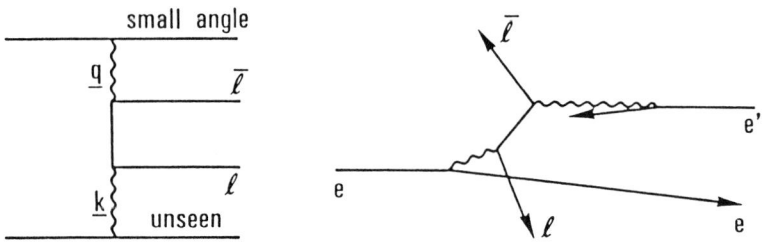

As far as multiperipherical graphs are concerned, there is another quite large contribution to a similar topology of events due to the folowing diagram and kinematic configuration where again k^2 is almost zero but $\underline{q}_e^2$ (instead of $\underline{q}^2$) is small.

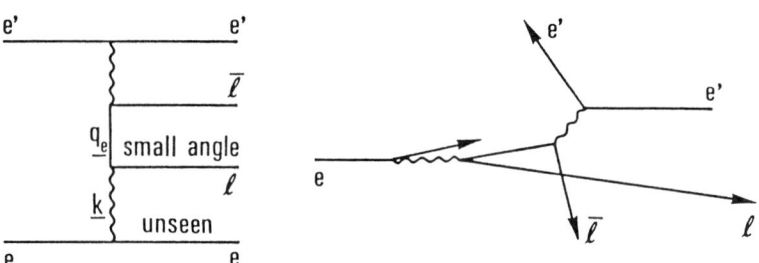

- This "Cabbibo Parisi" configuration is quite dangerous for e^+e^- production. Actually this contamination can be avoided to a large extent by means of cuts on the observed charge, tagged energy, acolineraly angles ...

As far as Conversion graphs are concerned, there is also other quite large contribution leading to a similar topology of events due to the folowing diagram and kinematic configuration.

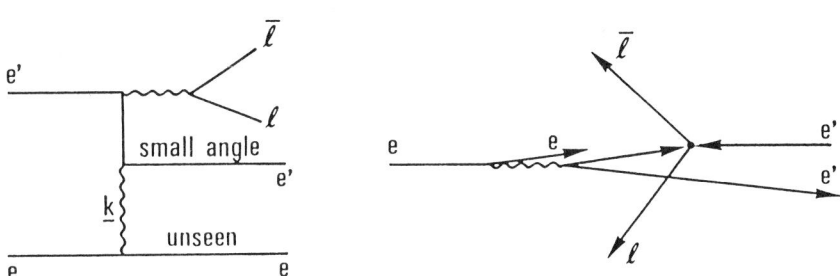

This process is quite similar to the previous one. It just involves an annihilation instead of a Möller diffusion. It leads to a contamination as well for µµ as for ee production. Actually it involves "back scattering" at small angle but since the charge is not generally observed it may also become dangerous. However such contamination can be again avoided to a large extent by various cuts.

Nevertheless as long as we want enough statistics within the cuts, there remain contaminations. Such contaminations were computed by F. Le Diberder [4]. The results can be sumarized as follows.

Conversion diagrams involve for $\ell\ell = \mu\mu$, a correction factor $|1+\Delta|^2$

$$\text{with } \Delta = \frac{y}{2} \sqrt{\frac{x(1-z)}{(1-x)(1-y)}} e^{i\varphi}$$

at y ~ .2 and x ~ .5, it leads to a correction of the order of 10%. This value of Δ decreases with x, but needs very stringent cuts on x or/and y to be negligible.

Exchange diagrams (only involved in $\ell\ell$ = ee) also involve a contamination which can be also expressed by a $(1+\delta)$ correction factor but the complete expression of δ is somewhat complicated. Actually the correction can be important. It depends not only on φ but also on Θ and is responsable for some charge assymetry.

As an example, within a realistic experimental situation for single tag measurement (CELLO) F. Le Diberder has shown that one gets the following contributions :

	Mult. ($\gamma^*\gamma$)	Conv.	Interference
ee → (e)e $\mu^+\mu^-$	89%	11%	~0
ee → (e)e e^+e^-	82%	33%	-15%

III AZIMUTAL CORRELATION IN DOUBLE TAG MEASUREMENTS

Let us consider the double tag experiment ee → ee$\ell^+\ell^-$ where both scattered electrons are detected at small angle and the produced leptons at large ones. Here both q^2 and k^2 (fig. 1) are small but finite.

The helicity structure of the completely diffirential cross-section can be written as :

$$\frac{d\sigma}{dL} = K \begin{bmatrix} I_{++,++} + I_{++,--} - 2\varepsilon\, I_{+-,--} \cos 2\varphi - 2\varepsilon'\, I_{++,+-} \cos 2\varphi' \\ + \varepsilon\varepsilon'\, I_{+-,+-} \cos 2(\varphi + \varphi') + \varepsilon\varepsilon' \cos 2(\varphi - \varphi') \\ + 15 \text{ Longitudinal terms} \end{bmatrix}$$

where

$$dL = \left(\frac{1}{2\pi}\right)^8 \delta^4 (p_o + p_o' - p - p' - p_e - p_e') \frac{d^3p}{2p_o} \frac{d^3p'}{2p_o'} \frac{d^3p_e}{2(p_e)_o} \frac{d^3p_e'}{2(p_e')_o}$$

is the Lorentz invariant phase space factor, and

$$K = \frac{(4\pi\alpha)^4}{8SQ^2Q'^2} (\cosh^2 \alpha + 1)(\cosh^2 \alpha' + 1)$$

with

$$\cosh \alpha = \frac{2S\, Z - \Sigma}{\sqrt{\Sigma^2 - Q^2Q'^2}} \quad ; \quad \cosh \alpha' = \frac{2S\, y - \Sigma}{\sqrt{Z^2 - Q^2Q'^2}}$$

defining :
$$W^2 = (q + q')^2, \quad S = (p_o + p'_o)^2, \quad \Sigma = W^2 + Q^2 + Q'^2$$
$$y = \frac{q\,p'_o}{p_o p'_o}, \quad z = \frac{q'\,p_o}{p_o p'_o}, \quad Q^2 = -\underline{q}^2, \quad Q'^2 = -\underline{q}'^2$$

and where :
$$\varepsilon = \frac{1-y}{1-y+y^2/2}, \quad \varepsilon' = \frac{1-z}{1-z+z^2/2}$$

are the polarization parameters of the photons.

In principle, it should be possible to extract all helicity terms from a double tag measurement using the 20 terms formula. In practice, this will be difficult as regards to the longitudinal terms, since they generally give only small contributions to the already small counting rate. Therefore it seems preferable to eliminate them by imposing Q and Q' to be very small with respect to W (Q, Q' « W/2).

Thus integrating the above 5 terms formula over all variables other than φ and φ', one gets :
$$\frac{d\sigma}{d\phi\,d\phi'} = \sigma_o + \sigma_1 \cos2\varphi + \sigma_2 \cos2\varphi' + \sigma_3 \cos2(\varphi+\varphi') + \sigma_4 \cos2(\varphi+\varphi')$$

with :
$$\sigma_o = \int d\chi\, K\,(I_{++,++} + I_{++,--}) \quad ; \quad \sigma_1 = (-2\varepsilon)\int d\chi\, K\,(I_{+-,++})$$

$$\sigma_2 = (-2\varepsilon')\int d\chi\, K\,(I_{++,+-}) \quad ; \quad \sigma_3 = (\varepsilon\varepsilon')\int d\chi\, K\,(I_{+-,+-})$$

$$\sigma_4 = (\varepsilon\varepsilon')\int d\chi\, K\,(I_{+-,-+}) \quad \text{where} \quad d\chi = \frac{dL}{d\phi\,d\phi'}$$

Those redefined terms can be, in principle, directly determined from a measurement, using the formulas :
$$\sigma_o = \frac{1}{4\pi^2}\int \frac{d\sigma}{d\phi\,d\phi'}\,d\phi\,d\phi'$$

$$\sigma_1 = \frac{1}{2\pi^2} \int \frac{d\sigma}{d\phi d\phi'} \cos 2\phi \, d\phi d\phi'$$

$$\sigma_2 = \frac{1}{2\pi^2} \int \frac{d\sigma}{d\phi d\phi'} \cos 2\phi' \, d\phi d\phi'$$

$$\sigma_3 = \frac{1}{2\pi^2} \int \frac{d\sigma}{d\phi d\phi'} \cos(\phi+\phi') \, d\phi d\phi'$$

$$\sigma_4 = \frac{1}{2\pi^2} \int \frac{d\sigma}{d\phi d\phi'} \cos(\phi-\phi') \, d\phi d\phi'$$

Actually, this is again not experimentally practicable so that it will be more convenient to determine the ratios looking at azimuthal correlations after integration over one angle. Setting $\bar{\phi} = \phi + \phi'$ and $\hat{\phi} = \phi - \phi'$, one has :

$$F_{(\phi)} = \frac{1}{2\pi\sigma_o} \frac{d\sigma}{d\phi} = 1 + \frac{\sigma_1}{\sigma_o} \cos 2\phi$$

$$F_{(\phi')} = \frac{1}{2\pi\sigma_o} \frac{d\sigma}{d\phi'} = 1 + \frac{\sigma_2}{\sigma_o} \cos 2\phi'$$

$$F_{(\bar{\phi})} = \frac{1}{2\pi\sigma_o} \frac{d\sigma}{d\bar{\phi}} = 1 + \frac{\sigma_3}{\sigma_o} \cos 2\bar{\phi}$$

$$F_{(\hat{\phi})} = \frac{1}{2\pi\sigma_o} \frac{d\sigma}{d\hat{\phi}} = 1 + \frac{\sigma_4}{\sigma_o} \cos 2\hat{\phi}$$

We notice that (in the C.of M frame) $\bar{\phi}$ is the azimutal angle between the 2 scattered electrons and, $\hat{\phi}$ is twice the angle between the bisectrix of the transverse momenta of both scattered electron and the transverse momentum of one produced lepton. Then the four above expressions respectively correspond to four different experimental angular distribution of the events.

I don't discuss, here again, the kinematic problems due to the Lorentz transformation. I just want to notice that analytic computation requires very strong constraints leading to very low counting rates. Anyway double tag events, as compared to single tag ones, have much lower counting rates. However they give a much higher analysing power, since all kinematic parameters are measured, which allows one, in principle, to check all constraints imposed by various dynamic models which could be very different. As an example, looking at pion pair production instead of lepton pair production, the "Born-term model" and the "Brodsky-Lepage's model" predict totally different azimutal correlations so that a few events could be enough to perform a significant test [3].

REFERENCES

(1)   CELLO Collaboration Phys. Lett. 126B (384) 1983
      PLUTO Collaboration Z. Phys. C27 (245) 1985
      PEP 4/9 Collaboration Phys. Lett. 174B (282) 1984

(2)   S. Ong, P. Kessler Mod. Phys. Lett. A2 863 (1987)

(3)   S. Ong, P. Kessler, A. Courau Mod. Phys. Lett. A4 909 (1989)

(4)   F. Le Diberder Ph. D. Thesis, Université Paris-Sud, Juin 1988.

Neutrino Counting at e^+e^- Colliders[*]

R.J. WILSON

*Boston University,
Boston, Massachusetts, 02215*

ABSTRACT

We discuss the experimental techniques required to make a measurement of the number of neutrinos at e^+e^- colliders. A summary of studies performed by the ASP and SLD collaborations at SLAC, is presented.

*Presented at the
Ann Arbor Workshop on QED Structure Functions, May 22-25, 1989.*

---

[*] Work supported by the Department of Energy, contract DE – AC02 – 86ER40284

© 1990 American Institute of Physics

1. Introduction

One of the outstanding puzzles in particle physics is the "generation problem". It began with the discovery of the muon as exemplified by I.I.Rabi's now famous "Who ordered that!". The mystery deepened with the discoveries of yet more leptons and quarks. Today, it is usual to present these fundamental particles in sets of weak iso-doublets, with an apparent symmetry of the lepton and quark sectors:

$$Leptons: \begin{pmatrix} e \\ \nu_e \end{pmatrix} \begin{pmatrix} \mu \\ \nu_\mu \end{pmatrix} \begin{pmatrix} \tau \\ \nu_\tau \end{pmatrix}$$

$$Quarks: \begin{pmatrix} u \\ d \end{pmatrix} \begin{pmatrix} c \\ s \end{pmatrix} \begin{pmatrix} t \\ b \end{pmatrix}$$

Only the top quark remains to be discovered to complete this pattern (although some would argue that the ν_τ has not actually been observed, only inferred from τ decay measurements). The Standard Model has no wisdom to offer on this empirical symmetry, in particular we are not told how many of these doublets or generations to expect. Attempts to produce "Grand Unified Theories" which might provide the answer by identifying a deeper all encompassing symmetry, have also yet to bear fruit. Some further measure of guidance from experiments is required.

One of the first efforts at any new accelerator is usually a search for new sequential charged leptons and quarks. This technique, however, is never likely to give a completely convincing answer to the puzzle. We already observe that the mass of each successive generation is significantly larger than the previous one, so a non-observation might simply mean that the next particle has a mass beyond the reach of the latest accelerator.

Fortunately, the Standard Model does gives us guidance concerning the neutrino masses: they are massless unless you add features not required by the model, such as right-handed neutrinos. Experimental limits on the neutrino masses support this view, or at least tell us that the neutrino has a mass very much lighter than its charged partner, even for the third generation tau-neutrino.[1] From this we may expect that the study of the decay of the Z^0 into neutrinos should settle the question (unless the next generation of neutrinos has a unreasonably large mass of more than about $M_{Z^0}/2$).

Unfortunately, it is not possible to observe the decay

$$Z^0 \to \nu_l \bar{\nu}_l \qquad (1.1)$$

directly due to the low interaction rate of the neutrinos with matter, but there are two methods by which the branching fraction may be determined indirectly. The first is to measure the width of the Z^0 resonance. This technique is applicable to any mode of production of the Z^0, and has been used to set limits on the number of neutrinos by the UA1 and UA2 groups at the CERN $Sp\bar{p}S$ collider.[2]

Another technique, suggested in 1978 by Ma and Okada,[3] is that e^+e^- colliding beam facilities could determine the total number of light-mass neutrinos directly by a measurement of the reaction:

$$e^+e^- \to \nu\bar{\nu}\gamma. \qquad (1.2)$$

Other groups have developed this idea further, improving the calculations and suggesting ways to perform the measurements.[4]

In this report we will discuss both of these techniques in the context of e^+e^- colliders. The results presented here are mainly based on the work of the neutrino counting study groups in the SLD[5,6] and Mark II[7,8] collaborations at the Stanford Linear Accelerator Center and from the ASP at SLC proposal.[9]

2. Z^0 Width

The partial width of the Z^0 boson for the decay to a fermion anti-fermion pair is given by the expression :

$$\Gamma(Z^0 \to f + \bar{f}) = C \frac{M_Z^3 G_F}{6\pi\sqrt{2}} \{v_f^2 + a_f^2 - \frac{2m_f^2}{M_Z^2}(2a_f^2 - v_f^2)\}\{1 - 4\frac{m_f^2}{M_Z^2}\}^{1/2} \qquad (2.1)$$

where the color factor C is 1 for leptons and 3 for quarks and the vector and axial vector weak coupling parameters are are listed in Table 1.

If the mass of the top quark is above threshold at the SLC (this now appears likely[11]) and we use the parameter values used in the SLD study shown in Table 2, then the total width of the Z can be conveniently expressed as:*

$$\Gamma_Z = \{2.58 + (N_\nu - 3) \times 0.171\} \text{ GeV} \qquad (2.2)$$

where N_ν is the number of neutrino species and all fermion masses have been set equal to zero. Table 3 shows the decay rate into each fermion type and includes a QCD correction term.

⋆ This expression is only slightly modified ($\approx$ 100 MeV on the total width) by recently reported new values for M_Z and $sin^2\theta_W$.[12]

The ultimate Z^0 width measurement error is dominated by the uncertainty in the beam energy. At the SLC, pulse-by-pulse beam energy monitors will provide beam energy measurements with an absolute error of ±20 MeV, which combined with an estimated error of ±30 MeV due to residual dispersion effects at the interaction point will yield an absolute measurement accuracy of the center of mass energy of ±40 MeV.[13] It is possible in principle to achieve a measurement error for Γ_Z comparable to the quadratic sum of the comparable smaller *relative* beam energy errors and the residual dispersion error, for a total systematic error of about ±30 MeV.

The Mark II collaboration has estimated that the systematic limit of $\delta\Gamma_Z = 30$ MeV will be reached with a data sample of approximately 5×10^4 Z^o events.[14] The interpretation of the width measurement in terms of neutrino generations does, however, also require that the radiative corrections to the Z^o line shape be well understood. These corrections are large, as can be seen in Fig. 1. In the SLD study we used the theoretical work of Karaev and Fadin,[15] and of Cahn.[16] The effect on Γ_Z of initial state radiation from the incident electron and positron, the dominant process contributing to the distortion of the Z^o line shape, was found to be an increase of about 55 MeV. This width increase is equivalent to $\approx 1/3$ of a neutrino generation. Such an effect can be corrected for provided that the beam energy spectrum is known. The systematic errors arising from radiative corrections should be relatively small, on the order of 10 MeV (see Ref. 14). It is also expected that QCD corrections will be manageable, with associated errors comparable to those due to radiative corrections (perhaps about 1%, or 25 MeV). A reasonable estimate of the total systematic error due to the above mentioned effects is about 40 MeV, corresponding to 0.23 generations (in our notation, 0.23 N_ν).

The effect of vacuum polarization corrections to the Z^o propagator has been included in our analysis, and is absorbed into a renormalization of the electromagnetic coupling constant, α. However, if the top quark mass is substantially above $M_Z/2$, relatively large effects on Γ_Z are expected due to the virtual $t\bar{t}$ loops. These loop effects have been calculated[17] and the results predict a width increase of about 0.09 N_ν for a top mass of 100 GeV, rising to 0.26 N_ν for a mass of 200 GeV (see Fig. 2). For the near future, a measurement of a heavy top mass by one of the hadron collider experiments is likely to be the only unambiguous means for reducing the systematic uncertainty of Γ_Z due to virtual top corrections.

A second method[18] determines the partial width to invisible final states from a measurement of the total width and the visible partial widths;

$$\Gamma_{invis} = \Gamma_{tot} - \Gamma_{ee} - \Gamma_{\mu\mu} - \Gamma_{\tau\tau} - \Gamma_{hadronic} \tag{2.3}$$

where the theoretical values are used for the leptonic widths. The observed muon-pair cross section is used to determine the total width from the equation:

$$\sigma_{\mu\mu} = \frac{12\pi}{M_Z^2} \frac{\Gamma_{\mu\mu}^2}{\Gamma_{tot}^2} \qquad (2.4)$$

The measurement of Γ_{invis} is particularly sensitive to systematic uncertainties in the muon pair and hadronic event efficiencies ($\epsilon_{\mu\mu}$ and $\epsilon_{hadronic}$) and to the error in the luminosity. However, assuming a 3% luminosity measurement, and systematic uncertainties in $\epsilon_{\mu\mu}$ and $\epsilon_{hadronic}$ of 3% and 1% respectively, Γ_{invis} can be measured to a precision of about 80 MeV with 2,000 events. The systematic limit of about 50 MeV will be reached for event samples of between 10^4 and 2×10^4 Z^0's, corresponding to $\delta N_\nu = 0.3$.

3. Single Photon Tag

An orthogonal method for determining the invisible width of the Z^0 is through the observation of the process:

$$e^+ + e^- \to Z^0 \to \gamma + Weakly\ Interacting\ Particles\ (WIP) \qquad (3.1)$$

The presence of a single unaccompanied photon in the detector indicates e^+e^- annihilation into particles which do not interact in the apparatus. The most well-known example of such a WIP is, of course, the neutrino. However, there are several "exotic" candidates outside of the Standard Model which can contribute to the invisible decay width (see Table 4).[19] We shall direct the discussion toward the "neutrino counting" process:

$$e^+ + e^- \to Z^0 \to \nu + \overline{\nu} + \gamma \qquad (3.2)$$

but return to comment on the exotics in the last section. In the rest of this section, we discuss this process and the possible sources of background affecting the measurement. In particular, we derive the statistical and systematic limitations of the measurement of N_ν by several different methods.

The basic tree level Feynman diagrams for the process are shown in Fig. 3. The major contributor is the annihilation via the Z^0 channel, Fig. 3(a), which depends on the number of neutrino species. Smaller contributions occur due to the W exchange term, Fig. 3(b), and the interference between these two diagrams.

The double differential cross section has been calculated at tree level (Ref. 4) and is given by

$$\frac{d^2\sigma}{dxdy} = \frac{G_F^2 \alpha}{6\pi^2} \frac{s(1-x)}{x(1-y^2)}[(1-x/2)^2 + x^2y^2/4]$$

$$\times \left\{ \frac{N_\nu(v_e^2 + a_e^2) + 2(v_e + a_e)[1 - s(1-x)/M_Z^2]}{[s(1-x)/M_Z^2 - 1]^2 + \Gamma_Z^2/M_Z^2} + 2 \right\} \quad (3.3)$$

where

$$x = E_\gamma/E_{beam}, \quad s = 4E_{beam}^2, \quad y = \cos\theta_\gamma$$

and

$$v_e = -\frac{1}{2}[1 - 4\sin^2\theta_W], \quad a_e = -\frac{1}{2}$$

The photon spectrum for different center of mass energies is shown in Fig. 5. The dashed line on the figure is an example of single photon source that is not due to intial state radiation through the Z^0 resonance ("neutralinos", see Fig. 4). Such processes may actually have the appearance of an unvetoed background and escape our notice!

Higher order radiative corrections[20] may reduce the cross section by as much as 25%, depending on experimental cuts. Fig. 6 shows schematically how the corrected cross section may be greater or less than the tree-level calculation depending on these cuts. A more detailed treatment of the W exchange diagram[21] gives corrections of no more than $2-3\%$ at $E_{cm} = M_Z$ and even less at center of mass energies away from the Z pole.

For the following discussion we use the simple tree-level expression (3.3) since it embodies all of the salient features, in the analysis of a real experiment a monte carlo which includes multiple photon emission is required.

In this section we discuss in detail the statistical and systematic errors of the single photon counting experiment. These depend on the choice of beam energy and on the fiducial region in which we choose to count signal photons. The experiment is also physically constrained by the machine components which determine the minimum angle at which certain backgrounds can be detected.

As an example, if a cut on the energy of observed photons is made at a fixed value, $E_{min}^\gamma = 1$ GeV, then the total accepted cross section depends on E_{cm} as shown in Fig. 7(a). The peak cross section of 195 pb occurs at $E_{cm} = 95$ GeV. As E_{min}^γ is increased, the position of the peak increases and its amplitude decreases. If we use a cut of $E_{min}^\gamma = 2$ GeV, rather than 1 GeV, the peak cross section would be 125pb and would occur at approximately 97 GeV. The decrease in the amplitude of the peak cross section as E_{min}^γ is increased is shown in Fig. 7(b). Running close to the Z^0 mass maximizes the peak cross section, however, the photon energy distribution at different center of mass energies (Fig. 5), becomes less steep with increasing beam energy. Indeed, when the center of mass energy is more than $\Gamma_Z \approx 2.5$ GeV above the mass of the Z^0, a clear reflection of the Z^0 excitation curve is visible. The systematic uncertainty in the accepted cross section caused by uncertainty in the energy calibration of the detector can largely be eliminated by choosing E_{cm} and E_{min}^γ large enough to avoid the steep photon spectra at values of E_{cm} near the mass of the Z^0. One must find a compromise between these competing effects.

In the next section we will discuss the uncertainty in our measurement that arises from uncertainty in the performance of the detector and the accelerator, and also from the experimental technique. We discuss running strategies which optimize the precision of the experimental result, and finally comment on the interpretation of the measurement.

4. Single Photon Measurement Errors and Running Strategy

4.1. BACKGROUNDS

The potentially most severe background to the single photon measurement is that from radiative bhabha scattering. The positioning of the final-focus quadrupoles, and the necessity to avoid placing detectors too close to the swath of synchrotron radiation produced by the quads and the bend magnets at the ends of the SLC arcs limits the smallest angle, θ_{veto}, at which high-energy electrons can be detected. At small scattering angles the $e^+e^- \to e^+e^-\gamma$ process is completely dominated by QED photon exchange; the Z^0 exchange contributions do not become significant until scattering angles of 20° or so. The veto angle required for a particular choice of p_{tmin}^γ is (by conservation of momentum):

$$\theta_{veto} < \frac{p_{tmin}^\gamma}{E_{beam}}. \tag{4.1}$$

We have calculated the radiative Bhabha cross section with the TEEGG[22] event generator, which uses matrix elements calculated by Berends and Kleiss but which was specifically designed to run efficiently in the small scattering-angle regime. These results are shown in Figs. 8 and 9. If the apparatus has a veto angle of 22 mr this corresponds to $p_{tmin}^\gamma = 0.72$ GeV at a beam energy of 48.0 GeV, we find that requiring p_t^γ of the signal photon to be greater than 1.1 GeV reduces the radiative Bhabha background to 2% of the signal. This process is completely dominated by the QED t-channel diagrams and a comparison of the event generator with data at PEP (using ASP) shows that we understand this background with good precision leading to an uncertainty in this background of less than 0.5% (Ref. 22). With these values of p^γ_{tmin} the backgrounds from all other "beam-beam" processes we have considered (τ production, two-photon processes, etc.) are negligible.[23]

Backgrounds from single-beam causes are more difficult to quantify. The most significant irreducible source in the lower energy experiments (PEP and PETRA) was due to beam-gas interactions. The effect was to produce essentially a line source of photo-produced π^0's with a number density given by

$$N(cm^{-1}) \approx 4.3 \times 10^{-7} \times P(10^{-9} torr) \times \sigma_{mol.}(\mu b)$$

$$\times \frac{100 pb^{-1}}{< L(10^{31} cm^{-2} s^{-1}) >} \times < I(mA/beam) > . \qquad (4.2)$$

A major design criteria for the ASP calorimeter at PEP was to allow for good reconstruction of photon candidates along the beam axis. This, combined with a p^γ_{tmin} requirement, was required to reduce the contamination to manageable levels. This background is not a problem at the SLC due to the low beam currents, but may require some consideration at LEP. Other sources such as off-axis electrons, beam halo etc. are less easily estimated. These require careful study using random and cosmic ray triggers. They can be the cause of such effects as energy deposition in the calorimeter which is then mistaken for a photon, or false vetos from the low angle calorimeter in a way which depends on the beam-conditions in an unpredictable manner.

4.2. ERRORS DUE TO DETECTOR PERFORMANCE

In the previous section we saw that p_t is a good variable to discriminate against the most significant backgrounds. Since backgrounds are more difficult to understand in detail than are questions of resolution and scale, it is preferable to define the signal fiducial cut in terms of transverse momentum rather than energy. However, comparing the signal energy and p^γ_t distributions shown in Figs. 5 and 10, we see that the reflection of the Z^0 line shape is not as clear in the latter distribution, although it does produce a broad spectrum for $E_{cm} > 95$ GeV.

The uncertainty in the p^γ_t calibration of the detector is a combination of energy and angle calibration and survey errors. To show the sensitivity of the cross section to uncertainties in the transverse momentum scale, we plot in

Fig. 11 the fractional change in the accepted $\gamma\nu\bar{\nu}$ cross section (with $N_\nu = 3$) as a function of p^γ_{tmin}. For example, with a $p^\gamma_{tmin} = 0.6~GeV/c$ and a 5% systematic uncertainty in $p^\gamma_t(30~MeV/c)$, the corresponding error in the cross section changes by more than a factor of two depending upon the center of mass energy chosen for the measurement.

The uncertainty in the measurement that we have discussed so far arises from uncertainty in the *calibration* of the detector. The detector also has finite *resolution* of the photon transverse momentum, this can produce a systematic error in the acceptance of the experiment. The effect of finite measurement resolution is to allow events with transverse momenta outside the nominal fiducial region to be counted, and to cause some photons that should be counted to be lost. If the resolution function is precisely known then we can properly compute the cross section that we are actually viewing with the experiment. However, uncertainty in the form of the resolution function will cause uncertainty in this computed value. This uncertainty will be less if the parent distribution is flat at the fiducial boundary than if the boundary occurs at a steep point in the distribution. This is likely to be a significant effect in the early life of large experiments. These often need a long running period during which they collect calibration events, such as muon and electron pairs, required to understand the systematics of the completed device.

4.3. Uncertainty in Machine Parameters

The effect of an uncertainty in the beam energy is equivalent to a scale change of the center of mass energy in the expression for the differential cross section. The fractional error of 0.05%, as determined by the SLC extraction-line spectrometer (see Ref. 13), leads to a 35 MeV shift in E_{cm}.

The effect of the beam energy spread is to average the center of mass energy of the collisions over the width of the distribution. We can fold the expected width into our cross section calculation, but uncertainty in this width results in an uncertainty in the calculated result. This error is greatest for experimental conditions with the center of mass energy chosen to be within one resonant width of the Z^0 peak. We have found that a non-gaussian beam energy profile (such as a low energy tail) can shift the apparent mass of the Z^0 by 0.15 GeV and affect the value of Γ_Z by approximately ± 0.15 GeV.

The nominal values and estimated errors for the three SLC machine parameters that we have considered are:

Parameter	Value	Uncertainty
Beam energy	48 GeV	$\pm 0.05\%$
Beam spread	$0.005 \cdot E_{beam}$	$\pm 10\%$
IP position	0.0	$\delta z \leq \pm 2mm$

None of these uncertainties will result in significant errors in the experiment (for example, the systematic error on the cross section for the ASP study parameters arising from the beam energy and spread uncertainties are approximately 0.2% and 0.4% respectively).

4.4. NORMALIZATION

To measure the coupling of the Z^0 to weakly interacting particles we must normalize the number of single photons that we observe to a well-understood final state. Experiments at e^+e^- colliding-beam machines have historically normalized physics measurements to Bhabha scattering,

$$e^+e^- \to e^+e^-(\gamma). \tag{4.3}$$

The outgoing electrons are usually measured at large polar angles to the beamline by the same apparatus used to detect the particles in the physics process under study. The (γ) in reaction (4.3) denotes the fact that careful allowance must be made for events in which a photon, radiated during the collision process, remains undetected by the experimental apparatus.[24] This reaction has been studied extensively at PETRA and PEP, and the systematic uncertainties that remain in our understanding of this process at energies below the Z^0 mass are about 1%.

Near the Z^0 pole the cross section is complicated by interference between QED and purely weak contributions, but we anticipate that precise studies of the differential cross section for this process will be of highest priority in the early stages of SLC and LEP experimental programs. Several groups[25] are working on Monte Carlo event generators of this, and other, processes for use in analysis of SLC and LEP experiments. In the studies discussed in this report, we have assumed that we will be able to compute the cross section with final-state electrons in the detector with a theoretical uncertainty of 1%, it is now likely that we will be able to do much better than that.

One such normalization procedure that we have considered is to normalize the single-photon sample to events in which there is at least one high-energy electron above some fiducial angle, $\theta = 20°$, and a second high-energy electron above some smaller angle, $\theta = 50mr$. From the lowest order matrix elements we find that, on the Z^0 peak, the cross section above $\theta = 20°$ is 1.8 nb with approximately equal contributions from t-channel photon exchange and the s-channel Z^0 propagator. For a center of mass energy one resonant width above the peak this falls to 1.0 nb, of which 80% is due to QED and is therefore sharply peaked in the forward direction. The total e^+e^- rate in this case is about six times larger than the radiative neutrino rate, so the normalization will not contribute substantially to the statistical error of the final result. The selection of these high-energy electron pairs will have no significant energy calibration error since there is no necessity to make a tight energy requirement. However the fractional error on the accepted cross section due to the polar angle cutoff is large due to the steep distribution. Since the angular distribution of the signal is also very steep ($1/sin^2\theta$) and cannot be optimized by choice of beam energy, the definition of this fiducial is potentially the largest source of systematic error.

Another approach is to attempt to remove such difficult to measure systematics by normalizing the cross section to a process with similar kinematics such as

$$e^+e^- \to \mu\bar{\mu}\gamma. \tag{4.4}$$

This process qualifies since the muon mass is very small compared to the center of mass energy and as it is large compared to the electron mass we expect that radiated photons will be mostly from the initial state. Another advantage is that muons do not easily shower in the detector and so are quite easy to track and identify, giving a high selection efficiency. We require the photon to satisfy the same cuts as are applied to the signal photon so that the ratio of the measured cross sections will cancel the detector systematics. However, we must again be careful to look at the effect of radiative corrections. Fig. 12 shows the dependence of the contributions from the initial and final state radiation on the center of mass energy. Close to the Z^0 peak the largest component is from final state radiation from the muons (Fig. 12(a)). This will have a different angular distribution than the signal which is entirely initial state radiation and there may be substantial differences in the systematic errors. Increasing the center of mass energy reduces the final state contribution (Fig. 12(b)). Another way to reduce this contribution is to apply a cut on the angle between the muon and the photon, this reduces the phase space for the final state radiation which is strongly peaked along the muon direction. However, Fig. 13 shows how a cut on the angle between the muon and the photon reduces the rate of the normalization process which is the limiting statistical error. As is often the case, one must carefully weigh the contribution from systematic and statistical errors.

4.5. SLD ANALYSIS SUMMARY

A summary of the analysis performed by the SLD neutrino counting group is given in Tables 5, 6, 7, and 8. The following assumptions were taken:

1. The effective average luminosity is $0.5 \times 10^{30} cm^{-2} sec^{-1}$.
2. The cross section for $e^+e^- \to \mu^+\mu^-$ at $E_{cm} = M_Z$ is 1.34 nb. The branching ratio is 3.27%.
3. The beam energy distribution is gaussian and the center of mass energy resolution is 50 MeV.
4. The calorimeter in the SLD detector has a photon energy resolution given by $\delta E = 9\sqrt{E}\%$.
5. The angular resolution of the calorimeter is good enough that it does not contribute to the overall error.
6. The ratio in the cross section for $e^+e^- \to \nu\bar{\nu}\gamma$ between the case for $4N_\nu$ and $3N_\nu$ is 1.30 at $E_{cm} = M_Z$ and 1.23 at $E_{cm} = 96.0$ GeV.
7. The theoretical uncertainty in the magnitude of the cross section for $e^+e^- \to \nu\bar{\nu}\gamma$ is 5%.
8. The theoretical uncertainty in the magnitude of the cross section for $e^+e^- \to e^+e^-\gamma$ (Radiative Bhabha) is 1%.
9. The theoretical uncertainty in the magnitude of the cross section for $e^+e^- \to \mu^+\mu^-\gamma$ is less than 1%.
10. The uncertainty in the luminosity measurement is 3%.
11. The veto angle for radiative Bhabhas is 22 mrad.

The results are obtained assuming that we run for the same integrated luminosity at a given run setting. From these tables we would conclude that the best determination is ultimately achieved by running at a center of mass energy near 96 GeV, requiring that the photon transverse momentum be greater than 1 GeV, and using the $\mu\mu\gamma$ ratio method, Table 8. However, we note that for shorter running periods that this is the worst method! The next best would appear to be that of Table 7. However, we can see from the number of $\mu\mu\gamma$ events that our assumption that the photons are solely from the initial state is not valid: since the cross-section for neutrinos is much larger than that for the muons, we would expect *fewer* such events than those from $\nu\nu\gamma$. This will invalidate the assumption of zero uncertainty contribution from radiative corrections and also introduce other detector systematic errors not accounted for here.

In Table 6 we see that the uncertainty from the radiative corrections is the dominating factor for long runs. Since this work was performed it has become clear that this contribution will be smaller. It is also clear that the SLD is unlikely to spend almost two years sitting above the Z^0 peak to perform this measurement. I believe that Table 5 represents the most likely analysis to be performed by the SLD: measuring single photons parasitically whilst collecting Z^0 data at the maximum rate. This limits the ultimate precision to around 0.25 species.

4.6. ASP AT SLC ANALYSIS SUMMARY

After its successful life at PEP it was proposed to upgrade the ASP apparatus to perform a precision measurement of the single photon cross section at the SLC (Ref. 9). As it was to be a single purpose experiment the design could be optimised in a manner not possible with a large general purpose device such as the Mark II, SLD or the LEP experiments. Its goal was to measure the cross section to a precision better than 5% with a contribution from systematics of less than 3%. Since the calorimeter had been calibrated at PEP with tens of thousands of kinematically constrained radiative bhabha events, it could lay claim to a low and precisely measured systematic error on the crucial parameters, giving it a considerable advantage over other devices. We estimate that photon transverse momenta are measured with a systematic uncertainty of less than 4% (24 Mev/c) in the range of energies and angles of interest, which (using Fig. 11) produces an uncertainty in the cross section at E_{cm}= 96 GeV of 1.2%.

Figure 14 shows a plan view of the device. The central tracking chamber used in the PEP experiment has been modified to allow the installation of precision "fiducial ring" calorimeters which determine the minimum polar angle for the photon. These eliminate any significant error in defining the fiducial volume. Due to the strong θ dependence of both the signal and the bhabha normalisation any fractional error in determining θ_{min} is magnified (by more than a factor of two for our cuts) in the cross section. The presence of the Fiducial Rings reduces the angular uncertainty by almost a factor of ten compared to what we are able to achieve with the lead-glass alone. In general purpose detectors it is difficult to optimize such considerations due to the competing needs of other analyses.

For this experiment it is optimal to run at 3 GeV above the mass of the Z^0 (e.g. 96 GeV/c^2 since we have used M_{Z^0}=93 GeV/c^2), this makes the uncertainty in the p_t^γ calibration the single largest contribution to the systematic uncertainty in the experiment.

The experimental parameters we have used are summarized in Table 9(a). The observed cross section includes the analysis efficiency and has been calculated to include the effect of finite resolution of the fiducial boundaries. The systematic errors for this set of experimental parameters are summarized in Table 9(b). As might be expected the major components are the momentum scale determination and the selection of the signal and normalization events. The uncertainty in the photon selection for the PEP analysis was determined to be 0.4% using the large sample of radiative Bhabha events but we have used a conservative 1% to account for changes in material in front of the calorimeter in the upgrade and any other effects that we may have neglected. The relative independence of these errors justifies the combination in quadrature, but even if there is some correlation or even a substantial underestimate of the errors (factors of two in any of the largest contributions) we would still keep the overall error less than the 3% goal.

5. Interpretation of the Single Photon Measurement

With this measurement we are making a general test of *any* deviation from the Standard Model prediction for this final state. We find it convenient to define a quantity,

$$R_\nu \equiv \frac{\Gamma(Z^0 \to \text{weakly interacting particles})}{\Gamma(Z^0 \to \nu\bar\nu)} \qquad (5.1)$$

where $\Gamma(Z^0 \to \nu\bar\nu)$ is the partial width of the Z^0 due to a single massless neutrino species. This quantity is the analog of the e^+e^- total cross section in that it "counts" the number of neutrino-like particles that exist with masses below M_{Z^0}. Since we expect there to be a contribution to reaction (3.1) from the three known neutrino species, it is useful to further define

$$\Delta R_\nu \equiv R_\nu - 3 \qquad (5.2)$$

The sensitivity of the experiment can now be characterized by the accuracy with which the experiment will determine ΔR_ν if $R_\nu = 3$, and all of the photons we see are produced by the standard processes in Fig. 3. For simplicity we combine all sources of error (including statistical and systematic contributions) in quadrature to define,

$$\delta(\Delta R_\nu) = R_\nu \cdot \left[\left(\frac{\delta(\sigma_\gamma/\sigma_{ee})}{(\sigma_\gamma/\sigma_{ee})}\right)^2 + \left(\frac{\delta\sigma_{ee}}{\sigma_{ee}}\right)^2 + \left(\frac{\delta\sigma_{\nu\bar\nu\gamma}}{\sigma_{\nu\bar\nu\gamma}}\right)^2\right]^{\frac{1}{2}}. \qquad (5.3)$$

The first term in (5.3) is the overall experimental error discussed in earlier sections, and the errors in the calculation of the cross sections are,

- $\frac{\delta\sigma_{ee}}{\sigma_{ee}}$ error in the e^+e^- cross section (1%)
- $\frac{\delta\sigma_{\nu\bar\nu\gamma}}{\sigma_{\nu\bar\nu\gamma}}$ error in $\nu\bar\nu$ (1% from radiative corrections and 1% from α_S.)

Using (5.3) with $R_\nu = 3$ and the results of the ASP analysis, we compute the results shown in Fig. 15. So that, for example, with an integrated luminosity of 12pb^{-1} we would reach $\delta(\Delta R_\nu) = 0.12$, or about one eighth of a neutrino generation.

It is clear that careful measurements of the Z^0 width will be performed at the SLC and LEP, and that the precision should reach 0.25 neutrino species, or better, from precision measurements of the Z^0 width. Although a precision on $\delta(\Delta R_\nu)$ in this range would be satisfactory to differentiate three from four generations in the Standard Model, we saw in Table 4 that this value need not be integer. We can rewrite the table in terms of ΔR_ν:

	ΔR_ν
Massive Dirac neutrinos	$\frac{\beta}{4} \cdot (3 + \beta^2)$
Massive Majorana neutrinos	β^3
Scalar neutrinos	$0.5 \cdot \beta^3$
Neutralinos	0 to > 1

Hence, the magnitude of their contributions is likely to be smaller than that for the usual massless Dirac neutrino.

If we discover a value of ΔR_ν not consistent with zero, then the significance of the result is given by the curve in Fig. 15 rescaled according to Eqn. (5.3). For example, a light scalar neutrino would produce $\Delta R_\nu = 0.5$, so that an integrated luminosity of 12pb^{-1} would yield $\delta(\Delta R_\nu) = 0.12$, and thus establish the existence of the new particle.

6. Conclusions

We conclude that the discovery potential of a "neutrino counting" experiment is quite general and covers a broad range of possible new physics. In particular a precision measurement made with a dedicated run using the single photon technique would either discover something new, or unambiguously rule out much of the parameter space of many theoretical ideas that try to go beyond the Standard Model.

It is now clear that the future of such a measurement lies with LEP and as such deserves high priority. It is entirely possible that the optimization required to make best use of the accelerator cannot be achieved with the current LEP detectors and that a new special purpose experiment is called for. It is amusing to note that, in preparation for possible installation at the SLC, the ASP calorimeter was carefully stored away mounted in a steel frame with a big hook attached to the top ....

Acknowledgements: The author wishes to thank the members of the SLD study group and the ASP at SLC group, in particular Peter Rowson for providing much of the material for the Z^0 width discussion. He also wishes to acknowledge many useful conversations with David Burke and Giovanni Bonvicini, and Jim Shank for his careful reading of the manuscript.

REFERENCES

1. M. Perl, Review talk at the International Lepton-Photon Symposium, Stanford, California, USA, 1989

2. S. Geer *et al.*, UA1 Collaboration, CERN-EEP/86-115 (1986); J. Appel *et al.*, UA2 Collaboration, ZPhys. $\underline{C30}$, 1 (1986).

3. E. Ma and J. Okada, Phys. Rev. Lett. $\underline{41}$, 287 (1978). Errata, Phys. Rev. Lett. $\underline{41}$, 1759 (1978).

4. K.J.F. Gaemers, R. Gastmans, and F.M. Renard, Phys. Rev. Comments $\underline{19D}$, 1605 (1979).
   G. Barbiellini, B. Richter, and J.L. Siegrist, Phys. Lett. $\underline{106B}$, 414 (1981). Proceedings of the SLC Workshop on Experimental Use of the SLAC Linear Collider, SLAC-Report-247, p.122 and 237 (1982).

5. SLD Physics Report-1, December 1988, unpublished.

6. SLD Design Report, SLAC-Report-273 (1988).

7. Mark II/SLC, Physics Working Group Note #10-3 (1987).

8. Mark II at SLC, SLAC-Report-273 (1989).

9. R.J. Wilson *et al.*, "Proposal for a Precision Measurement of the Process $e^+e^- \rightarrow$ Photon plus Weakly Interacting Particles at the SLC", SLAC-PROPOSAL-SLC-10 (1986).

10. J. Ellis, M. K. Gaillard, Annual Reviews of Nuclear and Particle Science, $\underline{32}$, 443 (1982).

11. See, for example, talk by S. Pekka at the International Lepton-Photon Symposium, Stanford, California, USA, 1989

12. See, for example, talks by G. Feldman (MarkII) and M. Campbell (CDF) at the International Lepton-Photon Symposium, Stanford, California, USA, 1989

13. J. Kent *et al.*, Precision Measurements of the SLC Beam Energy, SLAC-PUB-4922 (1989).

14. P. Rankin, Mark II/SLC Physics Working Group #1-9, Proceedings of the Third Mark II Workshop on SLC Physics, SLAC-Report-315 (1987).

15. E.A. Kuraev and V.S. Fadin, Sov. J. Nucl. Phys. $\underline{41}$, 466 (1985).

16. R.N. Cahn, Phys. Rev. $\underline{D36}$, 2666 (1987). We have used Cahn's method, but not his approximations.

17. B.W. Lynn, M.E. Peskin, and R.G. Stuart, SLAC-PUB-3725 (1985)

18. G. Feldman, MARK II/SLC-Physics Working Group Note #2-24, Proceedings of the Third Mark II Workshop on SLC Physics, SLAC-Report-315 (1987).

19. This final state has also been used at PEP and PETRA to set limits on the production of supersymmetric particles but where photinos or scalar-neutrinos are the WIPs and the exchange particle is not the Z^0 boson. At SLC/LEP energies the contribution from such SUSY particles is expected to be small.
    ASP Collaboration, C. Hearty et al., Phys. Rev. D39, 3207 (1989).
    CELLO Collaboration, H.J. Behrend et al., Phys. Lett. 215B, 186 (1988).
    MAC Collaboration, W.T. Ford et al., Phys. Rev. D33, 3472 (1986).
    Mark J Collaboration, B. Adeva et al., Phys. Lett. 194B, 167 (1987).

20. M. Igarashi and N. Nakazawa, TKU-HEP 86/01 or KUDP 86/02 (1986).

21. F.A. Berends, G.J.H. Burgers, C. Mana, M. Martinez, and W.L. Van Neerven, Nuc. Phys. B301, 583 (1988).

22. D. Karlen, Nuc. Phys. B289, 23 (1987). See also C. Mana and M. Martinez, Nuc. Phys. B287, 601 (1987).

23. See Ref. 7 and the paper by K. Riles in these proceedings

24. F.A. Berends and R. Kleiss, Nuc. Phys. B228, 537 (1983)

25. See, for example, papers by Bonvicini, Kuraev, Miquel, Ward in these proceedings.

TABLE CAPTIONS

1. Couplings for Z^0 to fermions

2. Electroweak Parameters used for the SLD study

3. Z^0 Decay rates using the SLD study parameters of Table 2 and the coupling coefficients of Table 1 (the $Z^o \to \bar{q}_i q_i g$ QCD correction is taken from Ref. 10).

4. Partial Width of the Z^0 into "exotic" particles (where $\beta = (1-4m^2/s)^{1/2}$)

5. SLD Analysis Summary, Method 1: Compare the observed number of $\nu\nu\gamma$ events with the number expected from the Standard Model.
   Run Conditions: $E_{cm} = M_Z$, $p_t^\gamma \geq 1\,\text{GeV}$, $\theta_{\gamma beam} \geq 10°$. For one additional neutrino $\delta\sigma/\sigma = 30\%$. All errors stated in units of N_ν.

6. SLD Analysis Summary, Method 2: Compare the observed number of $\nu\nu\gamma$ events with the number expected from the Standard Model.
   Run Conditions: $E_{cm} = 96.0$ GeV, $p_t^\gamma \geq 1$ GeV, $\theta_{\gamma beam} \geq 10°$. For one additional neutrino $\delta\sigma/\sigma = 23\%$. All errors stated in units of N_ν.

7. SLD Analysis Summary, Method 3: Compare the observed number of $\nu\nu\gamma$ events with the observed number of $\mu\mu\gamma$ events.
   Run Conditions: $E_{cm} = M_Z$, $p_t^\gamma \geq 1$ GeV, $\theta_{\gamma beam} \geq 10°$. $R = N(\nu\bar{\nu}\gamma)/N(\mu^+\mu^-\gamma)$, For one additional neutrino $\delta R/R = 48\%$. All errors stated in units of N_ν.

8. SLD Analysis Summary, Method 4: Compare the observed number of $\nu\nu\gamma$ events with the observed number of $\mu\mu\gamma$ events.
   Run Conditions: $E_{cm} = 96.0$ GeV, $p_t^\gamma \geq 1$ GeV, $\theta_{\gamma beam} \geq 10°$. $R = N(\nu\bar{\nu}\gamma)/N(\mu^+\mu^-\gamma)$, For one additional neutrino $\delta R/R = 28\%$. All errors stated in units of N_ν.

9. ASP at SLC Study: (a) Experiment Parameters (b) Systematic Error Contributions.

Table 1.

Final State	v_f	a_f
$e^+e^-, \mu^+\mu^-, \tau^+\tau^-$	$-\frac{1}{2}[1 - 4\sin^2\theta_W]$	$-\frac{1}{2}$
$\bar{\nu}_e\nu_e, \bar{\nu}_\mu\nu_\mu, \bar{\nu}_\tau\nu_\tau$	$\frac{1}{2}$	$\frac{1}{2}$
$\bar{u}u, \bar{c}c$	$\frac{1}{2}[1 - \frac{8}{3}\sin^2\theta_W]$	$\frac{1}{2}$
$\bar{d}d, \bar{s}s, \bar{b}b$	$-\frac{1}{2}[1 - \frac{4}{3}\sin^2\theta_W]$	$-\frac{1}{2}$

Table 2.

$$\sin^2\theta_W = 0.226$$
$$M_Z = \{\frac{\pi\alpha_R}{\sqrt{2}G_F}\}^{1/2} \times \frac{1}{\sin\theta_W \cos\theta_W} = 92.2 \text{ GeV}$$
$$\alpha_R^{-1} = 137.036 - 171/6\pi$$
$$G_F = 1.166 \times 10^{-5} GeV^{-2}$$
$$\Gamma(Z^0 \to \bar{q}_i q_i g) = 0.04 \times \Gamma(Z^0 \to \bar{q}_i q_i)$$

Table 3.

Final State	Γ(GeV)
$e^+e^-, \mu^+\mu^-, \tau^+\tau^-$	0.087
$\bar{\nu}_e\nu_e, \bar{\nu}_\mu\nu_\mu, \bar{\nu}_\tau\nu_\tau$	0.171
$\bar{u}u, \bar{c}c$	0.297
$\bar{d}d, \bar{s}s, \bar{b}b$	0.382
$\sum_i \bar{q}_i q_i g$	0.072
Total	2.58

Table 4.

Particle	$\Gamma(Z^0 \to X)$
Massive Dirac Neutrinos	$1.5\% \cdot \beta(3+\beta^2)$
Massive Majorana neutrinos	$6\% \cdot \beta^3$
Scalar neutrinos	$3\% \cdot \beta^3$
Neutralinos	0% to $\approx$10%

Table 5.

Running Time (days)	0.6	6	30	60	600
# of Z^0	1×10^3	1×10^4	5×10^4	1×10^5	1×10^6
# of $\nu\nu\gamma$	1	9	44	88	879
# of $ee\gamma$	0.1	1	7	13	134
Corresponding error in σ due to ...					
Statistics	3.5	1.0	0.50	0.35	0.12
Beam energy resolution	0.13	0.13	0.13	0.13	0.13
Photon energy resolution	0.02	0.02	0.02	0.02	0.02
$\delta \sin^2\theta_W$	0.34	0.10	0.10	0.07	0.03
Rad. Corr. Uncertainty	0.17	0.17	0.17	0.17	0.17
Luminosity Uncertainty	0.10	0.10	0.10	0.10	0.10
Total	3.5	1.0	0.57	0.42	0.27

Table 6.

Running Time (days)	0.6	6	30	60	600
# of Z^0	186	1860	9290	18600	1.86×10^5
# of $\nu\nu\gamma$	3	30	150	300	3000
# of $ee\gamma$	0.1	1	7	13	130
Corresponding error in σ due to ...					
Statistics	2.6	0.81	0.36	0.25	0.080
Beam energy resolution	0.01	0.01	0.01	0.01	0.01
Photon energy resolution	0.01	0.01	0.01	0.01	0.01
$\delta \sin^2\theta_W$	0.03	0.009	0.009	0.006	0.003
Rad. Corr. Uncertainty	0.22	0.22	0.22	0.22	0.22
Luminosity Uncertainty	0.13	0.13	0.13	0.13	0.13
Total	2.6	0.85	0.45	0.36	0.27

Table 7.

Running Time (days)	0.6	6	30	60	600
# of Z^0	1×10^3	1×10^4	5×10^4	1×10^5	1×10^6
# of $\nu\nu\gamma$	1	9	44	88	879
# of $\mu\mu\gamma$	4	37	185	370	3700
Corresponding error in R due to ...					
Statistics	2.4	0.82	0.37	0.26	0.08
Beam energy resolution	0.06	0.06	0.06	0.06	0.06
Photon energy resolution	0.02	0.02	0.02	0.02	0.02
$\delta sin^2\theta_W$	0.94	0.28	0.28	0.19	0.09
Rad. Corr. Uncertainty	0.0	0.0	0.0	0.0	0.0
Luminosity Uncertainty	0.0	0.0	0.0	0.0	0.0
Total	2.6	0.87	0.47	0.33	0.14

Table 8.

Running Time (days)	0.6	6	30	60	600
# of Z^0	186	1860	9290	18600	1.86×10^5
# of $\nu\nu\gamma$	3	30	150	300	3000
# of $\mu\mu\gamma$	1	9	45	91	914
Corresponding error in R due to ...					
Statistics	4.6	1.4	0.60	0.42	0.13
Beam energy resolution	0.03	0.03	0.03	0.03	0.03
Photon energy resolution	0.01	0.01	0.01	0.01	0.01
$\delta sin^2\theta_W$	0.14	0.14	0.04	0.04	0.01
Rad. Corr. Uncertainty	0.0	0.0	0.0	0.0	0.0
Luminosity Uncertainty	0.0	0.0	0.0	0.0	0.0
Total	4.6	1.4	0.60	0.42	0.13

Table 9.

(a)

Parameters	
Center of mass energy	$M_Z + 3.0$ GeV
Tranverse Momentum	0.6 GeV/c
Polar Angle	$\geq 20°$
Analysis Efficiency	78%
Observed cross section	$\sigma_{\nu\bar{\nu}\gamma} = 120 pb$

(b)

Systematic Errors	
Accelerator	0.5%
Momentum Calibration	1.2%
Momentum Resolution	0.1%
Angle Calibration	0.2%
Photon Selection	1.0%
e^+e^- Selection	1.2%
Combined (quadrature)	2.0%

FIGURE CAPTIONS

1) Effect of radiative corrections on the total cross section

2) Effect of the top quark mass on the Z^0 decay rate

3) Feynman diagrams for Single Photon Tagging: (a) neutral and (b) charged weak interactions.

4) Feynman diagram for a supersymmetric source (Higgsinos) of single photons

5) Photon Spectrum for different center of mass energies. The dashed line shows the effect due to the neutralino process of Fig. 4.

6) Schematic example of the effect of experimental parameters (center of mass energy and photon energy) on the magnitude of radiative corrections for photons radiated from the initial state.

7) (a) Cross-section for $\nu\bar{\nu}\gamma$ ($N_\nu = 3$) as a function of the center of mass energy for photons with $\theta_\gamma > 20°$ and $E_\gamma > 1.0$ GeV (the dashed line shows the effect due to the neutralino process of Fig. 4).
(b) Maximum cross-section for $\nu\bar{\nu}\gamma$ as a function of E^γ_{min}. The value of E_{cm} that produces the maximum accepted cross section increases as E^γ_{min} is increased.

8) Cross section as a function of p^γ_{tmin} for fiducial cut of $\theta_\gamma > 10°$ and veto angle > 22 mrad. Solid lines are the signal from $\nu\bar{\nu}\gamma$, dashed line is the radiative bhabha background.

9) Total radiative bhabha background as a function of the minimum angle at which the beam-energy electrons are detected with the requirements of $p^\gamma_t > 0.5$ GeV/c and $\theta_\gamma > 20°$. The cross section for the signal in the same fiducial volume is also shown for comparison.

10) Single Photon spectrum expressed in term of the tranverse momentum of the photon

11) Variation in the measured cross section as a function of the minimum accepted transverse momentum of the photon. The arrows correspond to an uncertainty in the p_t cut of ±5%; the solid line corresponds to E_{cm}=96 GeV and the dashed line to E_{cm}=95 GeV.

12) Contributions to the photon energy spectrum from the intial and final state. (a) $E_{cm} = M_Z$ (b) $E_{cm} = M_Z + 3$ GeV (M_Z=92.2 GeV)

13) Ratio of the $\nu\bar{\nu}\gamma$ to $\mu\bar{\mu}\gamma$ cross sections as a function of the minimum photon transverse momentum for different requirements on the angle between the muon and the photon.

14) Plan view of the main components of the ASP at SLC apparatus

15) Overall experimental error in the determination of ΔR_ν as a function of the accumulated luminosity.

Fig.1

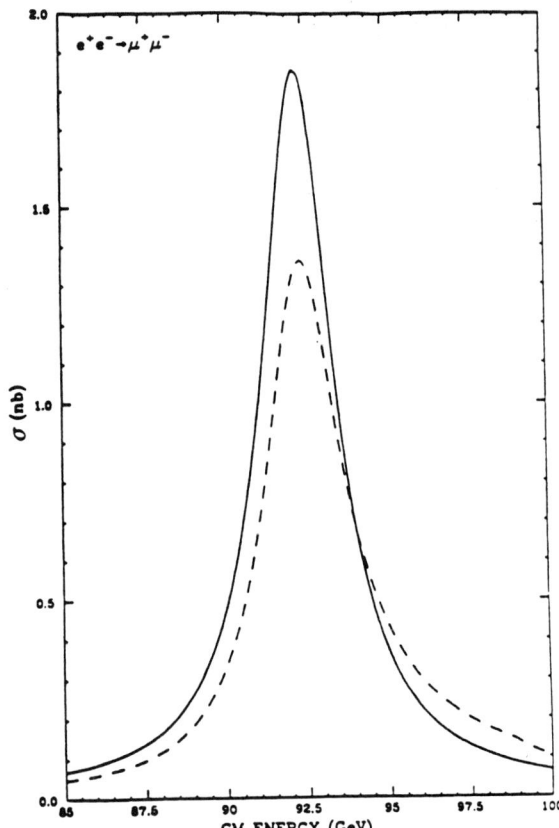

Fig.2

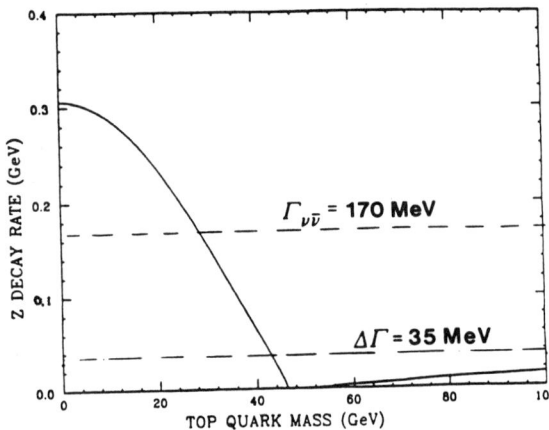

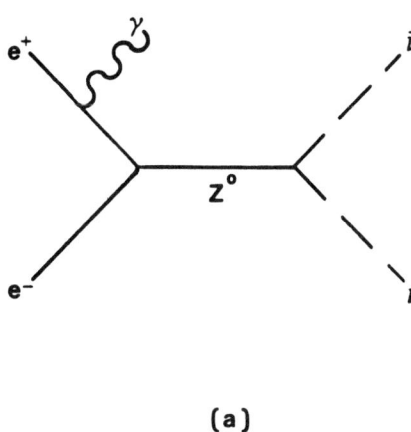

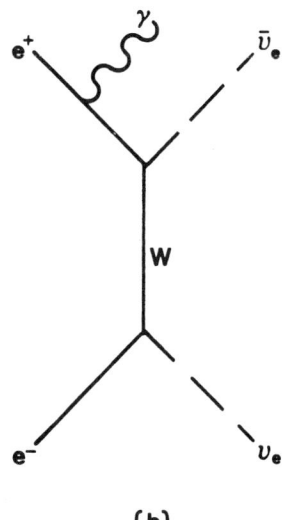

(a) (b)

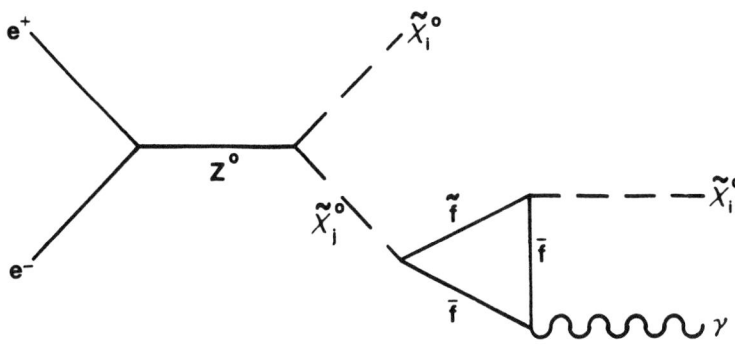

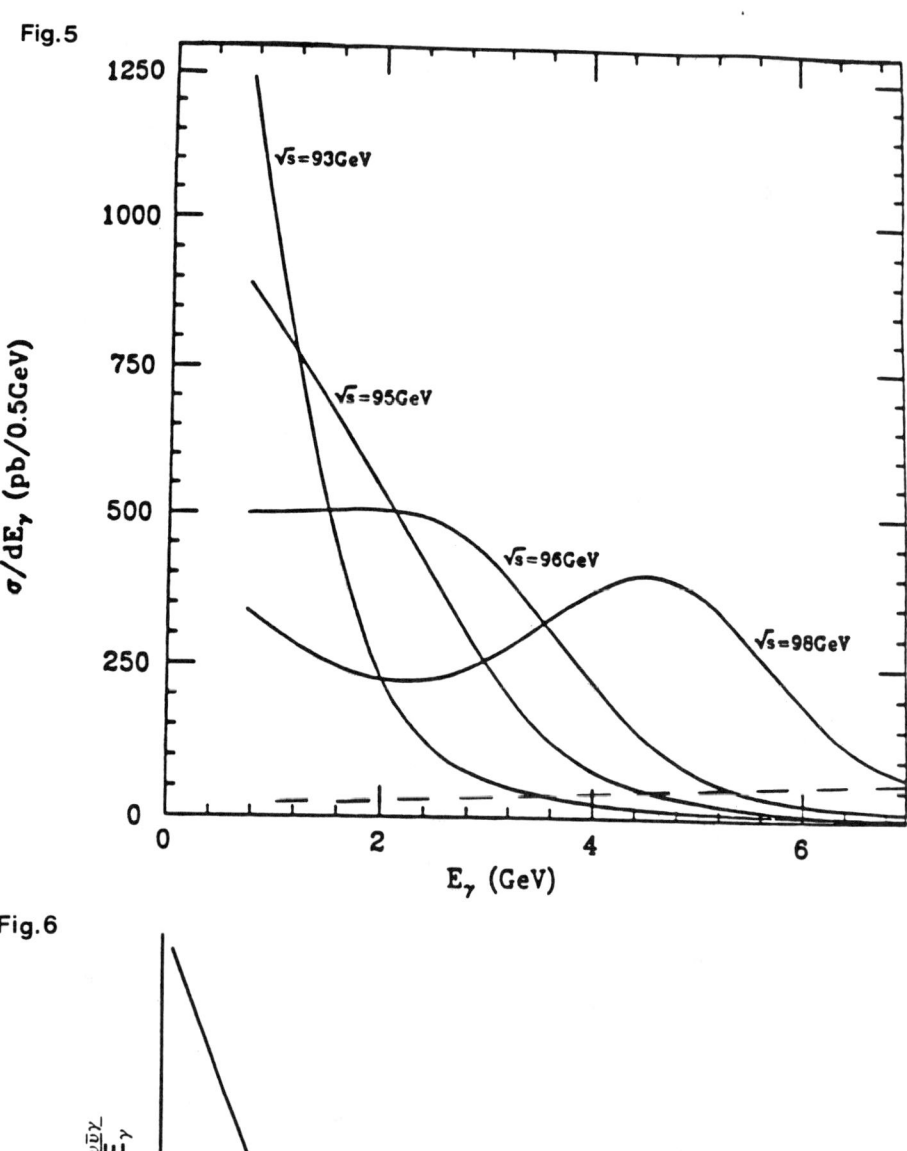

Fig. 5

Fig. 6

Fig.7

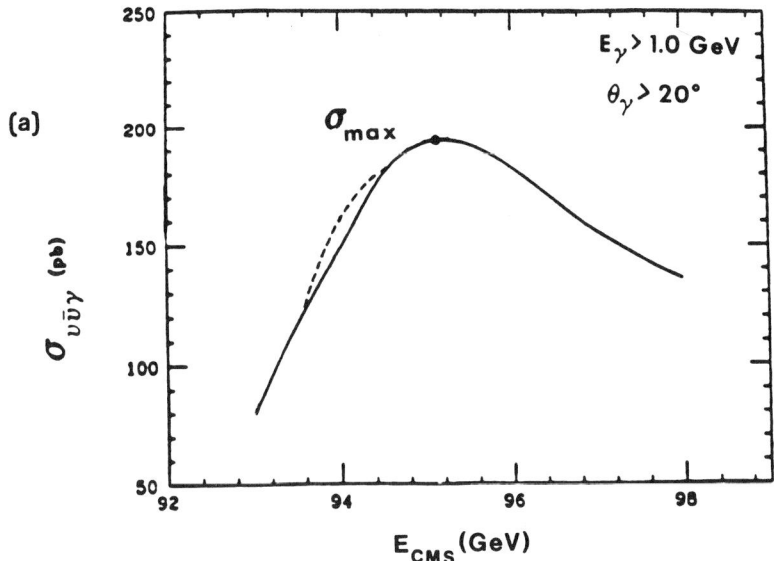

(a)

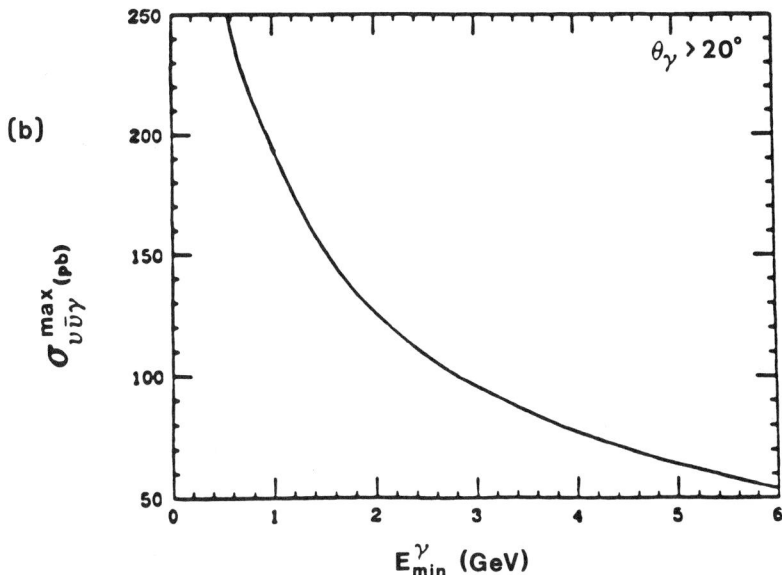

(b)

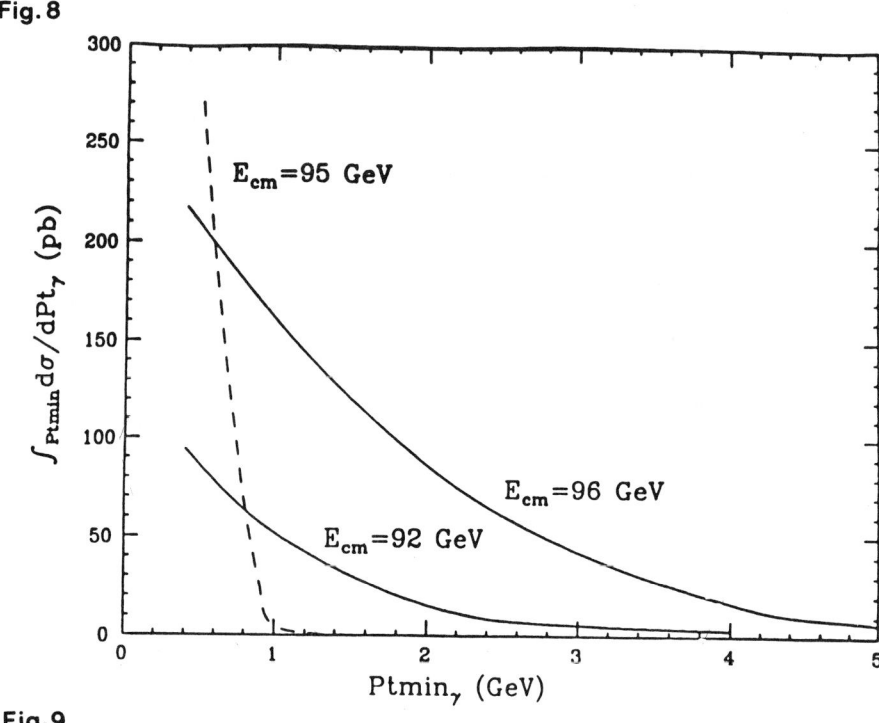

Fig. 8

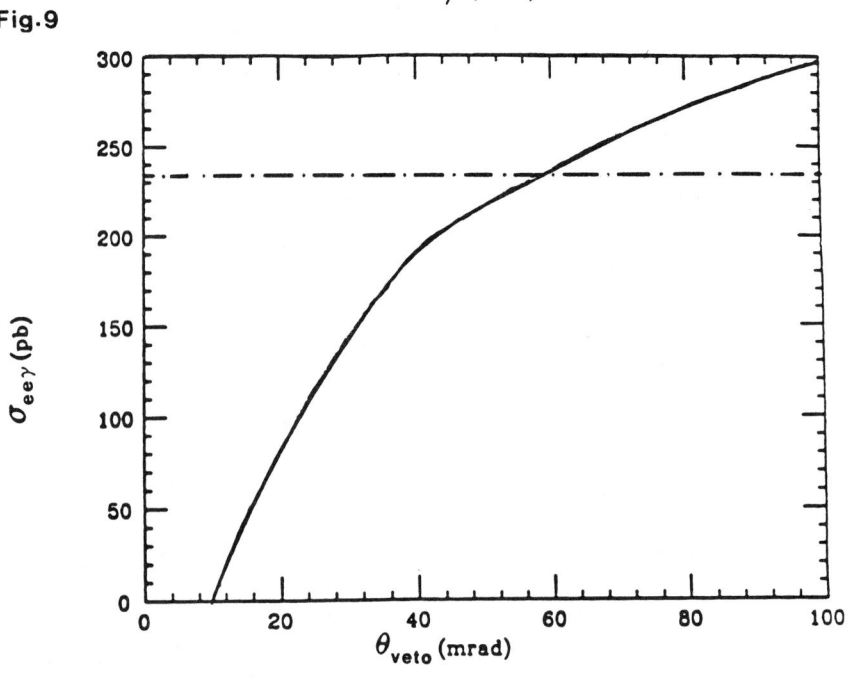

Fig. 9

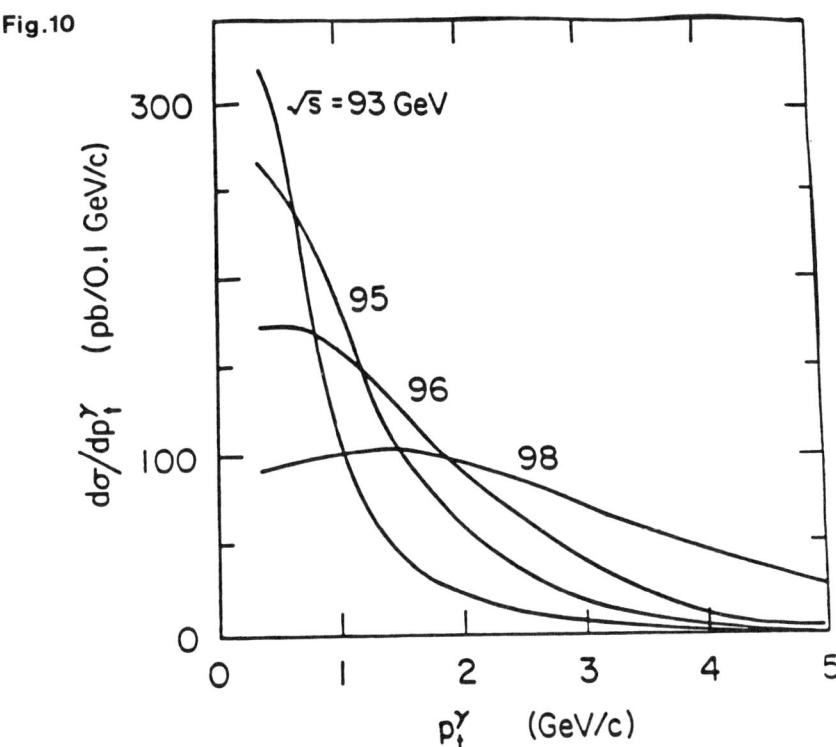

Fig.10

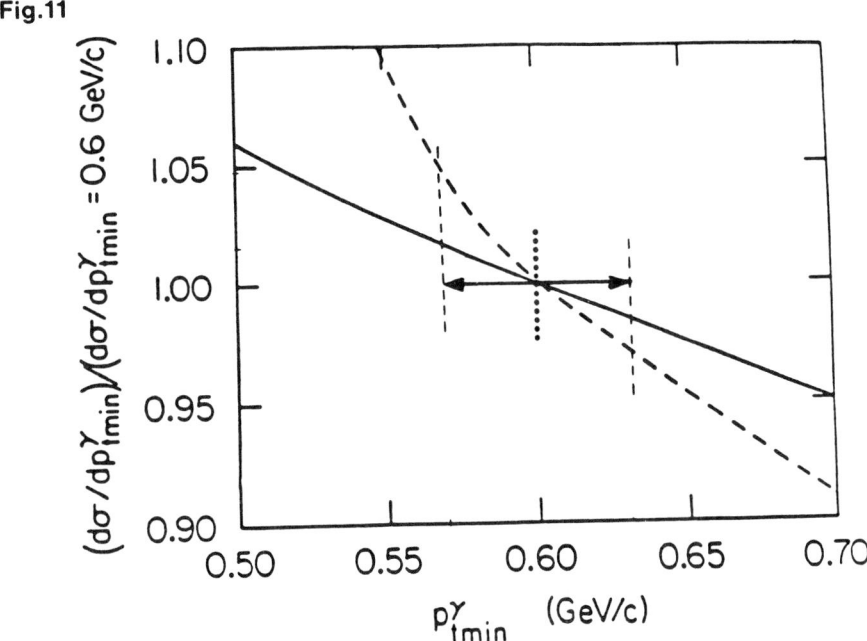

Fig.11

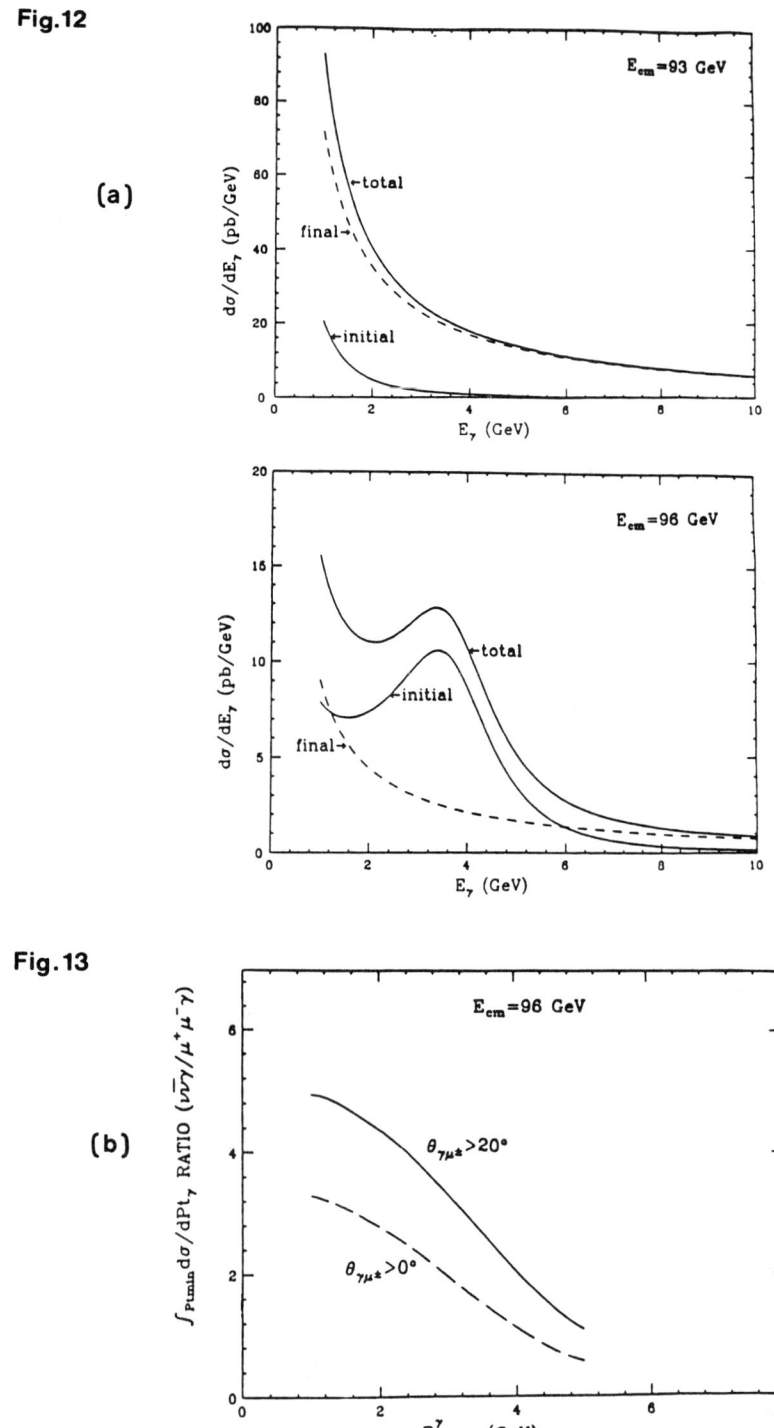

Fig.12

(a)

Fig.13

(b)

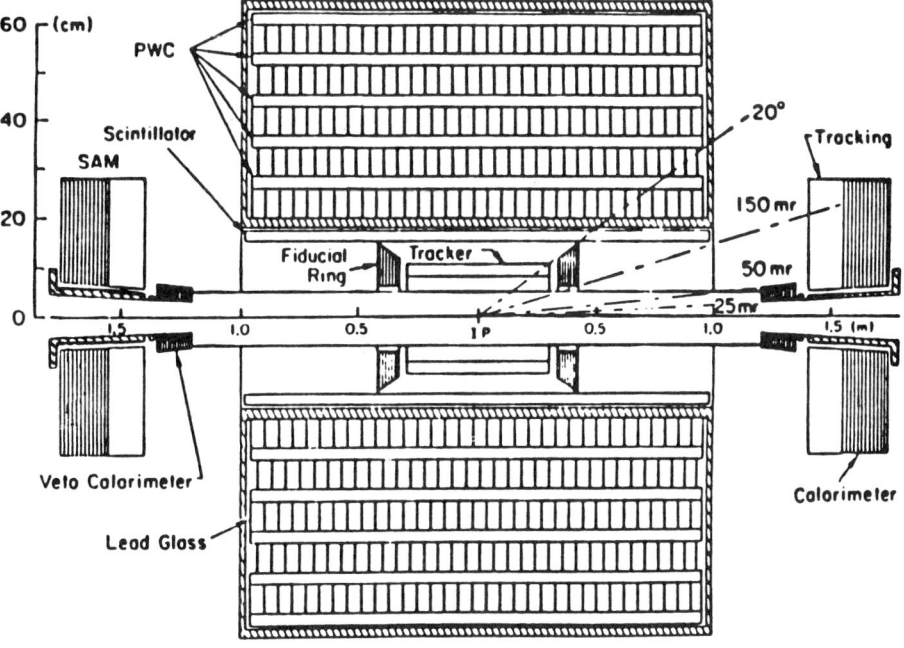

Fig.14

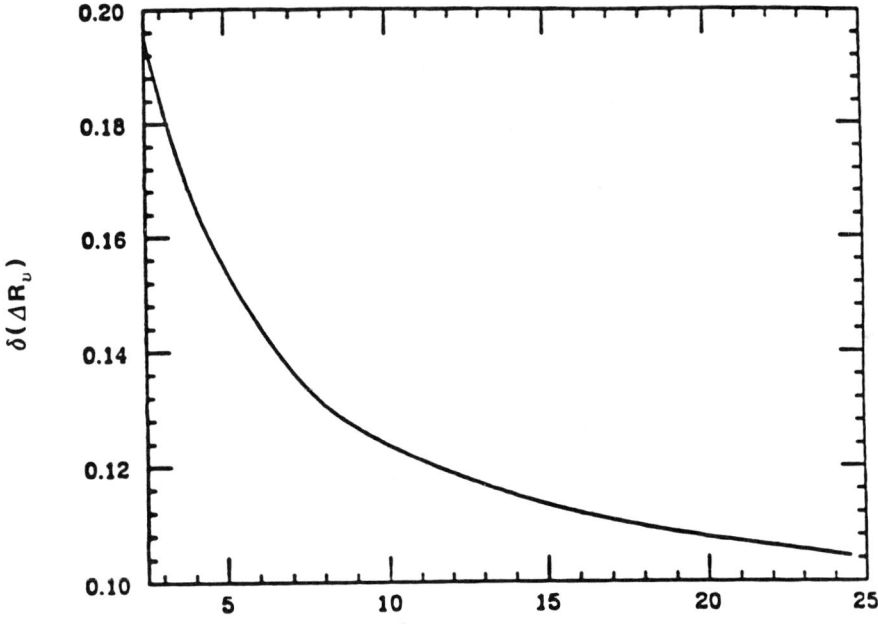

Fig.15

395

Radiative Corrections to the Neutrino Counting Experiment*

C. Mana
CERN, CH-1211 Geneva, Switzerland

M. Martinez, R. Miquel
Laboratori de Física d'Altes Energies
Universitat Autònoma de Barcelona, E-08193 Bellaterra (Barcelona) Spain

September 8, 1989

Abstract

Radiative corrections to radiative neutrino production and to radiative Bhabha scattering are discussed. Concerning the first process, one loop corrections are briefly reviewed. After emphasizing the importance of higher order QED corrections, their implementation using the Structure Functions approach is described. Next, the inclusion of the main weak corrections using the 'star' scheme is explained. Finally some results are presented, showing effects in the range 5-6% with respect to the pure one loop corrected cross-section. Also some comparisons with another calculation featuring multiphoton generation are shown, and good agreement is found between them, both for the total cross-section and the photonic distributions. Concerning the background process, hard photon corrections are computed. After comparing the results with the ones obtained using the EPA, this approximation is used for the virtual and soft part. Some results are presented showing corrections around 3%.

1 Introduction

Radiative neutrino production has been proposed long time ago ([1]) as a clean way of measuring the number of generations within the Standard Model framework. The first calculations were done in the lowest order approximation and using the contact approximation for the W diagrams ([2]). Not until recently complete tree level calculations have been performed ([3], [4]) and also one-loop radiative corrections have been taken into account ([4]–[6]). Since their effect has been found to be large (a common feature of radiative corrections to four fermion neutral current processes around the Z^0 peak), we have undertaken the task of including the main higher order (i.e., more than one-loop) radiative corrections.

Our goal is to achieve a precision better than 1% in the estimation of the total cross-section. This is of the order of the experimental needs, because, although the effect due to an extra generation would be much larger (around 25%), other reactions leading to single photon final states (radiative production of sneutrinos, photinos, etc.) have cross-sections in the per cent range. On the other hand, the error in the determination of the absolute luminosity will

---

*Presented by R. Miquel

be the limiting experimental error, and it will not be smaller than 1%. Therefore, this is a sensible choice for the planned accuracy.

Radiative Bhabha scattering was first pointed out to be the main background to the neutrino counting reaction in ref. [7]. The two final state electrons tend to be produced at very small angles, disappearing down the beam pipe undetected. Then, what remains is a single photon just like in the neutrino case. However, since the electrons go at small angle, they have a limited transverse momentum (p_T) which is compensated by the photon. Hence, due to momentum conservation, there is a limitation in the photon p_T ([7]). Therefore, a p_T cut results in an efficient signal over background ratio when running far above the Z peak because it eliminates most of the $e^+e^-\gamma$ events. However, when running close to or exactly at the Z^0 peak, it eliminates the signal events, too. Therefore, in this case, one has to soften the p_T cut and a large fraction of the background events remain. The only way to proceed then is subtracting the expected number of background events to get the signal cross-section. Then it is clear that a good estimation of the theoretical cross-section for the background process is needed at the few per cent level.

In section 2.1 we will review quickly the exact lowest order calculation for $e^+e^- \to \nu\bar\nu\gamma$. Next we explain with no technical details the one-loop corrections in section 2.2. After justifying the need for higher order corrections, we present in section 2.3 the two implementations we have done of the treatment of higher order QED corrections : one based in the 'inductive' approach ([8]) and another one based on the Structure Functions approach ([9]). Section 2.4 deals with the inclusion of the main higher order weak corrections by means of the 'star' scheme ([10]). Also the need for the inclusion of one loop corrections to the imaginary part of the Z^0 self energy is emphasized there. Finally, section 2.5 contains the results of a comparison done between our Monte Carlo program built along the lines explained in the previous sections (NNGG03) and another one based on the Yennie-Frautschi-Suura approach ([11]). These results have been taken from ref. [12].

In chapter 3 we present the calculation of radiative corrections to $e^+e^- \to e^+e^-\gamma$. Section 3.1 presents the results of the lowest order calculation. Then we will present our calculation of the O(α) QED corrections ([13]) through sections 3.2 and 3.3. Section 3.4 is devoted to the phase space integration and, finally, in section 3.5 we will discuss some results.

2 The process $e^+e^- \to \nu\bar\nu\gamma$

2.1 Lowest order calculation

The diagrams entering the tree level matrix element squared are those depicted in fig. 1. The result for the different helicity amplitudes can be found in ref. [4]. We have taken these results expressing the matrix element squared in terms of α, $\sin^2\theta_W$ and M_Z, i.e. not using the tree level relation between those quantities and G_μ to get an 'improved' tree level result which is neither the true tree level result nor the one obtained after the propagator (also called oblique) corrections are included. In Table 1 we can see the results for the integrated cross-section when making different approximations to the exact calculation. The meaning of the different cross-sections is as follows :

- σ_2^0 is the cross-section taking into account only the two Z^0 diagrams.

- σ_{GGR}^0 is the cross-section obtained neglecting the last diagram in fig. 1 and taking the limit $M_W^2 \to \infty$ in the other W diagrams. It corresponds to the classical result of

Figure 1: Feynman diagrams of the process $e^+e^- \to \nu\bar{\nu}\gamma$ in the Born approximation.

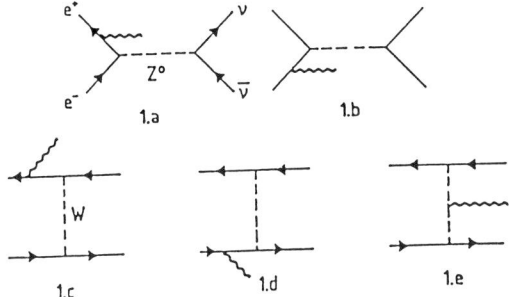

Table 1: Comparison between different approximations for the tree level cross-section.

$\sqrt{s}$(GeV)	σ_2^0(pb)	σ_{GGR}^0(pb)	σ_4^0(pb)	σ_5^0(pb)
98	126.5(3)	129.5(3)	128.7(3)	128.7(3)
150	5.80(3)	9.97(3)	7.74(5)	7.79(5)

Gaemers, Gastmans and Renard (GGR) presented in [2].

- σ_4^0 is the result obtained neglecting the last diagram but taking into account without any approximation the one-W diagrams.

- σ_5^0 is, finally, the total exact result without any approximation.

We have taken $M_Z = 93$ GeV, $\sin^2\theta_W = 0.230$, $E_\gamma > 1$ GeV and $|\cos\theta_\gamma| < \cos 15°$.
The main conclusions that we can extract from Table 1 are the following :

- The contribution from the diagram with two W propagators is negligibly small in all the energy range of LEP/SLC.

- The GGR calculation is essentially good enough at LEP1/SLC.

- At LEP1/SLC the bulk of the cross-section is due to the Z diagrams. Hence, we will only need to worry about radiative corrections to the Z diagrams.

In fig. 2 we show the W contribution, to be understood as the difference between the results of σ_5 and σ_2, as a function of the center of mass energy. We see that for energies between M_Z and 110 GeV it lies between 1% and 4%. This confirms our previous assumption. Therefore, in the following we will only compute radiative corrections to the Z diagrams.

2.2 One-loop corrections

As costumarily done, we divide the one-loop QED corrections into virtual ones and real ones. The first are due to the interference between the tree level diagrams and the diagrams shown

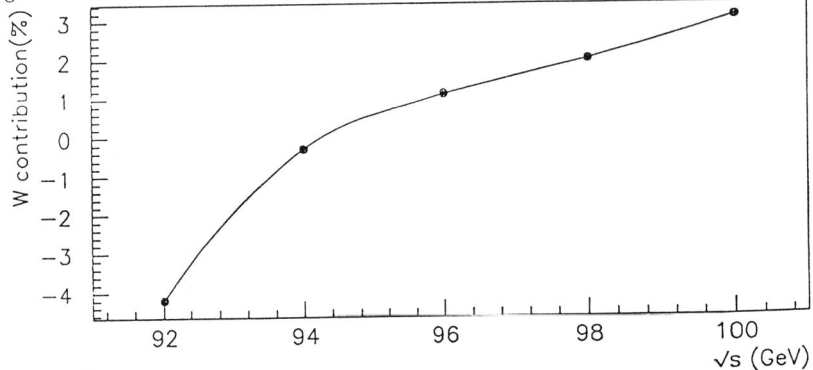

Figure 2: Contribution of the W diagrams squared and of their interference with the Z diagrams. $M_Z = 92$ GeV

in fig. 3. The calculation has been performed using the reduction techniques developped by Passarino and Veltman ([14]). The result can be found in [4]-[6].

Concerning the real photon corrections, we have to distinguish between the soft photon corrections and the hard photon ones. The first ones come from the diagrams of fig. 4 where one of the photons has an energy smaller than $x_0 E_b$, with x_0 an arbitrary (small : around $10^{-2} - 10^{-3}$) number. These corrections are computed analytically neglecting the soft photon kinematical effects and combined with the virtual photon ones to give an infrared finite cross-section depending on x_0.

The hard photon contribution cancels this dependence on x_0 by including events with may look like single photon events because of any of the following reasons:

- One of the photons have fractional energy larger than the minimum detectable photon energy, x_D, and the other one has $x_0 < x < x_D$.

- For this second photon, $x > x_D$ but it goes too low angle and, then, it cannot be detected.

- The two photons are almost collinear and are seen as a single object.

We have computed this cross-section with the spinor techniques explained, for instance, in ref. [6]. This has proved to be a powerful tool when dealing with calculations involving a lot of diagrams or a lot of particles.

Finally, we have also considered the more important weak corrections, namely (in the 'on-shell' renormalization scheme ([15])) the Z^0 self energy. This correction is taken into account just performing the following substitution wherever appears the Z propagator :

$$\frac{1}{q^2 - M_Z^2 + i\Gamma_Z M_Z} \longrightarrow \frac{1}{q^2 - M_Z^2 + Re\Pi_{ZZ}(q^2) + iIm\Pi_{ZZ}(q^2)} \quad (1)$$

where Π_{ZZ} is the Z^0 self-energy given by the diagram of figure 5. A discussion of the treatment

Figure 3: Diagrams contributing to the QED virtual corrections.

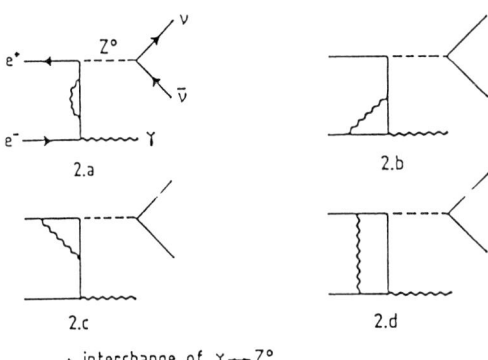

Figure 4: Real photon diagrams contributing to the one loop QED radiative corrections.

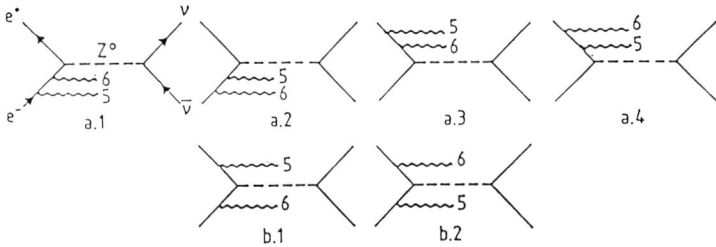

Figure 5: Z^0 self energy diagram.

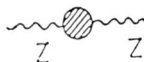

of the propagator corrections is given in section 2.4. At the moment, we just apply (1) to every propagator appearing both in the soft and hard parts. We neglect the non-leading weak corrections ([16]).

The total one-loop corrected cross-section can be expressed as :

$$\sigma_{\nu\bar{\nu}\gamma(\gamma)} = \sigma^{ew}_{V+S} + \sigma^{ew}_{\nu\bar{\nu}\gamma\gamma}$$
$$\equiv \sigma^{ew}_{\nu\bar{\nu}\gamma}(1 + \delta^{V+S}) + \sigma^{ew}_{\nu\bar{\nu}\gamma\gamma} \qquad (2)$$

where the ew superscript means that the Z^0 propagator has been corrected.

For the results given in this section we have used the following set of cuts to define the single photon event :

- Minimum photon energy : $E_D \equiv x_D E_b = 1$ GeV.
- Minimum detection angle : $\theta_D = 15°$.
- Veto angle : $\theta_V = 2.4°$.
- Two photon separation angle : $\theta_{RES} = 1°$.

and the following set of input parameters for the calculation :

- The fine structure function : $\alpha \simeq 1/137$.
- The muon decay constant : $G_\mu = 1.16637 \times 10^{-5}$ GeV^{-2}.
- The Z^0 mass : $M_Z = 93$ GeV.
- The Higgs boson mass : $M_H = 100$ GeV.
- The top quark mass : $m_t = 35$ GeV.

We use the value of G_μ to fix the value of M_W through the one-loop relation

$$\frac{G_\mu}{\sqrt{2}} = \frac{\pi\alpha}{2s_\theta^2 M_W^2 (1 - \Delta r)} \qquad (3)$$

We need the top and Higgs masses because they enter the calculations of the Z^0 self energy and of Δr. Finally, one has to choose a mass for the heavy charged leptons which are in the same $SU(2)$ doublet that the extra neutrinos, since they enter the calculation of the Z^0 self energy. We take for all of them $M_{HEAVY} = 100$ GeV. However, the numbers presented here are for $N_\nu = 3$. Furthermore, we have taken the soft-hard separation limit as $k_0 \equiv x_0 E_b = 0.2$ GeV.

Table 2 contains the results for the different corrections at two center of mass energies. In this table, we have defined $\delta_{Z^0} \equiv \frac{\sigma^{ew}_{\nu\bar{\nu}\gamma} - \sigma^0}{\sigma^0}$. We can see there that the QED corrections are extremely large at the Z peak : adding to δ^{V+S} the hard correction $\delta^H \equiv \frac{\sigma^{ew}_{\nu\bar{\nu}\gamma\gamma}}{\sigma_0}$, we find a total QED correction $\delta^{QED} \equiv \delta^{V+S} + \delta^H = -42.6\%$ Also the Z^0 propagator corrections are sizable. Comparing the first row (tree level calculation) with the last one (final result after the one-loop calculation) we see a huge effect when running at the Z^0 peak ($\sqrt{s} = 93$ GeV) : around -23%. This is almost the difference in the cross-section when adding a new generation (around 25%). At the higher energy, $\sqrt{s} = 100$ GeV, the total effect in the

Figure 6: Comparison between tree level and one-loop results :
a) Integrated cross-section.
b) Differential cross section with respect to the photon energy.
c) Differential cross section with respect to the photon polar angle.
d) Differential cross section with respect to the photon transverse momentum.
Solid line : corrected cross-section.
Dashed line : lowest order.

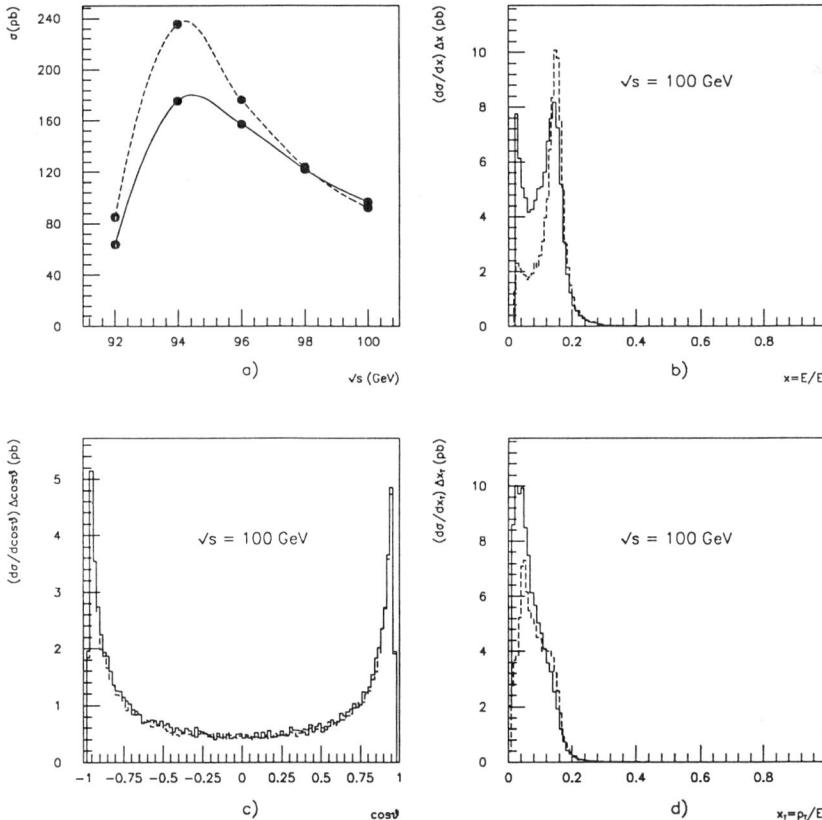

Table 2: Effect of the one-loop corrections. See in the main text the definitions of the quantities appearing in the table.

$\sqrt{s}(\text{GeV})$	93	100
$\sigma^0(\text{pb})$	77.9 ± 0.7	94.2 ± 0.5
δ^{V+S}	-62.5%	-57.3%
δ_{Z^0}	20.0%	15.3%
$\sigma^{ew}_{V+S}(\text{pb})$	44.8 ± 0.4	54.6 ± 0.3
$\sigma^{ew}_{\nu\bar{\nu}\gamma\gamma}(\text{pb})$	15.5 ± 0.3	42.8 ± 0.3
$\sigma_{\nu\bar{\nu}\gamma(\gamma)}(\text{pb})$	60.2 ± 0.4	97.5 ± 0.4

integrated cross-section is rather small ($\sim 3.5\%$). However, it is much more important in the photon distributions. We have shown in fig. 6 b), c) and d) a comparison between the distributions for the photon energy, the cosine of the polar angle and the tranverse momentum obtained with the tree level and the one-loop calculations for the two center of mass energies studied before. We can see that the energy distribution at $\sqrt{s} = 100$ GeV is greatly modified by the one-loop corrections, making the spectrum to be softer as a consequence of having the possibility of radiating an extra photon. Then, if an experimental cut is made in the photon energy in order to reduce the background, the actual cross-section will change a lot. So, even at this higher energy, radiative corrections are important. A similar effect can be seen in the tranverse momentum distribution.

2.3 Higher order QED corrections

The need for higher order corrections can be summarized in the following points which can be read off the one-loop results :

- The QED initial state corrections are really large in absolute value, around -40%. Therefore, we need higher order initial state QED corrections.

- The oblique corrections are large, around 15% : we need also higher order oblique corrections.

- From the previous two points we can extract that the effect of considering one-loop QED corrections combined with one-loop oblique corrections will be around -6%. Then we have to consider carefully the QED-oblique interplay.

Let's comment briefly why initial state QED corrections are large around the Z peak. We can write the cross-section for a general $e^+e^- \to f\bar{f}$ ($f \neq e$) process including one-loop initial state QED corrections as :

$$\sigma_1(s) = \sigma_0(s)(1 + \delta_1 + \beta \ln x_0) + \int_{x_0}^{1} \beta(\frac{1}{x} - 1 + \frac{x}{2})\sigma_0(s')dx \tag{4}$$

where the first part is the soft cross-section and the rest the hard one; x is the photon energy in units of the beam energy and $s' = s(1-x)$; x_0 is the soft-hard separation cut; δ_1 is the part of the soft radiative corrections which is independent of x_0, its value being

$$\delta_1 = \frac{\alpha}{\pi}(\frac{3}{2}\ln\frac{s}{m_e^2} + \frac{\pi^2}{3} - 2) = \frac{3}{4}\beta + \frac{\alpha}{\pi}(\frac{\pi^2}{3} - \frac{1}{2}) ; \tag{5}$$

β is defined as

$$\beta \equiv \frac{2\alpha}{\pi}(\ln\frac{s}{m_e^2} - 1) \tag{6}$$

and it can be regarded as an effective coupling constant for bremsstrahlung. At LEP energies it is large : $\beta \simeq 0.11$. This is one of the reasons causing the QED radiative corrections being so large at the Z_0 peak.

The other major reason has to do with the fact that we are close to a (relatively) narrow resonance as the Z^0 one. Since σ_0 is falling rapidly when far from the pole, the upper limit of the integral in (4) is effectively cut to

$$x_M \sim \frac{\Gamma_Z/2}{E_b} \tag{7}$$

where Γ_Z is the Z^0 width and E_b, the beam energy. And then we can approximate

$$\sigma_1(s) \sim \sigma_0(s)(1 + \delta_1 + \beta\ln\Gamma_Z/M_Z) \tag{8}$$

Putting some standard values for the Z width and mass, we find $\beta\ln\Gamma_Z/M_Z \sim -40\%$!

In the case of final state radiation or in absence of any resonance, the restriction in x_M would disappear and we would end up with no large logs.

There is still another, more technical reason for needing higher order QED corrections. We know that the soft-hard limit, x_0, has no physical content and that any physical result from a Monte-Carlo event generator has to be independent of the value of x_0 in a reasonable range. Since in the soft part we are neglecting the photon energy and the cross-section is a steep function of s, we cannot choose a large value for x_0 if we want a reasonable accuracy. For instance, at the Z peak and for $x_0 = 0.01$ (a standard value) we find

$$\frac{\sigma_0(M_Z^2) - \sigma_0(M_Z^2(1 - 0.01))}{\sigma_0(M_Z^2)} \sim 10\% \tag{9}$$

while for $x_0 = 0.001$ the difference is roughly one per mil. Then, we can conclude that we need to choose $x_0 \sim 0.001$ or lower. Actually, this can be taken into account easily modifying the soft-photon integral. However, choosing a large value for x_0 also affects the accuracy with which some distributions are reproduced. This is specially true for the acollinearity distribution.

On the other hand, if we look at equation (4) we can see that putting x_0 too small would cause the soft part to become negative. Since we have to interpret it as a probability density function, this is unacceptable. We have to put x_0 around 0.001 or higher. So, clearly, we have a problem also here, since there is almost no window left for x_0. We will see how this problem is also solved when the exponentiation is performed.

After seeing that one-loop corrections are not enough, the first idea has to be to try and compute two loop initial state QED corrections. This has been done in ref. [8] with the following result :

$$\sigma_2(s) = \sigma_0(s)(1 + \delta_1 + \delta_2 + \beta \ln x_0 + \delta_1 \beta \ln x_0 + \frac{1}{2}\beta^2 \ln^2 x_0)$$
$$+ \int_{x_0}^{1} dx \left[\beta\left(\frac{1}{x} - 1 + \frac{x}{2}\right)(1 + \delta_1 + \beta \ln x) + \delta_2^H\right] \sigma_0(s') \quad (10)$$

where δ_2 is the $O(\alpha^2)$ correction independent of x_0 and δ_2^H is the $O(\alpha^2)$ non-leading hard correction, the leading one being the one which is proportional to $1/x$. Now x is defined in such a way that $s' = s(1-x)$ is the q^2 in the boson propagator. The explicit expressions for δ_2 and δ_2^H can be found in [8].

If we look carefully at this last equation, it will become clear that it looks like an $O(\alpha^2)$ expansion of the following expression :

$$\sigma_{obs}(s) = \sigma_0(s)(1 + \delta_1 + \delta_2)x_0^\beta$$
$$+ \int_{x_0}^{1} dx \left[\beta x^{\beta-1}(1 + \delta_1 + \delta_2) - \frac{\beta}{2}(2-x)(1 + \delta_1 + \beta \ln x) + \delta_2^H\right] \sigma_0(s') \quad (11)$$

which we will take as our exponentiated formula.

With this treatment we have solved the two problems that we had mentioned before :

- The large corrections coming from the x_0-dependent terms in the initial state corrections have been summed up to all orders. Therefore, the overall precision of the calculation for the initial state QED corrections is around one per mil, the expected size of δ_3.

- The problem with the value of x_0 has also disappeared. Now we can choose x_0 much smaller than 0.001 without any problem, since the soft part never becomes negative. Actually, we could choose $x_0 = 0$, and this is indeed done in the Structure Functions approach, which will be presented later.

This approach can be implemented very easily in any one-loop Monte Carlo calculation. We have done it following these steps (it has to be noted that, in our problem, σ_0 will be the cross-section for radiative neutrino production, i.e., already with one hard photon) :

- In the virtual + soft photon corrections part : substitute $1 + \delta_1 + \beta \ln x_0$ by $(1 + \delta_1)x_0^\beta$.

- In the hard photon corrections part : substitute the term containing the piece $\beta \frac{1}{x}$ by $\beta \frac{1}{x} x^\beta (1 + \delta_1)$ Since it is not easy to isolate explicitely from our hard photon cross-section the $\frac{1}{x}$ term, we have just multiplyed everything by $x^\beta (1 + \delta_1)$. This makes us to include also the corrections to the hardest piece in an exponentiated way which may not be rigorous. However, we do know that the β^2 piece is taken into account correctly, so the difference with the rigorous treatment will be at most in the β^3 term. But in all the calculation we have neglected those terms, which could amount around 0.1%.

Since we are mainly interested in the total cross-section of the process as a measurement of the number of neutrino generations, this approximation will likely be enough. However, if we want to simulate our process accurately in order to take into account detector effects, for example, we have to worry also about how well do we reproduce the differential cross-sections. In this approach we will neglect all the effects of photons carrying a fractional energy smaller than x_0. This is not a serious problem since now, with the exponentiated formula (11), we can choose x_0 as small as we want, so that the effect will be really negligible. When the energy radiated is above this cut-off, we assume that is carried away by just one photon (besides the visible one). This is clearly an approximation, since what we are doing in (11) is adding up the contribution from the radiation of an infinite number of photons. However, when the radiated energy is large enough, it tends to be taken by just one hard photon or at most two. So our approximation is very good in this case. It has been shown in ref. [12] that the probability of having three photons or more of a noticeable energy is very small around the Z peak. Their effects in the total cross-section are accounted for, but their effects in the distributions are taken into account in an approximate way. This effect cannot produce differences in the observable cross-section at a level larger than some parts in 10^{-3}. Results in section 2.5 will confirm this statement.

The second approach follows the ideas of Kuraev and Fadin and Nicrosini and Trentadue ([9]) of using the Stucture Functions formalism. In this approach the initial state electrons are given a structure of electrons and photons. If $D_e(z,s)$ is the probability of finding an electron with fractional momentum z at center of mass energy s, we have

$$\sigma(s) = \int_{z_1^m}^1 dz_1 \int_{z_2^m}^1 dz_2 D_e(z_1,s) D_e(z_2,s) \sigma_0(sz_1z_2) \qquad (12)$$

The lower limit is the minimum fractional energy needed to create a pair of final state fermions.

The function $D_e(z,s)$ satisfies the Altarelli-Parisi equation ([17]). It can be solved to all order in the soft photon limit (large z) ([18]). And then the hard photon corrections up to order β^2 can be added by looking at the explicit iterative second order solution. Finally, the second order virtual corrections can be added by comparing with a complete second order calculation like that of [8]. The final result reads ([9]) :

$$\begin{aligned}D_e(z,s) &= \frac{1}{2}\beta(1-z)^{\frac{\beta}{2}-1}\Delta' - \frac{1}{4}(1+z)\beta \\ &+ \frac{1}{32}\beta^2\left[(1+z)[3\ln z - 4\ln(1-z)] - \frac{4}{1-z}\ln z - 5 - z\right]\end{aligned} \qquad (13)$$

where Δ' includes the virtual corrections up to $O(\alpha^2)$: $\Delta' = 1 + \frac{\delta_1}{2} + \cdots$. This leads to the following expression for the cross-section

$$\begin{aligned}\sigma(s) &= \int_0^1 dx H(x,s)\sigma_0(s(1-x)) \\ H(x,s) &= \Delta \cdot \beta x^{\beta-1} - \frac{1}{2}\beta(2-x) + \frac{1}{8}\beta^2\{(2-x)[3\ln(1-x) - 4\ln x] \\ &\quad - 4\frac{\ln(1-x)}{x} - 6 + x\} + O(\beta^3)\end{aligned} \qquad (14)$$

where $\Delta = \Delta'^2 - \frac{\beta^2\pi^2}{24}$ Comparing the result obtained for $H(x,s)$ with the integrand in the hard part of (11), we see that they are very similar, so that we would expect the two methods to give very similar results. We will see in the following that this is indeed the case.

We will use the expression with the Structure Functions (12) for the calculation of the total cross-section. However when the total amount of energy radiated is larger than a certain x_0 cut, we will use *only for the differential cross-section* the exact hard cross-section computed previously corrected with the replacement explained before for the $\frac{1}{x}$ piece. In this way, we are sure that we get the right answer for both the total cross-section and the main distributions. Let's see how we do it in practice :

- First, we generate the fractional energy of the electrons in the collision z_1 and z_2 according to an approximated distribution function $\tilde{D}_e(z,s) = \frac{\beta}{2}(1-x)^{\beta/2-1}$, and compute the new center of mass energy for the annihilation $s' = sz_1z_2$.

- Then we generate the phase-space variables for the two neutrinos and the visible photon according to the differential tree level cross-section at center of mass energy s' and boost them to the lab reference frame.

- Next, we compute the weight of the event as

$$W = \frac{D_e(z_1,s)}{\tilde{D}_e(z_1,s)} \frac{D_e(z_2,s)}{\tilde{D}_e(z_2,s)} \qquad (15)$$

Computing the mean of those weights and multiplying by the integral of the approximant we will find the total cross-section and with the help of a rejection algorithm we will have a sample of unweighted events.

- Once we have an unweighted event, we compute the amount of energy taken by the non-visible photons : $x = 1 - z_1z_2$. If $x < x_0$, we are through and go to next event; if $x > x_0$, we throw away this event and, instead, we generate an unweighted event according to the modified hard differential cross-section. Then, we go to next event.

This second implementation is supposed to give more accurate results than the first one for the distributions. As mentioned before, with the Structure Functions approach we have more kinematical information than with the 'inductive' approach : we know how the radiated energy is shared between the two initial state particles. Furthermore, all the effects of *collinear* radiation are taken into account in a precise way. With our approach of taking the exact second order matrix element for the hard photon radiation we are sure that we also take into account transverse radiation in a sensible way. For the total cross-sections, they are supposed to give equivalent results. Actually, we will see that this is indeed the case.

Now we are going to present the results of the contribution of the higher order QED corrections. Throughout this section we are going to use the following set of cuts to define a single photon event :

- Minimum energy : $E_D \equiv x_D E_b = 1$ GeV.

- Minimum polar angle : $\theta_D = 15°$.

- veto angle : $\theta_V = \theta_D = 15°$.

- Two photon separation angle : $\theta_{RES} = 1°$.

And the following input parameters for the calculation :

- $M_Z = 92$ GeV, $M_H = 100$ GeV, $m_t = 60$ GeV, $M_{HEAVY} = 100$ GeV.

Table 3: Comparison between one-loop and exponentiated cross-sections.

$\sqrt{s}$(GeV)	92	94	100
σ^1(pb)	63.7 ± 0.2	175.4 ± 0.6	96.7 ± 0.4
σ^I(pb)	67.5 ± 0.3	184.5 ± 0.7	95.6 ± 0.5
σ^{SF}(pb)	67.6 ± 0.4	185.1 ± 1.3	94.8 ± 0.7

And the standard values for α and G_μ. M_{HEAVY} is the mass of the leptons which are in the doublet with the extra neutrinos. It enters the calculation of the self energy of the Z, for instance. We have taken the soft-hard cut-off as $k_0 \equiv x_0 E_b = 0.2$ GeV for the pure one-loop calculation and 0.002 GeV for the other ones. Of course, the result is independent of the value chosen.

First of all in Table 3 we show a comparison between the pure one-loop result for the integrated cross-section (σ^1), the cross-section obtained after adding the higher order QED corrections via the implementation of the approach of Berends et al. (σ^I), and after adding the higher order QED corrections with the Structure Functions approach (σ^{SF}). In all the cases the only weak correction included is the modification of the Z^0 propagator as indicated by equation (1).

The main conclusions from this table are two : firstly, there is very good agreement between the two ways we have tried for including higher order QED corrections in this region. The results are always compatible and the differences of the central values are between 0.1 and 0.8%. On the other hand, the corrections with respect to the pure $O(\alpha)$ calculation are sizable, especially at the pole. They range from +6.0% at $\sqrt{s} = 92$ GeV to -1.1% at $\sqrt{s} = 100$ GeV. Comparing with table 2 where a comparison between tree level and one-loop results is made, we realize that higher order corrections go in the opposite way the order α ones go. They tend to compensate the overstimation of the radiative effects done by the first order corrections.

2.4 Higher order oblique corrections

We have chosen to use the 'star' scheme developped by Kennedy and Lynn ([10]) to include in our calculation the oblique corrections[1] Here we will give just a rough idea of the procedure. The details can be found in ref. [6].

We can write a general four-fermion neutral current process matrix element with the one-loop oblique corrections included through Dyson equations (which means that the leading-log terms are included to all orders) as :

$$M_{NC} = e_0^2 \frac{QQ'}{q^2 + \Pi_{AA}} + \frac{e_0^2}{s_0^2 c_0^2} \frac{\left[I_3 - Q(s_0^2 - s_0 c_0 \frac{\Pi_{ZA}}{q^2 + \Pi_{AA}})\right]\left[I_3' - Q'(s_0^2 - s_0 c_0 \frac{\Pi_{ZA}}{q^2 + \Pi_{AA}})\right]}{q^2 - \frac{e_0^2}{4\sqrt{2} s_0^2 c_0^2 G_\mu \rho_0} + Re\Pi_{ZZ} + iIm\Pi_{ZZ}} \quad (16)$$

---
[1]Recently, there has been some discussion about the gauge-invariance of the 'star' treatment([19]). The main conclusions seem to be that the treatment is gauge-dependent in a formal sense, but that this fact is completely irrelevat numerically, because the pieces missing to achieve gauge-invariance are very small in the t'Hooft-Feynman gauge in which the authors of ref. [10] work. For a more detailed discussion see, for instance, [6].

Now defining a set of finite functions (we do not write the imaginary parts)

$$
\begin{aligned}
e_*^2(q^2) &= \frac{e_0^2}{1 - \frac{\Pi_{AA}(q^2)}{q^2}} \\
s_*^2(q^2) &= s_0^2 - s_0 c_0 \frac{\Pi_{ZA}(q^2)}{q^2 + \Pi_{AA}(q^2)} \\
\frac{1}{4\sqrt{2}G_{\mu*}}(q^2) &= \frac{1}{4\sqrt{2}G_{\mu 0}} - \frac{s_0}{c_0}\Pi_{WW}(q^2) + \frac{s_0 c_0}{e_0}\Pi_{ZA}(q^2) + \frac{s_0^2}{e_0^2}\Pi_{AA}(q^2) \\
\frac{1}{\rho_*(q^2)} &= \frac{1}{\rho_0} - 4\sqrt{2}G_{\mu*}(q^2)\left(\Pi_{ZZ}(q^2) - \Pi_{WW}(q^2)\right)
\end{aligned}
\qquad (17)
$$

we obtain :

$$
M_{NC} = \frac{e_*^2(q^2)QQ'}{q^2} + \frac{e_*^2}{s_*^2 c_*^2} \frac{[I_3 - Q s_*^2(q^2)][I_3' - Q' s_*^2(q^2)]}{q^2 - \frac{e_*^2}{4\sqrt{2}s_*^2 c_*^2 G_{\mu*}\rho_*}(q^2) + iM_Z\Gamma_Z^*(q^2)}
\qquad (18)
$$

We have reabsorbed all the oblique corrections in four finite, universal functions using the Dyson equations. This treatment has some advantadges :

- The matrix element looks like the tree level one.

- All the leading-log terms of the oblique corrections are summed up to all orders and the corrections appear automatically in the Z width, for instance. The 'starred' functions being universal, the running coupling constants which appear in the numerator of the neutral current matrix element are the same that appear in the definition of Γ_Z^* and, thus, they cancel at the pole showing that the total cross-section is insensitive to the oblique corrections. Other schemes for computing radiative corrections would have obscured this fact.

- We can now take the new matrix element as an effective Born amplitude and apply to it QED corrections. In this way, we assure that the oblique corrections are not overstimated. This would happen if QED corrections were not factorized with respect to the oblique ones. Since, as we have seen, QED corrections are extremely large near the Z pole, this would have implied a large error in the oblique corrections.

One can also absorb the non-abelian vertex corrections (which are also universal) into the 'starred' functions ([6]). Then we will have also these corrections factorized with respect to the QED ones. However, the effect is almost negligible, as we will show later on.

Another important point concerning higher order oblique corrections is that of the corrections to $Im\Pi_{ZZ}$ around the pole, as pointed out for the first time by Wetzel ([20]). If we look at the expression for the matrix element for the Z^0 current, we see that at the Z peak, $q^2 = M_Z^2$, we just have

$$
M^{NC} \sim \frac{\alpha A}{iM_Z\Gamma_Z^*(M_Z^2)}
\qquad (19)
$$

where the A in the numerator means something without factors of α in the first approximation. Then the whole numerator is of order α. Since Γ_Z^* is also an $O(\alpha)$ quantity, the matrix element

Figure 7: Example of one loop corrections to $Im\Pi_{ZZ}$.

(19) is of order 1. This means that an order α correction to Γ_Z^*, which is equivalent to an order α^2 contribution to $Im\Pi_{ZZ}$, is just an order α correction to the matrix element, and, hence, it is probably needed to reach the accuracy we are seeking.

The fact that in front of the 'starred' width we have the 'starred' coupling constants means that we are including the oblique corrections also in the width, as we have pointed out previously. However, these are not the only corrections that have to be taken into account. There are also the direct corrections shown in the diagram of fig. 7. This kind of diagrams have not been computed at the moment. Nevertheless, we do not need to know their contribution at any q^2, but only when $q^2 \simeq M_Z^2$. We know that exactly at the Z pole the following equations hold

$$\alpha Im\Pi_{ZZ}^{(1)}(M_Z^2) = M_Z \Gamma_Z^{(0)} \qquad (20)$$

$$\alpha^2 Im\Pi_{ZZ}^{(2)}(M_Z^2) = M_Z(\Gamma_Z^{(1)} - \Gamma_Z^{(0)}) \qquad (21)$$

where $\alpha Im\Pi_{ZZ}^{(1)}$ is the one-loop self-energy, $\alpha^2 Im\Pi_{ZZ}^{(2)}$ is the second order contribution to this self energy, $\Gamma_Z^{(0)}$ is the tree level width and $\Gamma_Z^{(1)}$ is the one-loop corrected Z^0 width, computed in ref. [21]. It includes the complete electroweak corrections together with the QCD corrections to the hadronic branching ratios. This last effect turns out to be the one which is numerically the most important.

We also know that, to good accuracy, the behaviour of the imaginary part of the Z self energy around the peak is

$$Im\Pi_{ZZ}(s) \propto s. \qquad (22)$$

Then, a good approximation for the two-loop contributions to $Im\Pi_{ZZ}$ will be

$$Im\Pi_{ZZ}(s) \simeq \alpha Im\Pi_{ZZ}^{(1)}(s) + \frac{s}{M_Z^2} M_Z(\Gamma_Z^{(1)} - \Gamma_Z^{(0)}) \qquad (23)$$

This will be a very good approximation near the peak. Far away from it, it will not be so good, but then we do not need such a precision for the imaginary part, since the real part will no longer be small and will dominate the behaviour of the propagator.

Now, we are going to discuss the effect of putting the weak corrections using the 'star' scheme and the effect of the modification we have made to it. We see in Table 4 the cross-section with exponentiation, σ^I from Table 3, and the cross-sections with inductive exponentiation and 'star' scheme without (σ^{I*}) and with our modifications to include the non-abelian vertices ($\hat{\sigma}^I$). Note that since we are including in the Z width the QCD corrections, we need to fix the value of the strong coupling constant, α_s. We have taken $\alpha_s(M_Z^2) = 0.12$.

We see that the largest effect due to the inclusion of the 'star' treatment is at $\sqrt{s} = 94$ GeV, i.e., when we recover the Z peak due to the emission of the visible photon. This effect

Table 4: Comparison between the cross-sections without 'star' scheme (σ^I), with 'star' scheme unmodified (σ^{I*}) and with 'star' scheme modified by us ($\hat{\sigma}^I$).

$\sqrt{s}$(GeV)	92	94	100
σ^I(pb)	67.5 ± 0.3	184.5 ± 0.7	95.6 ± 0.5
σ^{I*}(pb)	66.4 ± 0.3	175.9 ± 0.8	93.3 ± 0.5
$\hat{\sigma}^I$(pb)	66.0 ± 0.3	175.6 ± 0.7	92.4 ± 0.5

Table 5: Percentual total corrections due to higher order effects.

$\sqrt{s}$(GeV)	92	94	100
δ(%)	3.6	0.1	-4.4

is important : around −4.6%. Below and above this energy the correction is decreasing up to around −2%. This kind of behaviour is reasonable since most of the effect is due to the inclusion of the one-loop corrections to the width and this is important when the Z^0 is on-shell. The correction due to the inclusion of the non-abelian vertices is almost negligible : it is always less that 1%.

Combining the higher order QED with the higher order weak corrections (with the 'star' scheme modified), we can find the total correction with respect to the one-loop cross-section. It is shown in Table 5 and in figure 8 a).

The QED correction dominates at the peak and hence the total correction is positive. Slightly above the peak, the two corrections are almost of the same size and opposite signs, so that the global correction is almost zero. We cannot see any particular reason for this and, therefore, we think it is completely casual. Finally, above the peak both corrections are negative giving a large total effect.

In fig. 8 b), c) and d) we can see the effect of all the higher order corrections altogether. As we see they do not modify very much the photon distributions, although some effects can be seen in the photon and transverse momentum distributions.

2.5 Comparison with other calculations

A thorough comparison has been performed in ref. [12] between our Monte Carlo calculation with the exponentiation a la Berends et al. and the version of KORL03 ([11]) modified by Colas, Mirabito and Wąs for the neutrino counting process. All the numbers in this section are taken from ref. [12].

The program in ref. [12] includes multiphoton radiation following the Yennie, Frautschi, Suura approach ([22]) and the oblique radiative corrections have been computed by Stuart ([23]). The neutrino version adds exactly the contributions of the one W diagrams and neglects the two W one.

The results have been obtained with the following set of parameters : $M_Z = 92$ GeV, $m_{top} = 60$ GeV, $M_H = 100$ GeV and $N_\nu = 3$. Furthermore only the events with at least one

Figure 8: Comparison between pure one-loop results and the ones obtained including higher order corrections :
a) Integrated cross-section.
b) Differential cross-section with respect to the photon energy.
c) Differential cross-section with respect to the photon polar angle.
d) Differential cross-section with respect to the photon transverse momentum.
Solid line : Higher order result with inductive exponentiation.
Dashed line : Higher order result with structure functions exponentiation.
Dotted line : One-loop result.

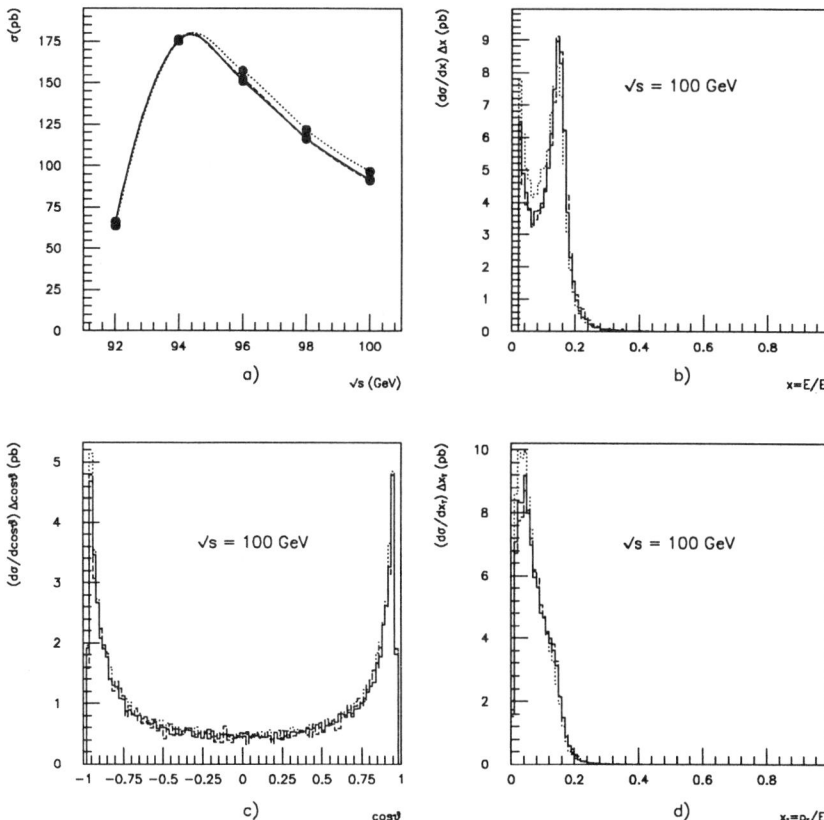

Table 6: Integrated cross-sections (pb) with at least one photon with $E_\gamma > 0.5$ GeV and $15° < \theta_\gamma < 165°$

$\sqrt{s}$(GeV)	90	92	94	99	110
KORL03	38.9 ± 0.2	143.7 ± 1.0	245.1 ± 1.2	117.0 ± 0.6	46.3 ± 0.3
NNGG03	38.4 ± 0.1	144.9 ± 0.2	245.4 ± 0.5	117.6 ± 0.1	46.0 ± 0.1

photon with $E_\gamma > 0.5$ GeV and $15° < \theta_\gamma < 165°$ are considered.

The agreement in the integrated cross-section has been found to be very good, better than 1% in the region of interest for the neutrino counting experiment, i.e., at and slightly above the Z peak. This is shown in Table 6 and in fig. 9 a) and b).

It has to be noted the fact that KORL03 has larger errors. The reason is that to have a large sample of events a lot of CPU time is needed, since most of the events will not have any hard photon. This is not the case with our program, which has always a photon from the tree level.

Also the main distributions have been compared in ref. [12] and the agreement has been found satisfactory. In fig. 10 a), b) and c) we show the distribution of energy, polar angle (defined in such a way that it is flat for a bremsstrahlung photon) and tranverse momentum for the hardest photon at a center of mass energy of 99 GeV. No noticeable differences between the two programs can be seen. The same has been checked at $\sqrt{s} = M_Z = 92$ GeV and at $\sqrt{s} = 110$ GeV.

3 The process $e^+e^- \to e^+e^-\gamma$

3.1 Lowest order calculation.

In this section we will just review some results found using the calculation of ref. [24]. This is an exact tree level calculation performed with the helicity amplitude method.

We will define a hypothetical experimental set-up with the following conditions :

- veto angle for electrons : $\theta_V = 2.5°$,
- minimum photon energy : $x_D E_b = 1$ GeV, and
- tagging angle for photons : $\theta_D = 15°$.

In fig. 11 a) we can see the integrated cross-section as a function of the center of mass energy compared with that obtained from the tree-level calculation of $e^+e^- \to \nu\bar{\nu}\gamma$. We have chosen $M_Z = 92$ GeV, $\sin^2\theta_W = 0.23$ and $N_\nu = 3$. The cross-section for the single photon final state coming from $e^+e^-\gamma$ is totally dominated by the diagrams with t-channel photons, and, hence, we do not see any enhancement of the cross-section near the Z pole. As we can see, in principle, the background is larger than the signal. However, if we look at the photon distributions above the Z peak, at $\sqrt{s} = 100$ GeV (fig. 11 b), 11 c) and 11 d)), we see that they are very different and that a clean selection of events can be made cutting on the transverse momentum (fig. 11 d)).

Figure 9: Comparison between our calculation and the one with KORALZ:
a) Integrated cross-section.
b) Ratio of cross-sections.

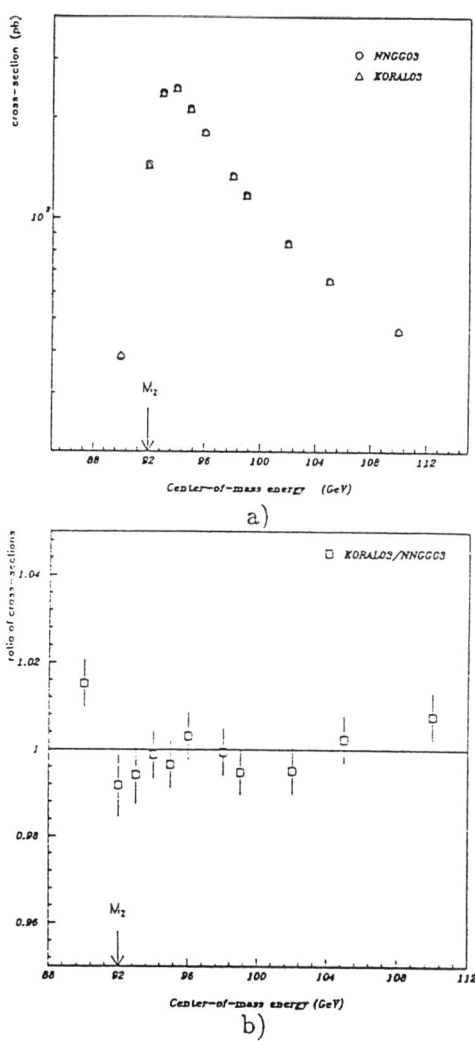

Figure 10 : Comparison between our calculation and the one with KORALZ ($\sqrt{s} = 99$ GeV):
a) E_γ distribution.
b) Angular distribution.
c) $p_{T\gamma}$ distribution.

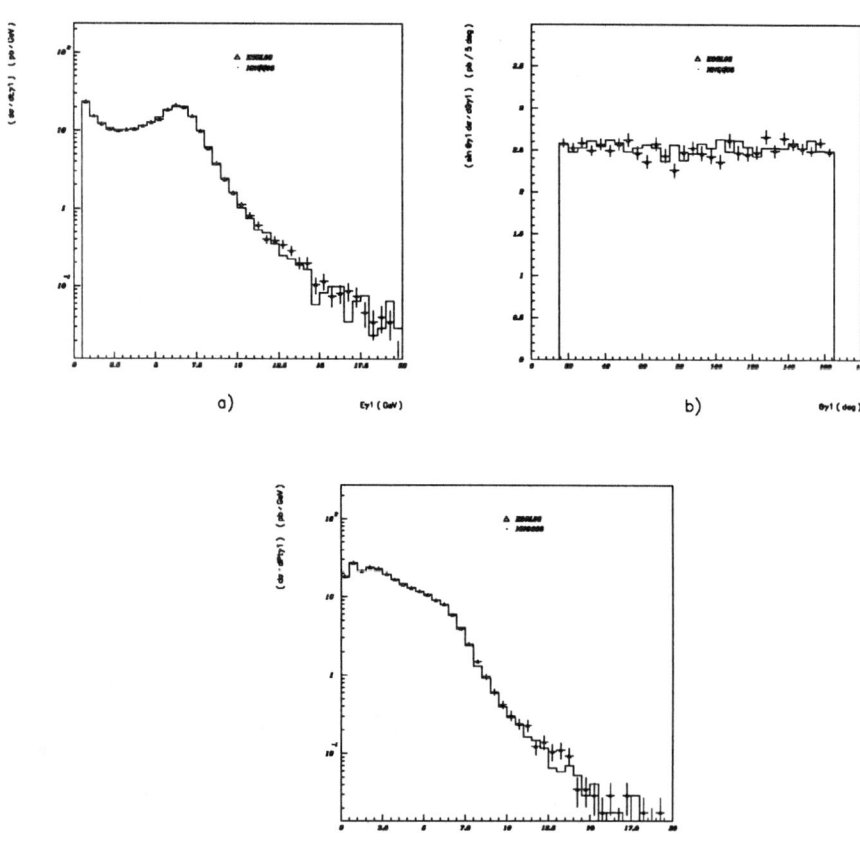

Figure 11: Results of the tree level calculation of radiative Bhabha scattering.
a) Integrated cross-section.
b) Energy distribution.
c) Angular distribution.
d) Transverse momentum distribution.
Solid line : $e^+e^- \to e^+e^-\gamma$.
Dashed line : $e^+e^- \to \nu\bar{\nu}\gamma$.

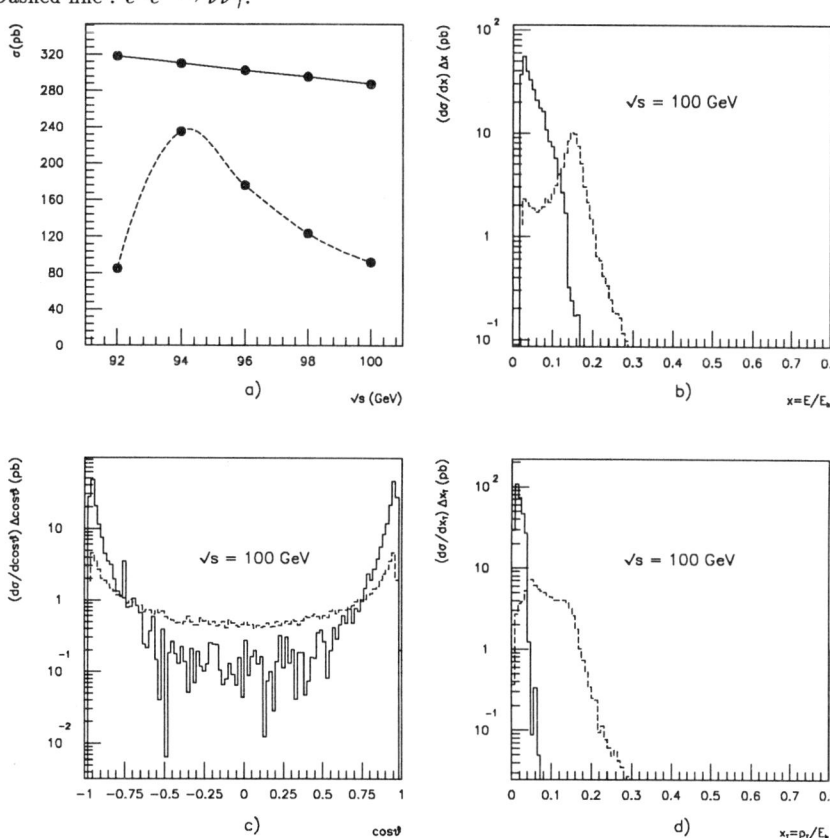

After obtaining the result for the tree level cross-section, we may worry about higher order corrections. It is clear that since this is a background process which can be mostly eliminated, we do not need as much precision as in the calculation of the signal cross-section. However, there are two reasons which justify the need for computing one-loop corrections to this process :

- The optimum neutrino counting experiment would be performed at an energy some GeV's above the Z peak. However, since most of the luminosity will be taken at the peak, certainly the experiment will also be done there. Then it is very difficult in this case to remove the background by means of kinematical cuts since the signal spectrum is also very soft and a cut on p_T would also cut most of the signal. Thus a background subtraction is needed and a precise calculation of the background cross-section has to be performed.

- The second reason is that a calculation of the one-loop corrections to $e^+e^- \to e^+e^-\gamma$ was indeed done, and the results (presented in the first version of the SLAC-PUB-4121 preprint, which (after some corrections) led to the article of ref. [25]) showed that the higher order correction was twice as large as the tree level cross-section! Since we were puzzled by this result we decided to start with this calculation.

We will begin the presentation of our one-loop calculation ([13]) by showing the contribution of the hard photon corrections.

3.2 Hard photon corrections.

3.2.1 Exact calculation : Helicity Amplitudes method.

The process

$$e^-(p_1) + e^+(p_2) \longrightarrow e^-(p_3) + e^+(p_4) + \gamma(p_5) + \gamma(p_6) \qquad (24)$$

is represented to order g^4 by as many as 80 diagrams. However, for the single photon configuration, the only non negligible contributions are those coming from the 10 diagrams shown in figure 12 plus the 10 additional ones obtained by exchanging the photon labels, i.e., all the diagrams containing a photon in t-channel. From now on, we will restrict ourselves to studying the contributions from those diagrams.

Clearly they form a gauge-invariant subset and we can group them in 4 topologically different groups (A,B,C,D in fig. 12) such that the amplitude for each diagram in one class can be obtained from any other belonging to the same class by an overall permutation of particle labels and helicities (and, occasionally, conjugation). They cannot be obtained with such a simple transformation from any one of the other classes. Therefore, we just have to calculate the matrix element, M, for one diagram of each class. We will use the helicity amplitudes approach. This method has proved to be very useful when dealing with a great number of diagrams and when large cancellations are expected to occur. When masses are to be kept (and this is indeed our case, due to the collinear peaks of the matrix element squared) the most efficient way to face the problem is the numerical implementation of the helicity amplitudes technique, as described in great detail in ref. [26], [27], [28], [29], among others. The amplitude for diagram A-1 reads:

$$-iM_{A1} = e^4 (b_2p_2 + b_4p_4)^{-2}((b_1p_1 + b_5p_5 + b_6p_6)^2 - m_e^2)^{-1}((b_1p_1 + b_5p_5)^2 - m_e^2)^{-1}$$

Figure 12: Hard photon corrections to the t-channel photon diagrams.

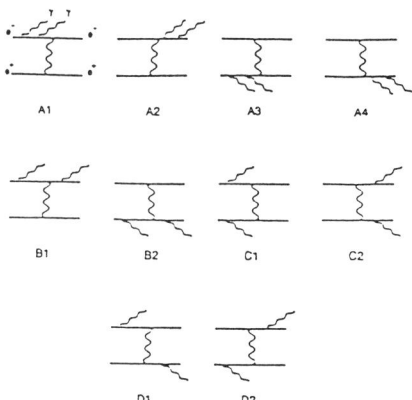

$$[\bar{u}(p_3)\gamma^\mu(b_1(\not{p}_1+m_1)+b_5\not{p}_5+b_6\not{p}_6)\not{\epsilon}_6^*(b_1(\not{p}_1+m_1)+b_5\not{p}_5)\not{\epsilon}_5^* u(p_1)]$$
$$[\bar{v}(p_2)\gamma_\mu v(p_4)] \quad (25)$$

with $b_i = 1$ for incoming particles and -1 for outgoing particles and $m_i = m_e$ for e^- and $-m_e$ for e^+. In this way, the characteristics of incoming (outgoing) and of particles (antiparticles) have been absorbed in b_i and m_i. Hence, for instance, we can substitute every $v(p_i)$ for an $u(p_i)$. Now the rest of diagrams of class A are obtained via:

$$\begin{aligned}
-iM_{A2} &= (-iM_{A1})^* \ (1 \leftrightarrow 3;\ 2 \leftrightarrow 4;\ \lambda_5 \leftrightarrow -\lambda_5;\ \lambda_6 \leftrightarrow -\lambda_6;\ 5 \leftrightarrow 6) \\
-iM_{A3} &= (-iM_{A1})^* \ (1 \leftrightarrow 2;\ 3 \leftrightarrow 4;\ \lambda_5 \leftrightarrow -\lambda_5;\ \lambda_6 \leftrightarrow -\lambda_6) \\
-iM_{A4} &= (-iM_{A1}) \ (1 \leftrightarrow 4;\ 2 \leftrightarrow 3;\ 5 \leftrightarrow 6)
\end{aligned} \quad (26)$$

and just changing $5 \leftrightarrow 6$ for the diagrams with the photon label exchanged. This procedure is repeated in a similar way for diagrams B, C and D.

Now, introducing the $Z(p_i, \lambda_i)$ functions as

$$Z(p_i, \lambda_i; p_j, \lambda_j; p_k, \lambda_k; p_l, \lambda_l; C_L, C_R, C'_L, C'_R) =$$
$$[\bar{u}(p_i, \lambda_i)\gamma^\mu(C_L P_L + C_R P_R)u(p_j, \lambda_j)][\bar{u}(p_k, \lambda_k)\gamma_\mu(C'_L P_L + C'_R P_R)u(p_l, \lambda_l)], \quad (27)$$

with $P_{\binom{R}{L}} \equiv \frac{1}{2}(1 \pm \gamma_5)$, and the polarization vector of the photons as

$$\epsilon^\mu(p_i, \lambda_i) = N_i \bar{u}(p_i, \lambda_i)\gamma^\mu u(p, \lambda_i), \quad (28)$$

being p any four-momentum occurring in the process different from p_5 and p_6, and N_i the normalization constant, we can express M_{A1} as

$$-iM_{A1} = e^4 \ (b_2 p_2 + b_4 p_4)^{-2} \ ((b_1 p_1 + b_5 p_5 + b_6 p_6)^2 - m_e^2)^{-1} \ ((b_1 p_1 + b_5 p_5)^2 - m_e^2)^{-1} \cdot T_{A1} \quad (29)$$

and

$$T_{A1} = N_5 N_6 \sum_{\lambda,\lambda'}(Z(p_2,\lambda_2;p_4,\lambda_4;p_3,\lambda_3;p_1,\lambda;1,1,1,1) \cdot b_1 \cdot$$
$$\cdot (Z(p_1,\lambda;p_5,\lambda';p,\lambda_6;p_6,\lambda_6;1,1,1,1) \cdot b_5 \cdot Z_5$$
$$+ Z(p_1,\lambda;p_1,\lambda';p,\lambda_6;p_6,\lambda_6;1,1,1,1) \cdot b_1 \cdot Z_1)$$
$$+ Z(p_2,\lambda_2;p_4,\lambda_4;p_3,\lambda_3;p_5,\lambda;1,1,1,1) \cdot b_5 \cdot$$
$$\cdot (Z(p_5,\lambda;p_5,\lambda';p,\lambda_6;p_6,\lambda_6;1,1,1,1) \cdot b_5 \cdot Z_5$$
$$\cdot Z(p_5,\lambda;p_1,\lambda';p,\lambda_6;p_6,\lambda_6;1,1,1,1) \cdot b_1 \cdot Z_1)$$
$$+ Z(p_2,\lambda_2;p_4,\lambda_4;p_3,\lambda_3;p_6,\lambda;1,1,1,1) \cdot b_6 \cdot$$
$$\cdot (Z(p_6,\lambda;p_5,\lambda';p,\lambda_6;p_6,\lambda_6;1,1,1,1) \cdot b_5 \cdot Z_5$$
$$\cdot Z(p_6,\lambda;p_1,\lambda';p,\lambda_6;p_6,\lambda_6;1,1,1,1) \cdot b_1 \cdot Z_1)) \qquad (30)$$

with

$$\begin{aligned} Z_1 &= Z(p_1,\lambda;p_1,\lambda_1;p,\lambda_5;p_5,\lambda_5;1,1,1,1), \\ Z_5 &= Z(p_5,\lambda;p_1,\lambda_1;p,\lambda_5;p_5,\lambda_5;1,1,1,1). \end{aligned} \qquad (31)$$

Given a particular set of four-momenta and helicities, one can evaluate the Z functions (tabulated in App. D) for diagrams A1, B1, C1 and D1, and, with the appropriate permutations and conjugations, the ones of all the remaining diagrams. Finally

$$\overline{|-iM|^2} = \frac{1}{8}\sum_{\lambda}|\sum_{j}(-iM_j)|^2 \qquad (32)$$

where j and λ run over all diagrams and helicity configurations respectively and the extra factor 1/2 accounts for the symmetrization of the final state photons.

Since we are evaluating a complete gauge-invariant set of diagrams and the effect of changing the choice of p is just a gauge transformation, we can check our calculation just by evaluating $\overline{|-iM|^2}$ with different choices of p. We see that the result is not affected at all by this choice. With the definition of $Z(p_i,\lambda_i)$ we have introduced, it is a straightforward exercise to include also the diagrams with an s-channel photon or with a Z^0, but their contributions are negligible in the single photon configuration.

3.2.2 Approximate calculation : Equivalent Photon Approximation.

In ref. [25] a different way has been chosen to compute the hard cross-section. We describe it here because it is the method we have used for the virtual and soft photon corrections. It is also relevant to try to compare the results obtained for the hard cross-section with both methods to justify the use of the approximation for the virtual and soft photon part.

This method is based on the so called Equivalent Photon Approximation (EPA). In short, it consists in considering the photon exchanged between the electron and the positron as quasi-real. Then the cross-section can be written as

$$d^8\sigma_{e^+e^- \to e^+e^-\gamma\gamma} = d^3n_{e^+(p_+) \to e^+(q_+)\gamma(\tilde{k})}d^5\sigma_{\gamma(\tilde{k})e^-(p_-) \to e^-(q_-)\gamma(k)\gamma(k_s)} \qquad (33)$$

where the first piece is the radiation spectrum of the positron and the second one is the differential cross-section for double Compton scattering. This kind of factorization is reasonable

Figure 13: Hard photon diagrams included in the EPA.

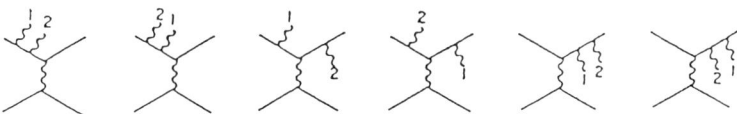

when $|t| \ll \tilde{s}$ where $t = (p_+ - q_+)^2$ and $\tilde{s} = (p_- + \tilde{k})^2$, that is the 'mass' of the virtual photon is much smaller than the center of mass energy of the γe collision.

With this approximation, it is clear that only the diagrams in fig. 13 will be considered. Of course, this could be sufficient when considering that the positron is going at very small angle. Then the charged conjugated diagrams are included just by symmetrization. Therefore, no interference between the set of diagrams in fig. 13 and their charged conjugated ones is included. Of course, no s-channel diagram or Z diagram is taken into account. It is clear then that the approximation will not be valid for general Bhabha scattering at large angles. We will see, however, that it is reasonable at very small angles, which is the region of interest for the evaluation of the background to $e^+e^- \to \nu\bar{\nu}\gamma$.

Expressions for $d^3n_{e^+ \to e^+\gamma}$ can be found in the literature (for instance, in ref. [30]). It can be written as:

$$d^3 n_{e^+ \to e^+ \gamma} = \frac{2\alpha}{\pi^2} \frac{1}{-t} \frac{s}{\tilde{s}} \left[1 - \frac{\tilde{s}}{s} + \frac{1}{2} \left(\frac{\tilde{s}}{s} \right)^2 + \frac{m_e^2}{t} \left(\frac{\tilde{s}}{s} \right)^2 \right] \frac{d^3 \vec{q}_+}{2 q_+^0} \tag{34}$$

The expression for the cross-section for the double Compton scattering

$$\gamma(\tilde{k}) e^-(p_-) \to e^-(q_-) \gamma(k) \gamma(k_s) \tag{35}$$

was first computed by Mandle and Skyrme ([31]). The result can be found in [32]. Combining it with the previous result for the radiation spectrum, we find finally the expression for the double radiative Bhabha cross-section with the EPA:

$$d^8 \sigma_{e^+e^- \to e^+e^-\gamma\gamma} = \frac{\alpha^4}{\pi^4 s} \frac{X_{MS}}{m_e^2} \frac{1}{-t} \left(\frac{s^2 + (s - \tilde{s})^2}{\tilde{s}^2} + \frac{2m_e^2}{t} \right) d^8 \Gamma$$

$$d^8 \Gamma = \delta^4(p_+ + p_- - q_+ - q_- - k - k_s) \frac{d^3 \vec{q}_+}{2q_+^0} \frac{d^3 \vec{q}_-}{2q_-^0} \frac{d^3 \vec{k}}{2k^0} \frac{d^3 \vec{k}_s}{2k_s^0}$$

(36)

with

$$\begin{aligned}
X_{MS} =& \; 2(ab - c)[(a+b)(x+2) - (ab-c) - 8] - 2x(a^2 + b^2) - 8c \\
&+ \frac{4x}{AB} \left[(A+B)(x+1) - (aA + bB)(2 + z\frac{1-x}{x}) + x^2(1-z) + 2z \right] \\
&- 2\rho[ab + c(1-x)]
\end{aligned}$$

$$a = \sum_1^3 \frac{1}{\kappa_i}, \quad b = \sum_1^3 \frac{1}{\kappa'_i}, \quad c = \sum_1^3 \frac{1}{\kappa_i \kappa'_i}$$

Figure 14: Virtual diagrams included in the EPA.

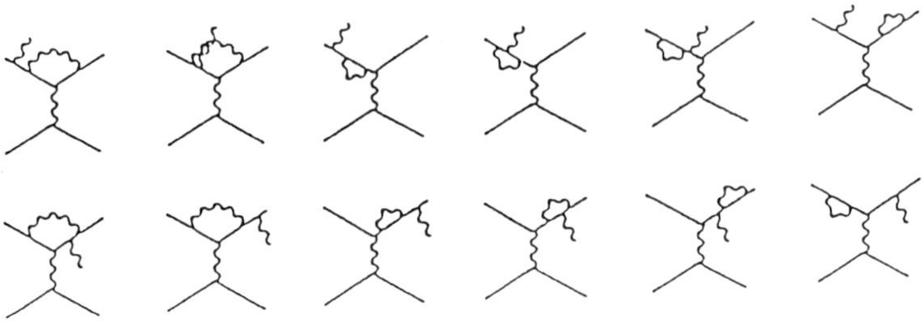

$$x = \sum_1^3 \kappa_i, \quad y = \sum_1^3 \kappa'_i, \quad z = \sum_1^3 \kappa_i \kappa'_i$$

$$A = \kappa_1 \kappa_2 \kappa_3, \quad B = \kappa'_1 \kappa'_2 \kappa'_3, \quad \rho = \sum_1^3 (\frac{\kappa_i}{\kappa'_i} + \frac{\kappa'_i}{\kappa_i})$$

$$m_e^2 \kappa_1 = p_- k, \quad m_e^2 \kappa_2 = p_- k_s, \quad m_e^2 \kappa_3 = -p_- \tilde{k},$$
$$m_e^2 \kappa'_1 = -q_- k, \quad m_e^2 \kappa'_2 = -q_- k_s, \quad m_e^2 \kappa'_3 = q_- \tilde{k}, \tag{37}$$

We have compared the results of this method with the exact tree level matrix element squared found in the previous section. The results of the comparison at some fixed phase space points relevant for the single photon configuration show us that this approximation gives essentially the right number. This reflects the fact that the contribution from diagrams C and D is very small and that the interference between the set formed by A1, A2, B1 and the diagrams with the photon labels exchanged, and their charge conjugated diagrams is also negligible for this configuration, as we have explicitly checked with the complete matrix element squared.

3.3 Virtual photon corrections.

Since no calculations of the virtual and soft photon corrections to the radiative Bhabha scattering exist in the literature at the moment, we have followed [25] in using the EPA for these corrections. Concerning the soft photon part, this approximation implies taking into account only the diagrams in figure 13 when one photon has an energy smaller than k_0, the soft-hard separation parameter. This seems to be a reasonable approach since, as we have mentioned previously, the EPA works very well for the hard photon part and, as we will show shortly, the total α^4 cross-section calculated in this way does not depend on the soft-hard separation parameter.

Concerning the virtual corrections, we follow the EPA approximation and just include the interference of the diagrams of fig. 14 with the ones in lowest order radiating a photon from the same fermionic line (and the same for the charge conjugated ones). These diagrams are sufficient to cancel out all the infrared divergences coming from the soft diagrams taken into account. We do not consider the interference of the diagrams in fig. 13 with their charge conjugated ones. This interference has an infrared divergence, cancelled by the box diagrams, which are also not considered. The remaining finite contribution, as well as the one coming from the rest of neglected terms, however, seems to be small ([25] and references therein).

The cross-section corresponding to the lowest order contribution and the virtual and soft photon corrections can be expressed as

$$d\sigma^{V+S} = (1+\delta) \cdot d\sigma^0, \tag{38}$$

where δ depends on k_0. We calculate δ from radiative corrections to Compton scattering. The EPA relates these two processes by

$$d^5\sigma_{e^-e^+ \to e^-e^+\gamma} = d^3n_{e^+ \to e^+\gamma^*} \cdot d^2\sigma_{\gamma^*e^- \to \gamma e^-}. \tag{39}$$

$d^3n_{e^+ \to e^+\gamma^*}$ is the equivalent photon spectrum, and, hence, δ is the same for the two processes. Then, we can write

$$d\sigma^{VS}_{e^-e^+ \to e^-e^+\gamma} = (1+\delta^{cm}_{\gamma^*e^- \to \gamma e^-}) \cdot d\sigma^0_{e^-e^+ \to e^-e^+\gamma} \tag{40}$$

$\delta^{cm}_{\gamma^*e^- \to \gamma e^-}$ is the virtual and soft photon correction to Compton scattering. It has the following expression ([33], [25])

$$\begin{aligned}
\delta = & -\frac{\alpha}{\pi U}\Big\{2(1-2y)U\ln\left(\frac{2k_0}{m_e}\right) + \frac{\pi^2}{6}\left(4 - 3t - \frac{1}{t} - \frac{2}{E^4 t^3}\right) \\
& + 4(2-U)y^2 - 4y + \frac{3}{2}U + \frac{2}{E^2 t^2} + 4\left(1 - \frac{1}{2t}\right)\ln^2 E \\
& + \left(2t + \frac{1}{t} - 2 + \frac{2}{E^4 t^3}\right)Li_2(1 - E^2 t) \\
& + \left[2 - 5t - \frac{2}{t} + 4y\left(\frac{2}{t} + t - 2\right)\right]\ln E \\
& - \frac{1}{2}U\ln^2(1-t) - U Li_2(t) \\
& + \left[1 - \frac{2}{t} - \frac{2}{E^2 t^2} - \frac{\frac{1}{2}E^2}{1 - E^2 t} + 4y\left(t - 1 + \frac{1}{2t}\right)\right]\ln(E^2 t)\Big\}
\end{aligned} \tag{41}$$

$$E = \frac{\sqrt{\hat{s}}}{m_e}, \quad t = \frac{1}{2}(1 + \beta_- \cos\theta_k), \quad y = \ln[E\sin(\theta_k/2)], \quad U = t + 1/t$$

and k_0 has to be specified in the $\gamma^* e^-$ center of mass frame.

3.4 Monte Carlo implementation and phase-space integration.

In this section we are going to describe the phase space variables we have used for the integration, together with the approximants chosen to handle the extremely sharp peaks showing up in the matrix element squared.

For the three body process including virtual and soft contribution, we have used the same phase space and approximants as those used in ref. [24] for the tree level calculation. This is the normal way to proceed since, in principle, virtual and soft photon corrections do not change the peaking structure of the matrix element squared.

In order to integrate the matrix element squared for the hard part over the allowed phase space of the single photon configuration, we have taken as independent variables:

- the final electron solid angle $d\Omega_3$,
- the final positron energy dE_4 and solid angle $d\Omega_4$,
- the missing photon energy dE_5 and solid angle $d\Omega_5$ with respect to either the initial electron or the final.

The choice of these variables is mainly required by the fact that the solid angle for the particles enumerated above can be very restricted due to the veto condition.

Now, the phase space density reads:

$$W_{ps} = \frac{1}{16} \frac{\beta_3^2 E_3 \, \beta_4 E_4 \, E_5}{E_{cm}\beta_3 + E_4(\beta_4 \cos\theta_{34} - \beta_3) + E_5(\cos\theta_{35} - \beta_3)} \qquad (42)$$

and therefore:

$$dR_4 = W_{ps} \, d\Omega_3 dE_4 \, d\Omega_4 \, dE_5 \, d\Omega_5, \qquad (43)$$

where the directions of the solid angles for particles 3, 4 and 5 can be referred to any defined direction.

After a careful study of the peaking structure of the total integrand (which is proportional to the product $\overline{|-iM|^2} W_{ps}$) when variables Ω_3, E_4, Ω_4, E_5 and Ω_5 are utilized, we apply Importance Sampling to eliminate the main peaks. The final integration variables η_i, $i = 1 \ldots 8$ are defined in the interval $[0, 1]$ and are related to the previous ones in the following way:

$$d\eta_1 = \kappa_1 \frac{d\cos\theta_3}{1 + \cos\theta_3 + \epsilon'} \qquad d\eta_2 = \kappa_2 \, d\phi_3$$

$$d\eta_3 = \kappa_3 \frac{dE_4}{(E_b - E_4)^2} \qquad d\eta_4 = \kappa_4 \frac{d\cos\theta_4}{1 - \cos\theta_4 + \delta} \qquad d\eta_5 = \kappa_5 \, d\phi_4$$

$$d\eta_6 = \kappa_6 \frac{dE_5}{E_5} \qquad d\eta_7 = \kappa_7 \frac{d\cos\theta_5}{1 + \cos\theta_5 + \epsilon} \qquad d\eta_8 = \kappa_8 \, d\phi_5, \qquad (44)$$

where κ_i are the transformation Jacobians and

$$\epsilon = \frac{2m_e^2}{s}$$

$$\delta = \epsilon \frac{(E_b - max(E_4))^2}{(max(E_4))^2}$$

$$\epsilon' = \epsilon^2, \qquad (45)$$

being actually the direction of particle 5 generated with a probability of 50 % with respect to the initial and final electron direction. For these integration variables the phase space density reads

$$W'_{ps} = W_{ps} \frac{1}{\prod \kappa_i} (1 + \cos\theta_3 + \epsilon') (E_b - E_4)^2 (1 - \cos\theta_4 + \delta) \, E_5 \, (1 + \cos\theta_5 + \epsilon) \qquad (46)$$

One proof of the fact that the behaviour of the function W'_{ps} compensates the peaking structure of the matrix element squared resulting in a rather smooth integrand is the fact that the generation efficiency (defined as the ratio between the average value of the total integrand and its absolute maximum) is bigger than 16 % even in extremely demanding cut conditions.

Table 7: Lowest order + virtual + soft photon corrections (σ^{V+S}), hard photon contribution (σ^H) and total $O(\alpha^4)$ cross-section (σ^4) for different values of $k_0 \equiv x_0 E_b$.

k_0 (GeV)	σ^{V+S} (pb)	σ^H (pb)	σ^4 (pb)
10^{-3}	21.8 ± 0.3	12.75 ± 0.11	34.6 ± 0.3
10^{-2}	26.5 ± 0.4	8.39 ± 0.07	34.9 ± 0.4
10^{-1}	31.0 ± 0.5	4.15 ± 0.09	35.1 ± 0.4

Table 8: Lowest order (σ^3) vs. one-loop corrected (σ^4) cross-sections. See the caption of the previous table.

	σ^3 (pb)	σ^{V+S} (pb)	σ^H (pb)	σ^4 (pb)	$\delta(\%)$
set-up 1	33.6 ± 0.5	26.5 ± 0.4	8.39 ± 0.07	34.9 ± 0.4	3.9
set-up 2	297.7 ± 0.7	208.4 ± 2.0	95.0 ± 0.9	303.4 ± 2.2	2.0

3.5 Results.

Following [25], we have defined a hypothetical experimental set-up such that we consider as single photon events those events fulfilling the following conditions :

- detected photon energy gretaer than $E_D = 0.5 GeV$

- detected photon θ angle between $\theta_D = 30°$ and $180° - \theta_D$

- undetected particles angle between $0°$ and $\theta_V = 15$ mrad or between $\pi - \theta_V$ and π.

Moreover, we take $\sqrt{s} = 94$ GeV and $\alpha = 1/137.036$.

In Table 7 we see the results we obtain for this configuration with the exact matrix element squared for the hard part (as described in section 3.2.1). We obtain very similar numbers with the EPA. Since, as Table 7 shows, the dependence in the soft-hard cutoff is negligible, we can assume that the approximation we use for the soft and virtual corrections is good enough. Table 8 shows the results of the lowest order (α^3) calculation and of our $O(\alpha^4)$ calculation. The result is that the hard photon corrections to $e^-e^+ \rightarrow e^-e^+\gamma$ happen to cancel almost exactly the soft + virtual corrections, giving an overall correction of about 3% for the single photon configuration. The result at $\sqrt{s} = 98$ GeV with another (maybe more realistic) definition of single photon configuration, namely $E_D = 1 GeV$, $\theta_D = 15°$, $\theta_V = 2.4°$, confirms this effect (Table 8). It is clear from Table 8 that we do not observe at all the enhancement that was described in the first version of [25] for the $O(\alpha^4)$ cross-section. Instead of this, we find a rather modest effect of considering α^4 corrections. Moreover, figures 15 a), 15 b) and 15 c) show that the distributions of the main variables of the detected photon (E_γ, $\cos\theta_\gamma$, p_{T_γ}) almost do not change from α^3 to α^4 calculation. The most important one in the single photon study, p_{T_γ}, does not change noticeably (fig. 15 c)) and hence a cut on p_{T_γ} reduces this background almost completely.

Figure 15 Comparison between tree level and one-loop corrected results for radiative Bhabha scattering :
a) $\cos\theta_\gamma$ distribution.
b) E_γ distribution.
c) $p_{T\gamma}$ distribution.

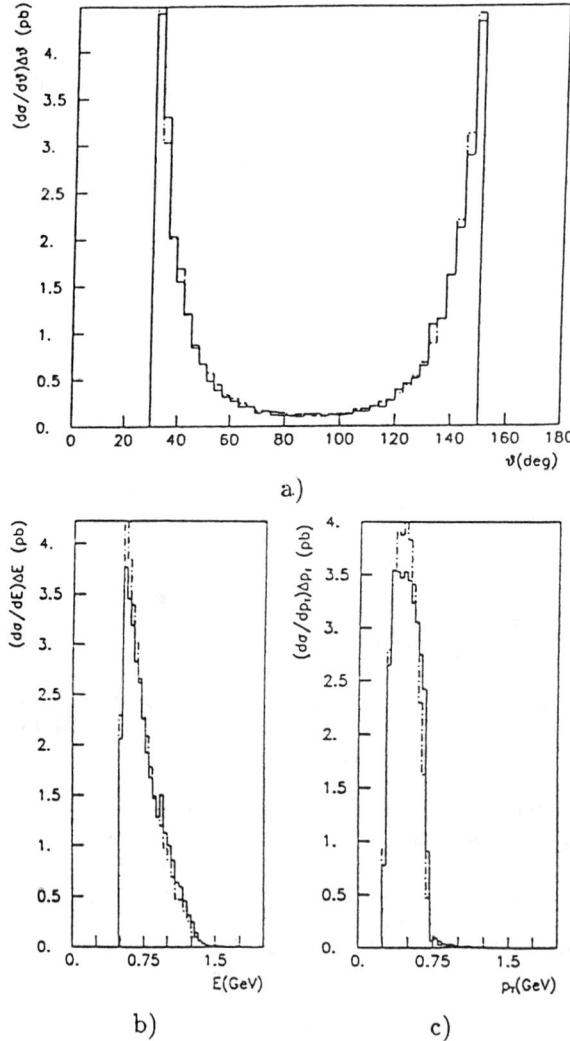

These results are in agreement with the result found in the definitive version of [25]. The differences between the two calculations are around 2%. This is also the kind of difference that we expect between the EPA result for the virtual and soft photon corrections and the exact one, which has not been computed as of now. Then, in our calculation, using the EPA for the virtual and soft part and our exact calculation for the hard photon part, we expect an accuracy of the order of few per cent. This is sufficient for experimental purposes.

4 Summary and conclusions

We have presented a calculation of $e^+e^- \to \nu\bar{\nu}\gamma$ including :

- Exact tree level calculation.
- Complete $O(\alpha)$ QED corrections to the Z diagrams.
- Leading QED corrections exponentiated.
- Oblique corrections and non-abelian vertices included via 'star' scheme.
- Corrections to $Im\Pi_{ZZ}$.

The calculation has been implemented as a Monte Carlo event generator of unweighted events. The one-loop corrections give a $\sim -25\%$ effect, while higher order corrections altogether amount something around $\pm 5\%$, depending on the center of mass energy.

Results from a comparison with KORL03 done in ref. [12] have been discussed showing a very good agreement (better than 1%) both in cross-section and main distributions in the region of experimental interest. This is a powerful check of both calculations.

We can conclude that the overall precision of our calculation is of the order of 1%, if not better. Hence, it is precise enough for the experimental needs.

Concerning the background process $e^+e^- \to e^+e^-\gamma$, we have explained the exact calculation of the hard photon corrections using the helicity amplitudes method. Then a comparison with the Equivalent Photon Approximation (EPA) calculation has been performed showing reasonable agreement between them. We then use the EPA for the virtual and soft photon corrections and after a complicated phase space integration, due to the presence of sharp peaks in the matrix element squared, we arrive at the final result for the one-loop QED corrections. The total correction to the cross-section lies between 2% and 4% depending on the cut conditions and is almost negligible in the photon distributions. Hence, we conclude that there is no need for higher order corrections.

Due to the uncertainty introduced by the EPA we can only claim a total accuracy of few per cent. However, this is also sufficient for the experiments, since radiative Bhabha scattering is a background and most of it (if not all) can be removed by kinematical cuts.

Acknowledgements

We thank very much the authors of ref. [12] for allowing us to take their results concerning the comparison between KORL03 and NNGG03.

References

[1] A.D. Dolgov, L.B. Okun and V.I. Zakharov, Nucl. Phys. **B41** (1972) 197
    E. Ma and J. Okada, Phys. Rev. Lett. **41** (1978) 287.

[2] K.J.F. Gaemers, R. Gastmans and F.M. Renard, Phys. Rev. **D19** (1979) 1605

[3] M. Martinez, unpublished (1985).
    L. Bento, J.C. Romao and A. Barroso, Phys. Rev. **D33** (1986) 148

[4] F.A. Berends, G.J.H. Burgers, C. Mana, M. Martinez and W.L. van Neerven, Nucl. Phys. **B301** (1988) 583.

[5] M. Igarashi and N. Nakazawa, Nucl. Phys. **B288** (1987) 301.

[6] R. Miquel, Ph. D. Thesis, Universitat Autònoma de Barcelona, UAB-LFAE 89-02 (1989).

[7] G. Barbiellini, B. Richter, J.L. Siegrist, Phys. Lett. **106B**, 414 (1981).

[8] F.A. Berends, G.J.H. Burgers and W.L. van Neerven, CERN-TH.4772/87 (1987).

[9] E.A. Kuraev and V.S. Fadin, Sov. J. Nucl. Phys. **41** (1985) 466
    O. Nicrosini and L. Trentadue, Phys. Lett. **B196** (1987) 551

[10] D.C. Kennedy and B.W. Lynn, SLAC–PUB–4039 (Rev) (1988)

[11] S. Jadach, R.G. Stuart, B.F.L. Ward and Z. Wąs, KORL03 Monte-Carlo, unpublished.

[12] P. Colas, L. Mirabito, Z. Wąs, preprint MPI-PAE/Exp.El 211 (1989).

[13] M. Martinez, R. Miquel, Universitat Autònoma de Barcelona preprint UAB-LFAE 87-01 (1987).

[14] G. Passarino and M. Veltman, Nucl. Phys. **B160** (1979) 151

[15] W.F.L. Hollik, DESY 88-188 (1988)

[16] M. Böhm and Th. Sack, Z. Phys. **C 35** (1987) 119

[17] G. Altarelli and G. Parisi, Nucl. Phys. **B126** (1977) 298

[18] V. Gribov and L. Lipatov, Sov. J. Nucl Phys. **15** (1972) 438, 675

[19] A. Sirlin, Contribution to the Brighton Workshop on radiative corrections, July 1989.

[20] W. Wetzel in CERN 86-02 (1986)

[21] W. Beenakker and W. Hollik, Z. Phys **C 40** (1988) 141

[22] D.R. Yennie, S.C. Frautschi and H. Suura, An. of Phys. **13** (1961) 379

[23] R.G. Stuart, unpublished.

[24] C. Mana, M. Martinez, Nucl. Phys. **B287**, 601 (1987).

[25] D. Karlen, Nucl. Phys. **B289**, 23 (1987).

[26] F.A. Berends, P.H. Daverveldt, R. Kleiss, Nucl. Phys. **B253**, 441 (1985).

[27] R. Kleiss, W.J. Stirling, Nucl. Phys. **B262**, 235 (1985).

[28] P.H. Daverveldt, Ph.D. Thesis, Univ. of Leiden (1985).

[29] M. Martinez, Ph.D. Thesis, Univ. of Barcelona (1986).

[30] G. Bonneau, F. Martin, Nucl. Phys. **B27**, 381 (1971).

[31] F. Mandle, T.H.R. Skyrme, Proc. Roy. Soc. **A215**, 497 (1952).

[32] J.M. Jauch and F. Rohrlich, *The Theory of Photons and Electrons* (Springer Verlag, New York, 1955), p. 237.

[33] K.J. Mork, Phys. Rev. **A4**, 917 (1971).

Study of the All Neutral Final State of $f_2(1270)$ Produced in Two–Photon Collisions as a Background to Radiative Neutrino Counting on the Z^0 Resonance[†]

M. Daoudi, J.G. Layter, K. Riles, and B.C. Shen
(Presented by K. Riles)

Physics Department
University of California
Riverside, California

Abstract

Upcoming experiments at LEP will measure the number of neutrino generations from the cross section for single-photon production in e^+e^- annihilation on or above the Z^0 resonance. We consider one potential background to this process, two-photon production of f_2, with subsequent decay into four photons. Monte Carlo studies of this reaction are presented and the results discussed. From analysis of a sample corresponding to 10 pb^{-1} of integrated luminosity, we conclude this reaction constitutes a small, and for certain event selection criteria, a negligible background.

[†] Work supported by the Department of Energy, contract DE–AM03–76SF00010

Introduction

In 1978 Ma and Okada[1] suggested that measurement of $e^+e^- \to \nu\bar{\nu}\gamma$ at new colliding beam facilities could determine directly the total number of light-mass neutrinos. Several other groups have developed this idea further.[2,3] Ideally, one would take data at a center-of-mass energy approximately 10 to 15 GeV above the peak of the Z^0 resonance, in order to obtain a clean peak in the photon energy spectrum. It is unlikely, however, that LEP will run at energies significantly higher than the Z^0 peak in the first few months after startup. Therefore we consider the measurement of the partial width of the Z^0 decaying into $\nu\bar{\nu}$ at a center-of-mass energy at or near the peak.

In this case, one has the advantage of a large cross section, but the low photon energies emitted in the radiative neutrino production imply a large background from radiative Bhabha production, where both the electron and positron escape detection at low angles. This background has been studied extensively by others[4] and will not be further discussed here.

A potential background that has received less attention comes from two-photon reactions leading to the production of neutral pions that subsequently decay into photons, one of which satisfies the single-photon tagging requirements. In this note, we present an analysis of Monte Carlo generated events of the reaction $\gamma\gamma \to f_2 \to \pi^0\pi^0 \to \gamma\gamma\gamma\gamma$. This reaction is chosen because of the large $\gamma\gamma$ coupling of f_2 and because the f_2 production and decay characteristics are well understood. We do not treat the contribution from continuum $\pi^0\pi^0$ production, but measurements from the Crystal Ball collaboration[5] indicate such a contribution should be negligible.

Two-Photon Monte Carlo Program

For event generation, we have used the Monte Carlo program PLEASURE_DOME, written by A.R. Barker,[6] where we generate events of the type:

$$e^+e^- \to e^+e^- f_2 \to e^+e^- \pi^0\pi^0 \to e^+e^- \gamma\gamma\gamma\gamma,$$

corresponding to an integrated luminosity of 10 pb^{-1} at a center-of-mass energy of 92 GeV. The parameters used for the f_2 are the following:

$M(f_2) = 1.274$ GeV/c^2 $\Gamma(f_2) = 185$ MeV
$J^P = 2^+$ Helicity $= \pm 2$
$\Gamma_{\gamma\gamma}(f_2) = 3$ KeV $B(f_2 \to \pi^0\pi^0) = 0.286$

Figure 1 shows the f_2 peak in the invariant mass of generated π^0 pairs. Events are generated using the ρ form factor for the virtual photon propagators. We consider an untagged sample, that is, only events with the electron and positron recoiling less than 25 mrad, assuming that recoils at larger angles would allow detection and identification by forward calorimetry.

Analysis

We first examine the energy and angular distributions of the photons in the final state. Figure 2a shows a scatter plot of photon polar angle θ_γ vs the photon energy E_γ for all four photons in each event. Figures 2b and 2c are the projections in E_γ and θ_γ, respectively. Most of the photons have energies below 1 GeV and are nearly parallel to one beam direction.

We assume in this analysis that events trigger if a photon with energy greater than 1 GeV is detected in the polar angular interval between 45° and 135°. Figure 3a shows the distribution of the number of photons with energy greater than 1 GeV in each event, and fig. 3b shows the same distribution when only the angular interval of 45° to 135° is considered. Most of the events have no photon satisfying the trigger requirements. If the 0–photon bin is suppressed, as in fig. 3c, we see that there are 49 events with one and only one photon satisfying the trigger requirements and 7 events with two photons. We emphasize that this sample corresponds to a run with 10 pb^{-1} of accumulated luminosity.

Since we trigger on one photon, we can safely discard the 7 events with two tagged photons. We now consider the energies and directions of the additional photons in the remaining events, in the hope of determining useful and realistic veto conditions. Figure 4a shows the energy-angle correlation of the *other* three photons. It is interesting to note that these photons, whose energies are typically less than 1 GeV(see fig. 4b), are distributed roughly isotropically (fig. 4c). Assuming we can detect photons with high efficiency above a certain energy threshold E_{th} for angles as low as 8.5°(150 mrad)with respect to each beam direction, we determine the number of events that can be rejected by detection of a second photon in that angular range with energy greater than E_{th}. The number of events rejected is plotted vs E_{th} in fig. 5, assuming 100% photon detection efficiency for energies greater than E_{th}. For example, if we can detect all photons with energy greater 0.6 GeV in the angular veto range, we can reject 37 of the 49 events, that is, about 75%. This corresponds to a cross section of about 1 pb, which is much smaller than that for radiative Bhabhas satisfying the same tagging criteria. If we can lower our veto energy threshold to 0.4 GeV, maintaining 100% detection efficiency, we can reject 92% of the events, leaving a negligible background. The minimum opening angle between the triggering photon and the remaining three photons is about 10°, as seen in fig. 6, allowing easy separation of photons from the same π^0 decay.

Double-Radiative Neutrino Pair Production

In vetoing events with a second detected photon, there is the danger of discarding events due to neutrino production with multiple initial-state radiation. To address this concern, we have used a Monte Carlo program[7] that simulates fermion-antifermion production with arbitrary numbers of initial state photons, using a structure function treatment of the incoming electron and positron beams.[8]

Requiring that one such initial-state photon satisfy the tagging cuts described above and discarding events with any additional photon satisfying the veto cuts, we find the numbers given in the table below for the radiative neutrino signal and for the f_2 background. All numbers shown are expectations in 10 pb^{-1} of data. The first column gives the number of events expected from the radiative neutrino signal when running on the Z^0 peak (assumed 92 GeV), while the second column gives the expectation when running at a center-of-mass energy 1 GeV above the peak. The third column gives the expectation from the f_2 background when running on the peak, which for practical purposes, is the same as that expected 1 GeV above the peak. All errors shown are statistical. The first row of the table shows the number of events expected with at least one tagged photon and no veto condidion on extra detected photons. The second row shows the number expected with a veto cut(A) on additional photons with energy greater than 0.6 GeV in the polar angular range $8.5° < \theta_\gamma < 171.5°$. The third row shows the number expected with a veto energy cut(B) of 0.4 GeV and the same veto angular range.

$\int \ell dt = 10\text{pb}^{-1}$	$e^+e^- \to \nu\bar{\nu}\gamma$		$e^+e^- \to e^+e^- f_2$
	$\sqrt{s} = 92$ GeV	$\sqrt{s} = 93$ GeV	$\sqrt{s} = 92$ GeV
$N_{TAG} \geq 1$, no veto	348 ± 19	653 ± 26	56 ± 7
$N_{TAG} = 1$, veto A	334 ± 18	629 ± 25	12 ± 3
$N_{TAG} = 1$, veto B	329 ± 18	620 ± 25	4 ± 2

One sees from these numbers that the $\gamma\gamma \to f_2$ background is a significant fraction of a single neutrino generation, but that imposing veto conditions on additional detected photons substantially removes the background, while sacrificing little of the radiative neutrino signal.

Conclusions

From analysis of Monte Carlo events of reaction (1), corresponding to an integrated e^+e^- luminosity of 10 pb^{-1}, we have studied the possibility of event rejection based upon detection of additional photons. We have also studied the effect of such event rejection on the radiative neutrino signal, allowing for multiple initial-state radiation. We conclude that the f_2 contributes only a small background to the reaction $e^+e^- \to \nu\bar{\nu}\gamma$, and in the case of high tagging-energy threshold with sensitivity to low-energy photons at forward angles, the f_2 background should be negligible.

REFERENCES

1. E. Ma and J. Okada, Phys. Rev. Lett. **41,** 287 (1978).
2. K.J.F. Gaemers, R. Gastmans, and F.M. Renard, Phys. Rev. Comments **19D,** 1605 (1979).
3. G. Barbiellini, B. Richter, and J.L. Siegrist, Phys. Lett. **106B,** 414 (1981).
4. C. Mana and M. Martinez, Nuc. Phys. **B287,** 601 (1987). D. Karlen, Nuc. Phys. **B289,** 23 (1987).
5. D.A. Williams, in *Proceedings of the XXIII International Conference on High Energy Physics*, p. 1223, Berkeley, California, 1986.
6. A.R. Barker, Ph.D. Thesis, University of California, Santa Barbara, December, 1988 (unpublished).
7. G. Bonvicini and L. Trentadue, submitted to Nucl. Phys. B, UM–HE–88–36. See also contributions by Bonvicini and Trentadue to this workshop.
8. E.A. Kuraev and V.S. Fadin, Sov. Journ. Nucl. Phys. **41,** 466 (1985). See also contributions by Kuraev and Fadin to this workshop.

FIGURE CAPTIONS

1) Generated f_2 invariant mass distribution.

2) a) Scatter plot of energy *vs* polar angle for all photons in $\gamma\gamma \to f_2$ production. b) Projection of a) onto photon energy axis. c) Projection of a) onto polar angle axis.

3) a) Multiplicity of photons with energy greater than 1 GeV. b) Multiplicity of photons with energy greater than 1 GeV in the angular range $45° < \theta_\gamma < 135°$.

4) a) Scatter plot of energy *vs* polar angle for the *other* three photons in an event with exactly one photon satisfying the tagging requirements. b) Projection of a) onto energy axis. Projection of a) onto polar angle axis.

5) Number of tagged events(exactly one photon satisfies tagging cuts) rejected because an additional photon is detected with energy greater than E_{th} in the angular range $8.5° < \theta_\gamma < 171.5°$.

6) Opening angle between triggering photon and other three photons for f_2 background events with exactly one photon satisfying tagging criteria.

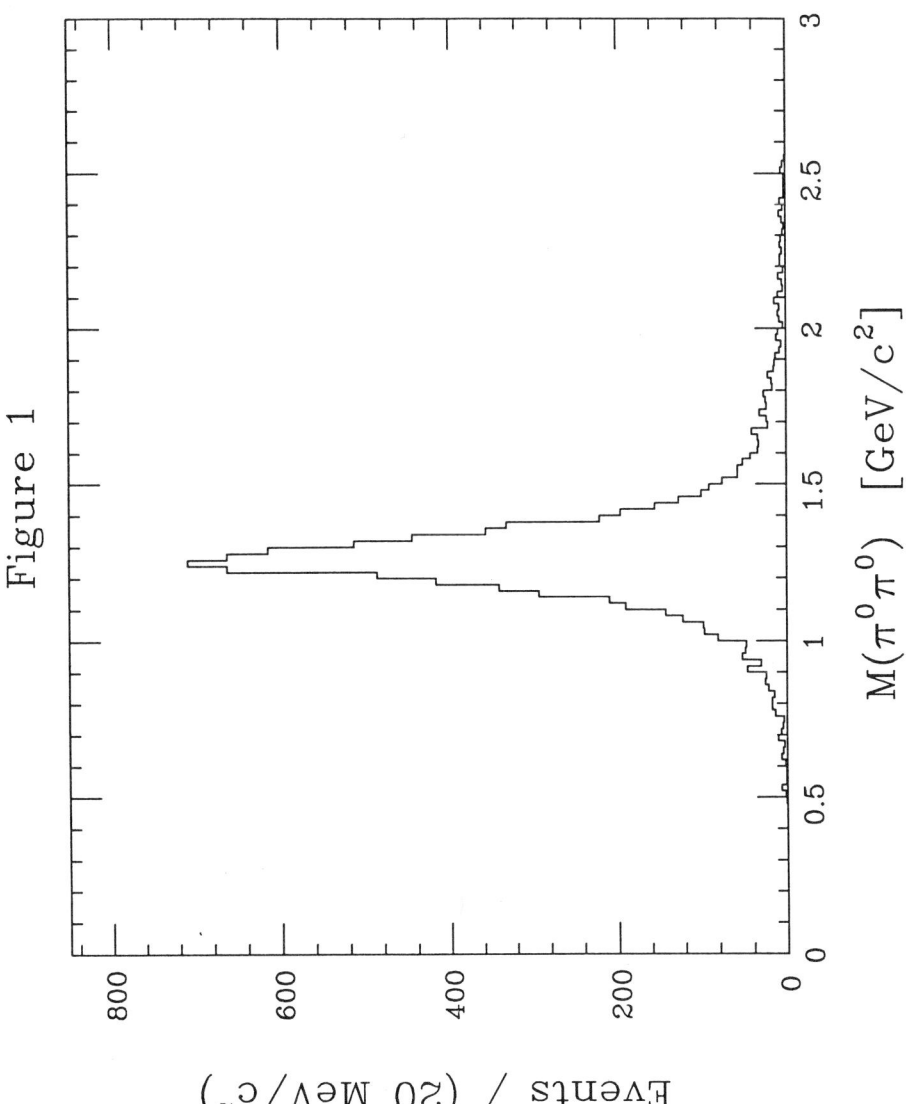

Figure 1

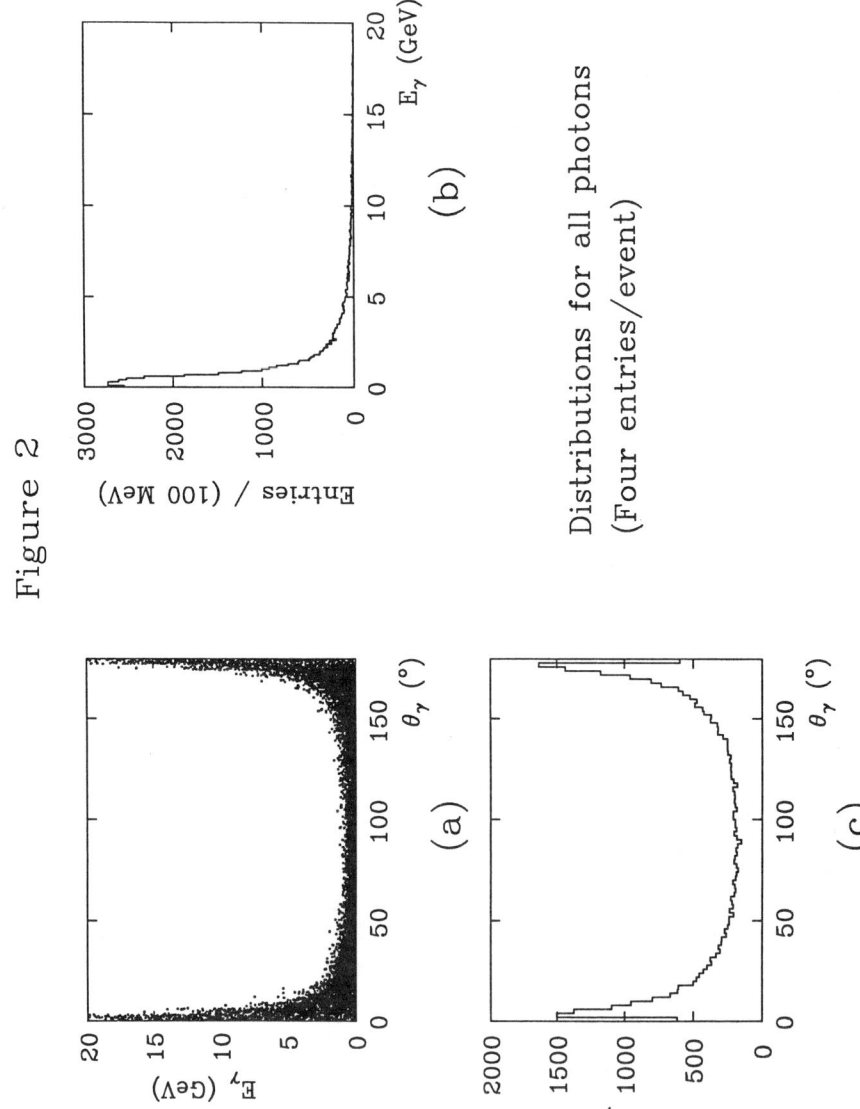

Figure 2
Distributions for all photons
(Four entries/event)

Figure 3

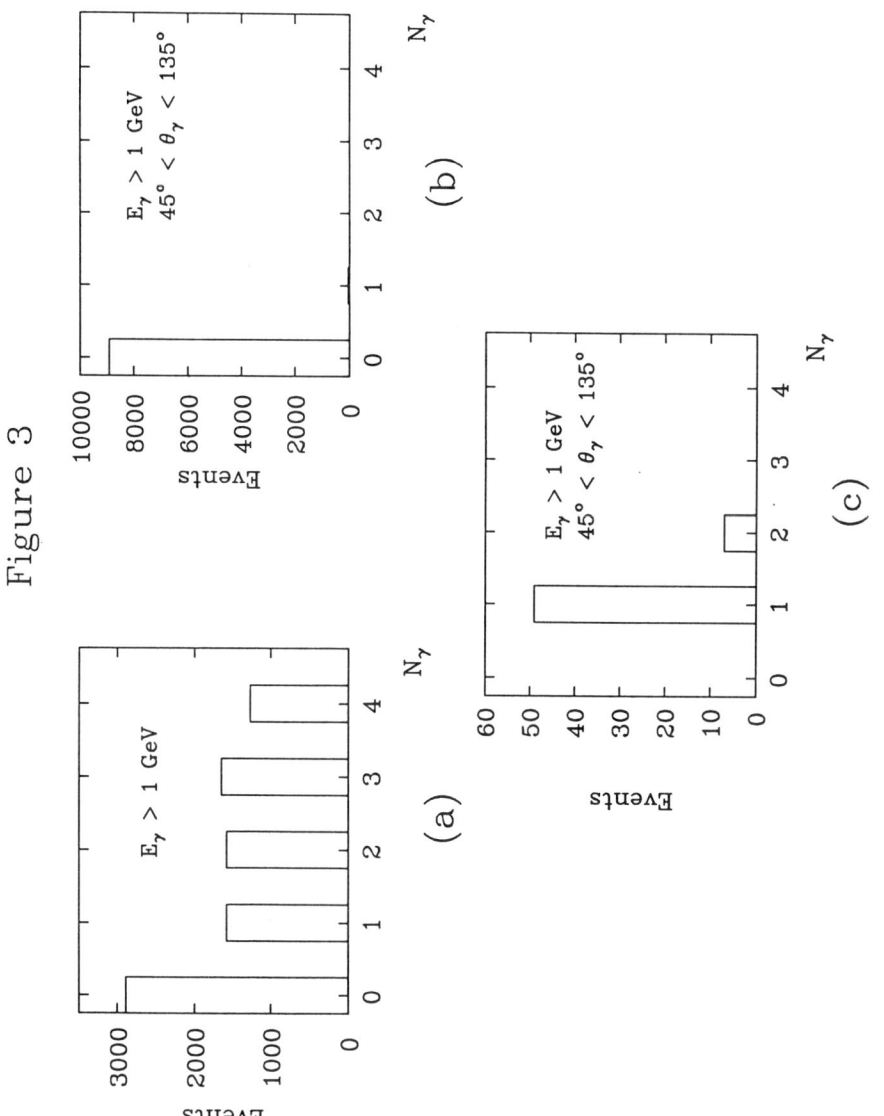

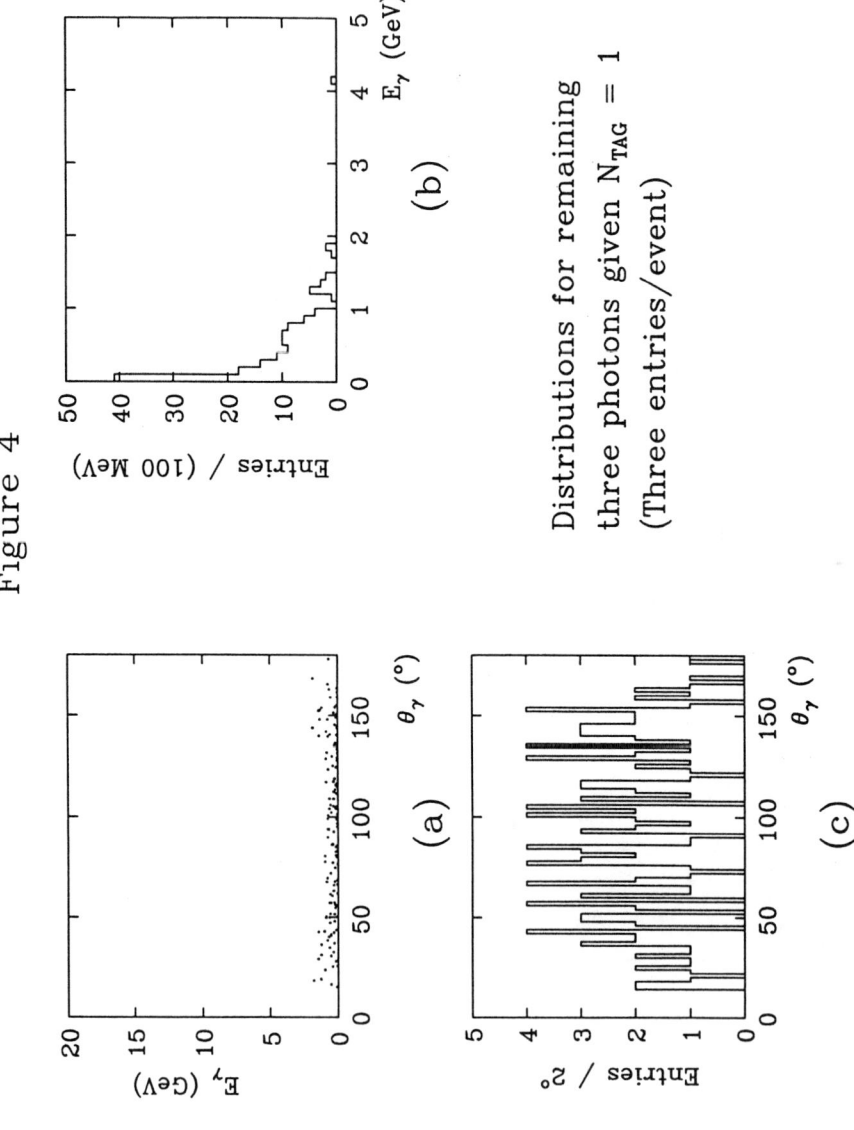

Figure 4

Distributions for remaining three photons given $N_{TAG} = 1$ (Three entries/event)

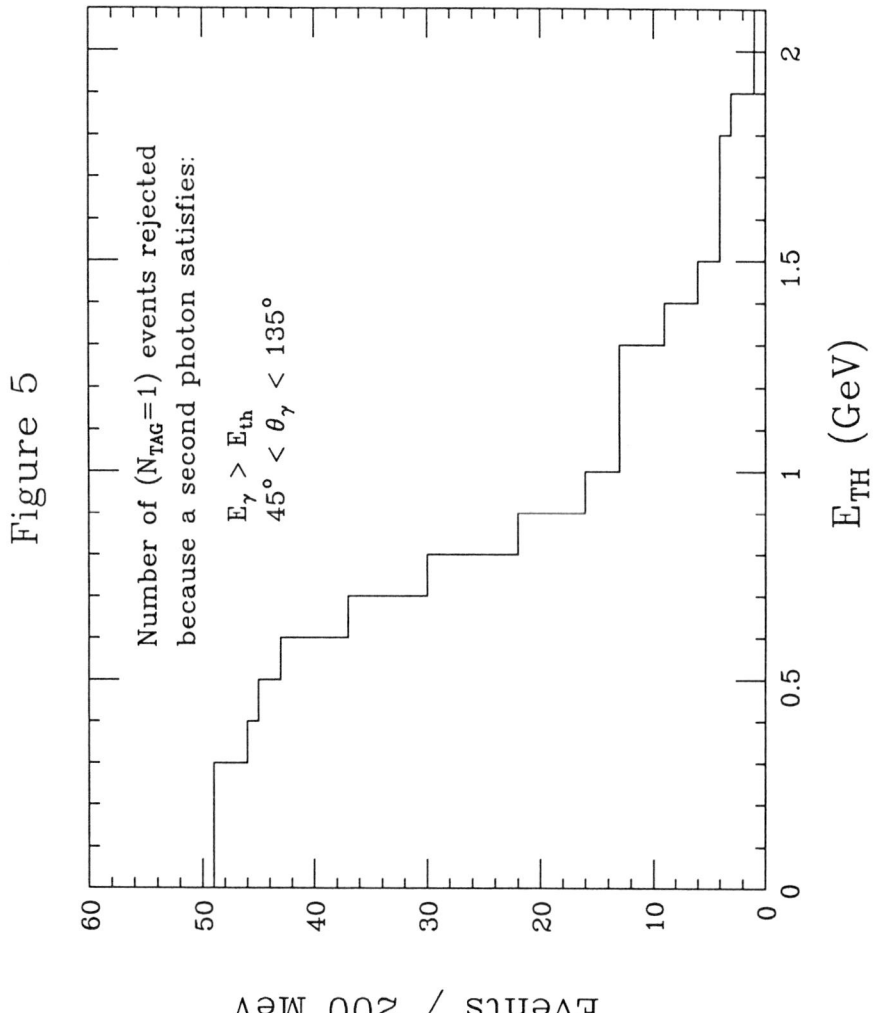

Figure 5

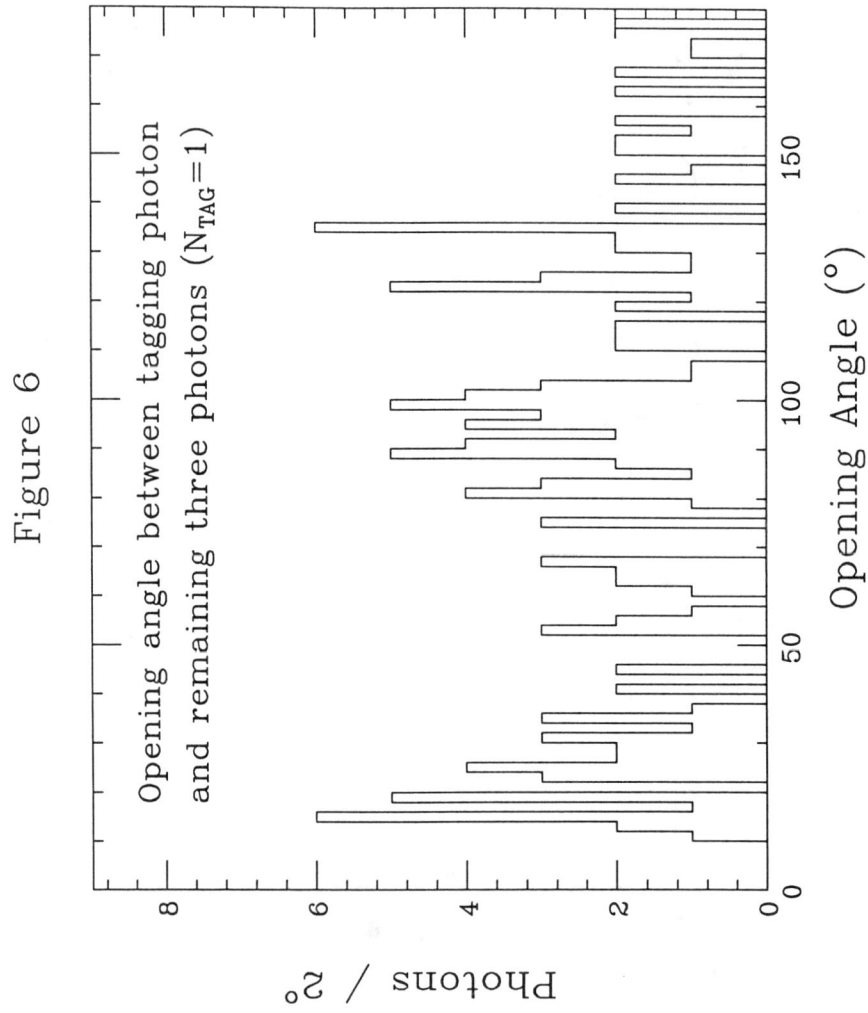

Figure 6

RADIATIVE CORRECTIONS TO POLARIZED COMPTON SCATTERING

Helene Veltman
University of California, Berkeley, CA

ABSTRACT

The left-right asymmetry for Compton scattering is calculated including lowest order corrections. The asymmetry is given as a function of the energy of the scattered electron.

1. INTRODUCTION

It is expected that some time in the future, the SLC will conduct experiments using a longitudinally polarized electron beam (the positron beam remains unpolarized)[1]. As is well known, by evaluating the left-right asymmetry, for example, Standard Model parameters can be measured with even greater accuracy then when using unpolarized beams[2]. Furthermore, one-loop effects are expected to be observable. In order to achieve this desired accuracy, a precise measurement of the degree of polarization of the incoming electron beam is required. The errors in the asymmetry and in the polarization P of the electron beam are related as follows, neglecting statistical errors

$$\frac{\Delta A_{LR}}{A_{LR}} = \frac{\Delta P}{P} \text{ where } A_{LR} = \frac{\sigma_L - \sigma_R}{\sigma_L + \sigma_R} \qquad (1)$$

Here σ_L and σ_R denote the cross-sections for the process $e^-e^+ \to f\bar{f}$ (f is a fermion) with left and right hand polarized incoming electrons.

P can be measured through Compton scattering. To this purpose the longitudinally polarized electron beam is scattered with left and right circular polarized photons and the energy of the scattered electron is measured. From this, one determines the asymmetry $A_{LR}^{\text{exp}}(e^-\gamma \to e^-\gamma)$ as a function of energy (from now on A_{LR} stands for $A_{LR}(e^-\gamma \to e^-\gamma)$ and L and R for left and right circular polarized photons). The relation between A_{LR}^{exp} and A_{LR}^{th}, where A_{LR}^{th} is the value for the asymmetry when the electron beam is 100% polarized, is

$$A_{LR}^{\text{exp}} = P \cdot A_{LR}^{\text{th}} \qquad (2)$$

A_{LR}^{th} can be theoretically evaluated and thus P can be determined. It is this A_{LR}^{th} that is calculated including lowest order corrections.

The metric is such that $p^2 = -m^2$ for a particle on mass-shell with mass m and momentum p.

This paper contains a brief description of the calculation, followed by the results and conclusion. For the more elaborate and detailed evaluation see ref.[3].

2. EXPERIMENTAL SET-UP

The experimental set-up at the SLC leads to 2 important consequences for the calculation.
- Although the energy of the incoming electron beam is 50 GeV, the electron mass cannot be neglected, since the energy of the incoming photon beam is in the electronvolt range. This also explains why the electron scattering angle will not be measured. It is very close to zero.
- The scattered photon will not be measured at all and thus besides soft bremsstrahlung also hard bremsstrahlung must be included.

© 1990 American Institute of Physics

Of course there is always a background process. At the SLC the degree of polarization of the electron beam will be measured along the beam line a certain distance away from the interaction point. It is only the energy of the scattered electron that will be measured. This means that if the polarization is measured when there is scattering at the interaction point, radiative Bhabha-scattering is a source of background for small scattering angle. The energy of the scattered electron will be different from 50 GeV for the process $e^-e^+ \to e^-e^+\gamma$. This background has been calculated[4] and it is concluded that it is not too large and it does not influence the accuracy of the polarization measurement[1]. We will not go into this any further.

3. THE ASYMMETRY IN LOWEST NON ZERO ORDER AND THE ONE LOOP CORRECTIONS

In section 3.1 we define the kinematics and the asymmetry for the process $e^-\gamma \to e^-\gamma$. The asymmetry is evaluated in lowest non zeroth order. In section 3.2 we include the one loop corrections.

3.1 KINEMATICS AND THE DEFINITION OF THE ASYMMETRY

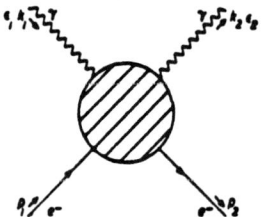

Fig.1 Compton scattering

The process is defined in Fig.1. p_1 and p_2 are the momentum 4-vectors of the incoming and outgoing electron. k_1, ϵ_1 and k_2, ϵ_2 are the momentum 4-vectors and polarization 4-vectors of the incoming and outgoing photon.

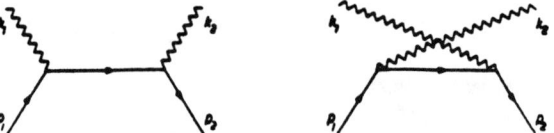

Fig.2 Lowest order diagrams for Compton scattering

Let A be the amplitude of the process considered. In lowest order the corresponding Feynman diagrams are shown in Fig.2. The differential cross-section as a function of the energy of the scattered electron in the lab frame is given by

$$\frac{d\sigma}{dE} = \frac{1}{32\pi} \frac{1}{|p_1 k_1|} \cdot \frac{1}{(|p_1| - E_{1\gamma})} \cdot |A|^2 \qquad (3)$$

where $|p_1|$ = momentum of the incoming electron beam, $\vec{p}_1 = |p_1|\hat{z}$

$E_{1\gamma}$ = energy of the incoming photon, $\vec{k}_1 = -E_{1\gamma}\hat{z}$

$|p_1 k_1|$ = absolute value of the dot product of p_1 and k_1

The amplitude A can be defined as follows

$$A = \epsilon_1^\mu \epsilon_2^\alpha \cdot M(\mu, \alpha) \qquad (4)$$

with
$$M(\mu, \alpha) = \bar{u}(p_2) \cdot M_u(\mu, \alpha) \cdot \frac{1}{2}(1 - i\gamma^5 \rlap{/}{s})u(p_1) \qquad (5)$$

This expression includes the projection operator corresponding to a longitudinally polarized electron. In lowest non zero order the expression for $M_u(\mu, \alpha)$ is given by

$$M_u^0(\mu, \alpha) = (-ie)^2 \cdot \left\{ \frac{[-i(\rlap{/}{p}_1 + \rlap{/}{k}_1) + m]}{2(p_1 k_1)} + \frac{[-i(\rlap{/}{p}_1 - \rlap{/}{k}_2) + m]}{-2(p_1 k_2)} \right\} \qquad (6)$$

The scattered electron and photon are unpolarized, thus the amplitude squared is

$$|A|^2 = \epsilon_1^\mu \, \epsilon_1^\nu \cdot \left\{ M_0(\mu, \alpha) \, M_0(\nu, \alpha)^* \right\} \qquad (7)$$

since $\Sigma \epsilon_2^\alpha \epsilon_2^\beta = \delta^{\alpha\beta}$. For Left + Right and Left - Right scattering one has

$$\epsilon_1^\mu \, \epsilon_1^\nu (\text{Left} + \text{Right}) = \frac{1}{2} \begin{pmatrix} 1 & 0 & 0 & 0 \\ 0 & 1 & 0 & 0 \\ 0 & 0 & 0 & 0 \\ 0 & 0 & 0 & 0 \end{pmatrix}$$

$$\epsilon_1^\mu \, \epsilon_1^\nu (\text{Left} - \text{Right}) = \frac{1}{2} \begin{pmatrix} 0 & -i & 0 & 0 \\ i & 0 & 0 & 0 \\ 0 & 0 & 0 & 0 \\ 0 & 0 & 0 & 0 \end{pmatrix}$$

where Left and Right stand for left and right circular polarized photons which travel along the z-axis.

The asymmetry as a function of the energy E of the scattered electron is

$$A_{LR}(E) = \frac{\frac{d\sigma}{dE}\big|_{\text{Left}} - \frac{d\sigma}{dE}\big|_{\text{Right}}}{\frac{d\sigma}{dE}\big|_{\text{Left}} + \frac{d\sigma}{dE}\big|_{\text{Right}}} \qquad (8)$$

Fig.3 (left) Unpolarized differential cross-section in lowest order in $mbarn/GeV$ as a function of the energy in GeV of the scattered electron for E(incoming electron)=50 GeV and E(incoming photon)=2.34 eV. Fig.4 (right) Same as Fig.3, but here E(incoming photon)=4 eV.

Fig. 3 and 4 show the unpolarized differential cross-section in lowest non zero order as a function of the energy of the scattered electron with the energy of the incoming electron equal to 50 GeV and the energy of the incoming photon equal to 2.34 eV. Fig. 5 and 6 show the asymmetry in lowest non zero order for the same values of the energy of the incoming electron and photon. It has been checked that the Left + Right cross-section equals the unpolarized cross-section. Thus in lowest non zero order

$$|A_0|^2 = \epsilon_1^\mu \epsilon_1^\nu \text{(Left + Right)} \cdot \{M_0(\mu\alpha)M_0(\nu\alpha)^*\} = \frac{1}{2}M_0(\mu\alpha)M_0(\mu\alpha)^* \qquad (9)$$

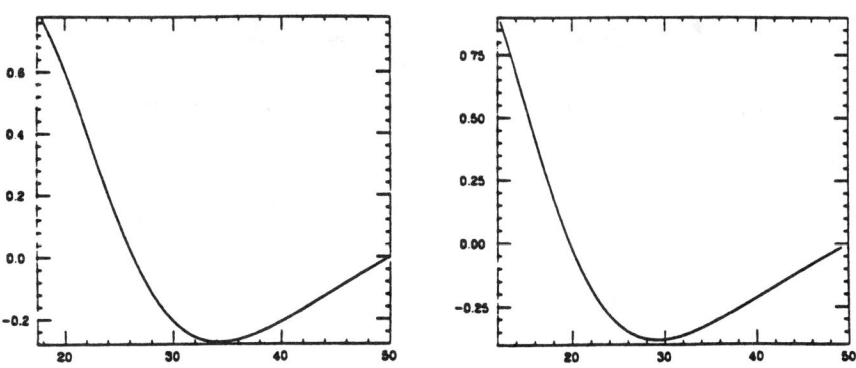

Fig.5 (left) The asymmetry as a function of the energy of the scattered electron in lowest order for E(incoming electron)=50 GeV and E(incoming photon)=2.34 eV. Fig.6 (right) Same as Fig.5 but here E(incoming photon)=4 eV.

3.2 THE ONE LOOP CORRECTIONS

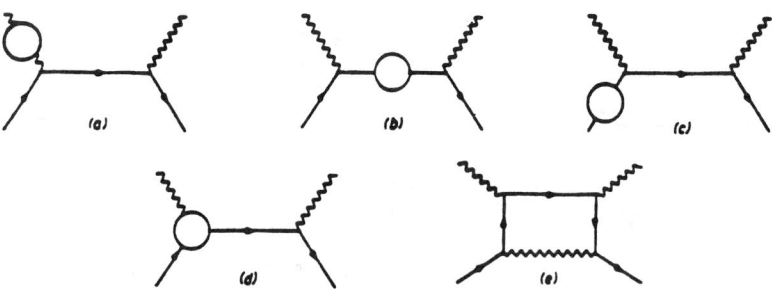

Fig.7 One loop diagrams

Now consider the one loop corrections (see ref.[3] for the complete discussion). The topologies of the corresponding Feynman diagrams are displayed in Fig.7. We use the on-shell renormalization scheme and the occuring one loop integrals are evaluated using dimensional regularization. After on shell photon wave function renormalization the external photon line correction (fig. 7(a)) does not contribute. The amplitude squared of the lowest order plus the one-loop corrections is

$$|A|^2 = |A_0|^2 + A_0 \cdot \{A^P + A^{W.R.} + A^V + A^B\}^* + \{A^P + A^{W.R.} + A^V + A^B\} \cdot A_0^* \qquad (10)$$

and the differential cross-section is again given by equation (3). $|A_0|^2$ is the lowest order amplitude squared. A^P is the amplitude due to the electron propagator correction (Fig.7(b)). External electron wave function correction (Fig.7(c)), the vertex correction (Fig.7(d)) and the box diagrams (Fig.7(e)) give rise to $A^{W.R.}$, A^V and A^B. $A^{W.R.}$ and A^B are both infrared divergent. The one loop corrected amplitude is calculated as follows. First the one loop integrals occuring in the expressions for the Feynman diagrams of fig.7 are expressed in terms of the form-factors[5]. Then the amplitude squared is algebraically evaluated by using the computer program SCHOONSCHIP[6]. Finally the form-factors are numerically evaluated by using the computer program FORMF[7].

$|A|^2$ indeed does not contain any ultraviolet divergencies. Furthermore, it has been verified that the $L + R$ cross-section equals the unpolarized cross-section;

$$|A|^2 = \epsilon_1^\mu \epsilon_1^\nu \text{ (Left + Right)} \{M(\mu\alpha)M(\nu\alpha)^*\} = \frac{1}{2}M(\mu\alpha)M^*(\mu\alpha) \qquad (11)$$

Since the $L + R$ cross-section was evaluated in the same manner as the $L - R$ cross-section, it follows that the above equation also functions as a good check for the expression of the $L - R$ cross-section. It is obvious that $|A|^2$ is still infrared divergent.

4. BREMSSTRAHLUNG

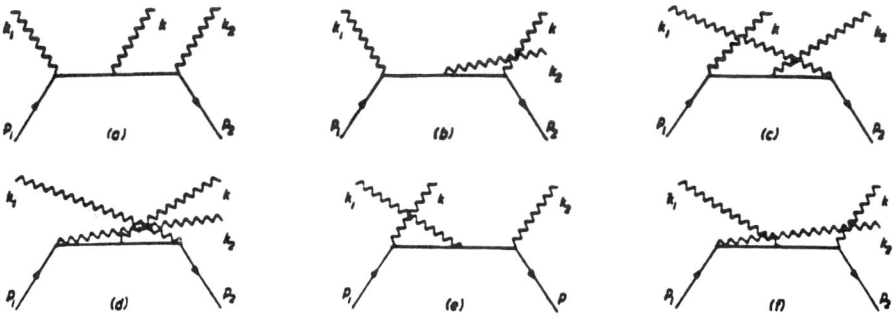

Fig.8 Diagrams for Bremsstrahlung

The diagrams are shown in Fig.8. The momentum of the extra photon is denoted by k. Since only the scattered electron will be seen, total bremsstrahlung has to be considered. The total bremsstrahlung is defined as soft bremsstrahlung, which contains the infrared divergence, plus the hard bremsstrahlung. Thus we have

$$\int_\lambda^{max} dk_0 \frac{d\sigma}{dk_0}(\text{total}) = \int_\lambda^{max} dk_0 \left\{ \frac{d\sigma}{dk_0}(\text{soft}) + \frac{d\sigma}{dk_0}(\text{hard}) \right\}$$
$$= \int_\lambda^{max} dk_0 \frac{d\sigma}{dk_0}(\text{soft}) + \int_0^{max} dk_0 \left\{ \frac{d\sigma}{dk_0}(\text{tot}) - \frac{d\sigma}{dk_0}(\text{soft}) \right\} \qquad (12)$$

Since the soft bremsstrahlung piece is being subtracted in the second part of the integral, it follows that the hard bremsstrahlung is infrared finite and the lower limit of that integration maybe taken to be zero.

The scattered photon will not be seen at all and the upper limit of the integration is the maximum photon energy kinematically allowed. Because of Bose statistics, the integral must be multiplied by a factor 1/2. This factor 1/2 can also be achieved by requiring that the energy k_0 of this photon is always smaller than the energy of the other outgoing photon (the one with momentum k_2).

4.1 SOFT BREMSSTRAHLUNG

The soft bremsstrahlung approximation is defined as follows (the photon with momentum k is allowed to go soft).
- Neglect those amplitudes which do not contain k in the propagator.
- Neglect k in the numerator of the amplitudes.
- Neglect k in the delta function.

This means that we may throw away diagrams (a) and (d) of Fig.8. We so arrive at

$$A_{\text{soft}} = A_0 \cdot e \cdot \left\{ \frac{(p_1 \epsilon)}{(p_1 k)} - \frac{(p_2 \epsilon)}{(p_2 k)} \right\} \tag{13}$$

where ϵ is the polarization vector of the photon with momentum k. The amplitude squared is

$$A_{\text{soft}}^2 = A_0^2 e^2 \left\{ \frac{-m^2}{(p_1 k)^2} - \frac{m^2}{(p_2 k)^2} - \frac{2 p_1 p_2}{(p_1 k)(p_2 k)} \right\} \tag{14}$$

The differential cross-section can thus be expressed in terms of the lowest order differential cross-section

$$\left(\frac{d\sigma_{\text{soft}}}{dE} \right) = \left(\frac{d\sigma}{dE} \right)_{\text{lowest order}} \{f_1 + f_2\} \tag{15}$$

where

$$f_1 = \frac{-m^2 e^2}{(2\pi)^3} \left\{ \int \frac{d^3k}{2k_0} \frac{1}{(p_1 k)^2} + \int \frac{d^3k}{2k_0} \frac{1}{(p_2 k)^2} \right\}$$

and

$$f_2 = -\frac{(2p_1 p_2) e^2}{(2\pi)^3} \int \frac{d^3k}{2k_0} \frac{1}{(p_1 k)(p_2 k)} \tag{16}$$

The infrared divergence contained in the f_1 part of the soft bremsstrahlung differential cross-section cancels against the infrared divergence due to the electron wave function correction. Similarly, the infrared divergence contained in the f_2 part cancels against the infrared divergence due to the box diagrams.

4.2 HARD BREMSSTRAHLUNG

For evaluating the total bremsstrahlung, all 6 diagrams of Fig.8 have to be taken into account. The expression for the amplitude squared has been obtained by using the computer program SCHOONSCHIP. Since it has not been evaluated by hand, the advantage of using the so-called spinor technique[8,9] for massive spinors[10] is not clear. The cross-section for the hard bremsstrahlung is defined as

$$\sigma_{\text{hard}} = \frac{1}{(2\pi)^5} \int \frac{d^3k}{2k_0} \cdot \frac{1}{16(p_1 k_1)} \cdot \left\{ \frac{A_{\text{tot}}^2}{X_2} - \frac{A_{\text{soft}}^2}{X_1} \right\} \tag{17}$$

with

$$X_1 = p_{1z} + k_{1z} = |p_1| - E_{1\gamma}$$

and

$$X_2 = X_1 - k_z + k_x \cdot \left(\frac{p_{2z}}{p_{2x}} \right) = X_1 - E_\gamma \left\{ \cos\theta_\gamma - \sin\theta_\gamma \cos\varphi_\gamma \left(\frac{\cos\theta_p}{\sin\theta_p} \right) \right\} \tag{18}$$

The momentum 4-vector k is defined as

$$k = \begin{pmatrix} E_\gamma \sin\theta_\gamma \sin\varphi_\gamma \\ E_\gamma \sin\theta_\gamma \cos\varphi_\gamma \\ E_\gamma \cos\theta_\gamma \\ iE_\gamma \end{pmatrix}$$

Note that in the limit of small photon energy, X_2 becomes equal to X_1. There are 3 integration variables left, namely E_γ, $\cos\theta_\gamma$ and φ_γ of the photon with momentum k. This integration has been performed in the center of mass frame by using the Monte Carlo programs VEGAS[11] and RIWIAD[12]. The results of these two programs were in agreement within 2%. It has been checked that the infrared divergent pieces cancel precisely and the programs behave very properly even when taking the lower limit of integration of E_γ exactly zero;

$$\lim_{E_\gamma \to 0} \left\{ \frac{A^2_{tot}(L \pm R)}{X_2} - \frac{A^2_{soft}(L \pm R)}{X_1} \right\} = 0 \qquad (19)$$

5. RESULTS AND CONCLUSION

The results can be found in tables 1–4. The first 2 tables show the correction to the unpolarized differential cross section, the last 2 tables show the correction to the asymmetry. The corrections are evaluated for 2 different values of the energy of the incoming photon beam, while the energy of the incoming electron beam remains the same, namely 50 GEV. The correction to the unpolarized differential cross section is very small and varies between 0.20% and 0.50%. The correction to the asymmetry is totally negligible at the lower end of the spectrum. At the other end of the spectrum it seems that the correction becomes quite large, around −4%. The reason for this is that the lowest order value is already extremely small and approaching zero for a scattering energy of the electron of 50 GEV. Such a correction would thus not be observable and is not very meaningfull.

TABLE 1

Energy in GeV	Lowest order	Correction in %
17.90	17.548	0.28
19.90	13.267	0.28
21.90	10.619	0.27
23.90	9.008	0.26
25.90	8.074	0.24
27.90	7.594	0.23
29.90	7.423	0.22
31.90	7.464	0.21
33.90	7.653	0.20
35.90	7.945	0.20
37.90	8.310	0.19
39.90	8.724	0.19
41.90	9.172	0.19
43.90	9.643	0.20
45.90	10.127	0.20
47.90	10.620	0.21

TABLE 2

Energy in GeV	Lowest order	Correction in %
12.30	14.043	0.45
14.30	10.353	0.44
16.30	8.194	0.42
18.30	6.891	0.39
20.30	6.096	0.37
22.30	5.617	0.35
24.30	5.342	0.33
26.30	5.203	0.32
28.30	5.156	0.30
30.30	5.173	0.29
32.30	5.235	0.29
34.30	5.329	0.28
36.30	5.446	0.28
38.30	5.580	0.28
40.30	5.726	0.28
42.30	5.880	0.29
44.30	6.041	0.30
46.30	6.206	0.32
48.30	6.373	0.34

Table 1 (left) Correction to the unpolarized differential cross section for E(incoming electron) $=50$ GeV and E(incoming photon)$=2.34$ eV. Column 1: Energy of the scattered electron in GeV. Column 2: Differential cross section in $mbarn/GeV$ in lowest order. Column 3: Lowest order correction in %. Table 2 (right) Same as Table 1, but here E(incoming photon)$=4$ eV.

TABLE 3		
Energy in GeV	Lowest order	Correction in %
17.90	0.7726	0.09
19.90	0.6096	0.09
21.90	0.4170	0.11
23.90	0.2175	0.16
25.90	0.0368	0.77
27.90	−0.1054	−0.22
29.90	−0.2016	−0.10
31.90	−0.2545	−0.08
33.90	−0.2727	−0.09
35.90	−0.2659	−0.11
37.90	−0.2421	−0.14
39.90	−0.2084	−0.20
41.90	−0.1691	−0.29
43.90	−0.1272	−0.45
45.90	−0.0847	−0.76
47.90	−0.0427	−1.69

TABLE 4		
Energy in GeV	Lowest order	Correction in %
12.30	0.8858	0.04
14.30	0.6354	0.04
16.30	0.3805	0.05
18.30	0.1459	0.10
20.30	−0.0497	−0.26
22.30	−0.1973	−0.08
24.30	−0.2969	−0.07
26.30	−0.3547	−0.07
28.30	−0.3791	−0.09
30.30	−0.3786	−0.11
32.30	−0.3606	−0.14
34.30	−0.3308	−0.18
36.30	−0.2936	−0.24
38.30	−0.2521	−0.31
40.30	−0.2084	−0.43
42.30	−0.1640	−0.60
44.30	−0.1198	−0.90
46.30	−0.0765	−1.55
48.30	−0.0345	−3.74

Table 3 (left) Correction to the asymmetry for E(incoming electron)=50 GeV and E(incoming photon)=2.34 eV. Column 1: Energy of the scattered electron in GeV. Column 2: Asymmetry in lowest order. Column 3: lowest order correction in %. Table 4 (right) Same as Table 3, but here E(incoming photon)=4 eV.

ACKNOWLEDGEMENTS

The author would like to especially thank her advisor M.K. Gaillard for her invaluable assistance and helpful discussions. The author also thanks R. Cahn for suggesting this problem.

REFERENCES

(1) Proposal for polarization at the SLC.
(2) J.M. Dorfan, TASI lectures in elementary particle physics (1984) 540.
    D. Williams, Ed.
(3) H. Veltman, Berkeley preprint LBL-27163, UCB-PTH-89/8.
    To be published in Phys. Rev. D.
(4) G. Altarelli and B. Stella, Lett. Nuovo Cimento 9 N10(1974) 416.
(5) G. Passarino and M. Veltman, Nucl. Phys. B160(1979) 151.
(6) M. Veltman, SCHOONSCHIP.
(7) M. Veltman, FORMF.
(8) P. de Causmaecker, R. Gastmans and W. Troost, Nucl. Phys. B206(1982) 53.
(9) F.A. Berends, R. Kleiss, P. de Causmaecker, R. Gastmans, W. Troost and
    T.T. Wu, Nucl. Phys. B206(1982) 61.
(10) A. Gongora-T. and R.G. Stuart, MPI-PAE/PTH 55/87, July 1987.
(11) G. Lepage, VEGAS.
(12) B. Lautrup, RIWIAD, CERN-DD. Long writeup D114.
(13) G. 't Hooft and M. Veltman, Nucl. Phys. B153(1979) 365.